Azure Data Engineering Cookbook

Second Edition

Get well versed in various data engineering techniques in
Azure using this recipe-based guide

Nagaraj Venkatesan

Ahmad Osama

BIRMINGHAM—MUMBAI

Azure Data Engineering Cookbook

Second Edition

Publishing Product Manager: Reshma Raman

Senior Editor: Nazia Shaikh

Content Development Editor: Manikandan Kurup

Technical Editor: Sweety Pagaria

Copy Editor: Safis Editing

Project Coordinator: Farheen Fathima

Proofreader: Safis Editing

Indexer: Sejal Dsilva

Production Designer: Joshua Misquitta

Marketing Coordinators: Priyanka Mhatre and Nivedita Singh

First published: March 2021

Second edition: September 2022

Production reference: 2070922

Published by Packt Publishing Ltd.

Livery Place

35 Livery Street

Birmingham

B3 2PB, UK.

ISBN 978-1-80324-678-9

www.packt.com

Contributors

About the authors

Nagaraj Venkatesan works as a cloud solution architect at Microsoft. At Microsoft, he works with some of the largest companies in the world, solving their complex data engineering problems and helping them build effective solutions using cutting-edge technologies based on Azure. Nagaraj, based out of Singapore, is a popular member of the data and AI community and is a regular speaker at several international data and AI conferences. He is a two-time **Microsoft Most Valuable Professional (MVP)** award winner, in 2016 and 2017. Nagaraj shares his technical expertise through his blog and on his YouTube channel called *DataChannel*. He also holds a master's degree in computing from the National University of Singapore.

Ahmad Osama works for Pitney Bowes Pvt. Ltd. as a technical architect and is a former Microsoft Data Platform MVP. In his day job, he works on developing and maintaining high performant, on-premises and cloud SQL Server OLTP environments as well as deployment and automating tasks using PowerShell. When not working, Ahmad blogs at DataPlatformLabs and can be found glued to his Xbox.

About the reviewers

Milind Kumar Chaudhari is an experienced cloud data engineer/architect who specializes in designing, implementing, and managing complex data pipelines. He is a data enthusiast and passionate about leveraging the best data engineering practices to solve critical business problems. He is also currently a technical reviewer with O'Reilly Media.

Vijay Kumar Suda is a senior cloud big data architect with over 20 years of experience working with global clients. He has worked in Switzerland, Belgium, Mexico, Bahrain, India, and Canada and helped customers across multiple industries. He has been based out of the USA since 2008. His expertise includes data engineering, architecture, the cloud, AI, and machine learning.

Firstly, I'd like to thank my parents, Sri Koteswara Rao and Rajyalakshmi, for their love and support in every step of my life. I'd like to thank my wife, Radhika, my son, Chandra, and my daughter, Akshaya, for their daily support and patience. I'd like to thank my siblings, Rama, Swarana, and Dr. SriKumar, for their support, and finally, I'd like to thank Packt for the opportunity to review this book.

Table of Contents

3

Building Data Ingestion Pipelines Using Azure Data Factory   59

4

8

Processing Data Using Azure Synapse Analytics   311

11

Monitoring Synapse SQL and Spark Pools   443

Preface

Data is the new oil and probably the most valuable resource. Data engineering covers how one can gain insights out of data. This book will introduce the key processes in data engineering (ingesting, storing, processing, and consuming) and share a few common recipes that can help us develop data engineering pipelines to gain insights into our data.

The book follows the logical data engineering process by beginning with Azure Data Lake and covering data ingestion using Azure Data Factory into Azure Data Lake and Azure SQL Database in the first few chapters. In these chapters, the book also covers the management of common storage layers such as Azure Data Lake and Azure SQL Database, focusing on topics such as security, high availability, and performance monitoring. The middle chapters focus on data processing using Azure Databricks, Azure Synapse Analytics Spark pools, and Synapse dataflows, and data exploration using Synapse serverless SQL pools. The final few chapters focus on the consumption of the data using Synapse dedicated SQL pool and Synapse Spark lake databases, covering the tips and tricks to optimize and maintain Synapse dedicated SQL pool databases and lake databases. Finally, the book also has a bonus chapter on managing the overall data engineering pipeline, which covers pipeline monitoring using Azure Log Analytics and tracking data lineage using Microsoft Purview.

While the book can be consumed in parts or any sequence, following along sequentially will help the readers experience building an end-to-end data engineering solution on Azure.

Who this book is for

The book is for anyone working on data engineering projects in Azure. Azure data engineers, data architects, developers, and database administrators working on Azure will find the book extremely useful.

What this book covers

Chapter 1, Creating and Managing Data in Azure Data Lake, focuses on provisioning, uploading, and managing the data life cycle in Azure Data Lake accounts.

Chapter 2, Securing and Monitoring Data in Azure Data Lake, covers securing an Azure Data Lake account using firewall and private links, accessing data lake accounts using managed identities, and monitoring an Azure Data Lake account using Azure Monitor.

Chapter 3, Building Data Ingestion Pipelines Using Azure Data Factory, covers ingesting data using Azure Data Factory and copying data between Azure SQL Database and Azure Data Lake.

Chapter 4, *Azure Data Factory Integration Runtime*, focuses on configuring and managing self-hosted integration runtimes and running SSIS packages in Azure using Azure-SSIS integration runtimes.

Chapter 5, *Configuring and Securing Azure SQL Database*, covers configuring a Serverless SQL database, Hyperscale SQL database, and securing Azure SQL Database using virtual networks and private links.

Chapter 6, *Implementing High Availability and Monitoring in Azure SQL Database*, explains configuring high availability to Azure SQL Database using auto-failover groups and read replicas, monitoring Azure SQL Database, and the automated scaling of Azure SQL Database during utilization spikes.

Chapter 7, *Processing Data Using Azure Databricks*, covers integrating Azure Databricks with Azure Data Lake and Azure Key Vault, processing data using Databricks notebooks, working with Delta tables, and visualizing Delta tables using Power BI.

Chapter 8, *Processing Data Using Azure Synapse Analytics* covers exploring data using Synapse Serverless SQL pool, processing data using Synapse Spark Pools, Working with Synapse Lake database, and integrating Synapse Analytics with Power BI.

Chapter 9, *Transforming Data Using Azure Synapse Dataflows*, focuses on performing transformations using Synapse Dataflows, optimizing data flows using partitioning, and managing dynamic source schema changes using schema drifting.

Chapter 10, *Building the Serving Layer in Azure Synapse SQL Pools*, covers loading processed data into Synapse dedicated SQL pools, performing data archival using partitioning, managing table distributions, and optimizing performance using statistics and workload management.

Chapter 11, *Monitoring Synapse SQL and Spark Pools*, covers monitoring Synapse dedicated SQL and Spark pools using Azure Log Analytics workbooks, Kusto scripts, and Azure Monitor, and monitoring Synapse dedicated SQL pools using Dynamic Management Views (DMVs).

Chapter 12, *Optimizing and Maintaining Synapse SQL and Spark Pools*, offers techniques for tuning query performance by optimizing query plans, rebuilding replication caches and maintenance scripts to optimize Delta tables, and automatically pausing SQL pools during inactivity, among other things.

Chapter 13, *Monitoring and Maintaining Azure Data Engineering Pipelines*, covers monitoring and managing end-to-end data engineering pipelines, which includes tracking data lineage using Microsoft Purview and improving the observability of pipeline executions using log analytics and query labeling.

To get the most out of this book

Readers with exposure to Azure and a basic understanding of data engineering should easily be able to follow this book:

Software/hardware covered in the book	OS requirements
Azure subscription	Windows 10 or above
PowerShell 7 or above with Azure PowerShell installed	
SQL Server Management Studio installed	
Power BI Desktop installed	

If you are using the digital version of this book, we advise you to type the code yourself or access the code via the GitHub repository (link available in the next section). Doing so will help you avoid any potential errors related to the copying and pasting of code.

Download the example code files

You can download the example code files for this book from GitHub at `https://github.com/PacktPublishing/Azure-Data-Engineering-Cookbook-2nd-edition`. In case there's an update to the code, it will be updated in the existing GitHub repository.

We also have other code bundles from our rich catalog of books and videos available at `https://github.com/PacktPublishing/`. Check them out!

Download the color images

We also provide a PDF file that has color images of the screenshots and diagrams used in this book. You can download it here: `https://packt.link/CJshA`.

Conventions used

There are a number of text conventions used throughout this book.

`Code in text`: Indicates code words in text, database table names, folder names, filenames, file extensions, pathnames, dummy URLs, user input, and Twitter handles. Here is an example: "Observe that the `CopyFiles` package is now listed under the `AzureSSIS | Projects` folder."

A block of code is set as follows:

```
CREATE TABLE dbo.transaction_tbl WITH (DISTRIBUTION = ROUND_
ROBIN)
AS
Select * from dbo.ext_transaction_tbl;
GO
Select TOP 100 *  from dbo.transaction_tbl
GO
```

Any command-line input or output is written as follows:

```
Connect-AzAccount
```

Bold: Indicates a new term, an important word, or words that you see onscreen. For example, words in menus or dialog boxes appear in the text like this. Here is an example: "The **Configuration** section under the **Source** section of **Copy Data tool** can remain with defaults."

> **Tips or important notes**
> Appear like this.

Sections

In this book, you will find several headings that appear frequently (*Getting ready*, *How to do it...*, *How it works...*, *There's more...*, and *See also*).

To give clear instructions on how to complete a recipe, use these sections as follows.

Getting ready

This section tells you what to expect in the recipe and describes how to set up any software or any preliminary settings required for the recipe.

How to do it...

This section contains the steps required to follow the recipe.

How it works...

This section usually consists of a detailed explanation of what happened in the previous section.

There's more...

This section consists of additional information about the recipe in order to make you more knowledgeable about the recipe.

See also

This section provides helpful links to other useful information for the recipe.

Get in touch

Feedback from our readers is always welcome.

General feedback: If you have questions about any aspect of this book, mention the book title in the subject of your message and email us at customercare@packtpub.com.

Errata: Although we have taken every care to ensure the accuracy of our content, mistakes do happen. If you have found a mistake in this book, we would be grateful if you would report this to us. Please visit www.packtpub.com/support/errata, selecting your book, clicking on the Errata Submission Form link, and entering the details.

Piracy: If you come across any illegal copies of our works in any form on the Internet, we would be grateful if you would provide us with the location address or website name. Please contact us at copyright@packt.com with a link to the material.

If you are interested in becoming an author: If there is a topic that you have expertise in and you are interested in either writing or contributing to a book, please visit authors.packtpub.com.

Share your thoughts

Once you've read *Azure Data Engineering Cookbook, Second Edition*, we'd love to hear your thoughts! Scan the QR code below to go straight to the Amazon review page for this book and share your feedback.

https://packt.link/r/1-803-24678-2

Your review is important to us and the tech community and will help us make sure we're delivering excellent quality content.

1

Creating and Managing Data in Azure Data Lake

Azure Data Lake is a highly scalable and durable object-based cloud storage solution from Microsoft. It is optimized to store large amounts of structured and semi-structured data such as logs, application data, and documents.

Azure Data Lake can be used as a data source and destination in data engineering projects. As a source, it can be used to stage structured or semi-structured data. As a destination, it can be used to store the result of a data pipeline.

Azure Data Lake is provisioned as a storage account in Azure, capable of storing files (blobs), tables, or queues. This book will focus on Azure Data Lake storage accounts used for storing blobs/files

In this chapter, we will learn how to provision, manage, and upload data into Data Lake accounts and will cover the following recipes:

- Provisioning an Azure storage account using the Azure portal
- Provisioning an Azure storage account using PowerShell

- Creating containers and uploading files to Azure Blob storage using PowerShell
- Managing blobs in Azure Storage using PowerShell
- Configuring blob lifecycle management for blob objects using the Azure portal

Technical requirements

For this chapter, the following are required:

- An Azure subscription
- Azure PowerShell

The code samples can be found at `https://github.com/PacktPublishing/Azure-Data-Engineering-Cookbook-2nd-edition`.

Provisioning an Azure storage account using the Azure portal

In this recipe, we will provision an Azure storage account using the Azure portal. Azure Blob storage is one of the four storage services available in Azure Storage. The other storage services are **Table**, **Queue**, and **File Share**. Table storage is used to store non-relational structured data as key-value pairs, queue storage is used to store messages as queues, and file share is used for creating file share directories/mount points that can be accessed using the NFS/SMB protocols. This chapter will focus on storing data using the Blob storage service.

Getting ready

Before you start, open a web browser and go to the Azure portal at `https://portal.azure.com`. Ensure that you have an Azure subscription. Install Azure PowerShell on your machine; instructions for installing it can be found at `https://docs.microsoft.com/en-us/powershell/azure/install-az-ps?view=azps-6.6.00`.

How to do it...

The steps for this recipe are as follows:

1. In the Azure portal, select **Create a resource** and choose **Storage account – blob, file, table, queue** (or search for storage account in the search bar; do not choose **Storage accounts (classic)**).

2. A new page, **Create a storage account**, will open. There are six tabs on the **Create a storage account** page – **Basics**, **Advanced**, **Networking**, **Data protection**, **Tags**, and **Review + create**.

3. In the **Basics** tab, we need to provide the Azure **Subscription**, **Resource group**, **Storage account name**, **Region**, **Performance**, and **Redundancy** values, as shown in the following screenshot:

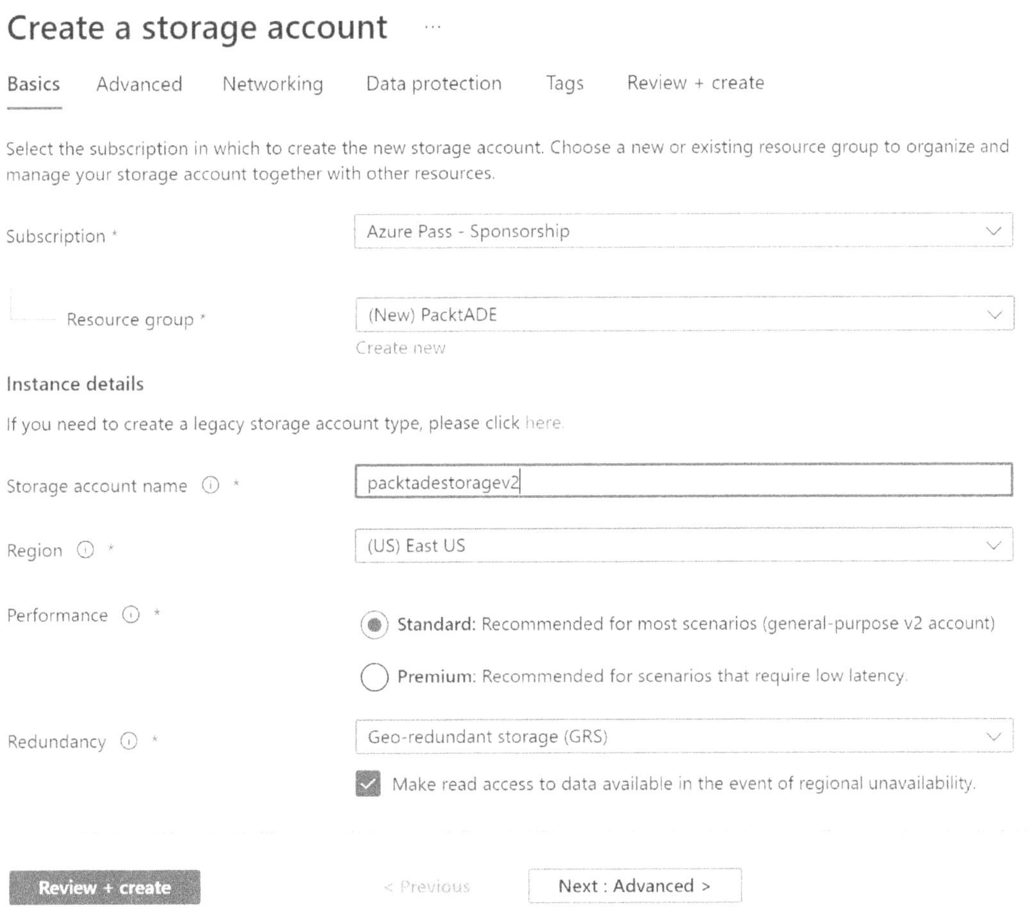

Create a storage account

Basics   Advanced   Networking   Data protection   Tags   Review + create

Select the subscription in which to create the new storage account. Choose a new or existing resource group to organize and manage your storage account together with other resources.

Subscription *   Azure Pass - Sponsorship

Resource group *   (New) PacktADE
                                  Create new

Instance details

If you need to create a legacy storage account type, please click here.

Storage account name ⓘ *   packtadestoragev2

Region ⓘ *   (US) East US

Performance ⓘ *   ⦿ Standard: Recommended for most scenarios (general-purpose v2 account)

                                  ○ Premium: Recommended for scenarios that require low latency.

Redundancy ⓘ *   Geo-redundant storage (GRS)

                                  ☑ Make read access to data available in the event of regional unavailability.

[Review + create]   < Previous   [Next : Advanced >]

Figure 1.1 – The Create a storage account Basics tab

4. In the **Advanced** tab, we need to select **Enable hierarchical namespace** under the **Data Lake Storage Gen2** settings:

Basics
Advanced
Networking
Data protection
Tags
Review + create
Require secure transfer for REST API operations
Enable infrastructure encryption
Enable blob public access
Enable storage account key access
Default to Azure Active Directory authorization in the Azure portal
Minimum TLS version
Version 1.2
Data Lake Storage Gen2
The Data Lake Storage Gen2 hierarchical namespace accelerates big data analytics workloads and enables file-level access control lists (ACLs). Learn more
Enable hierarchical namespace

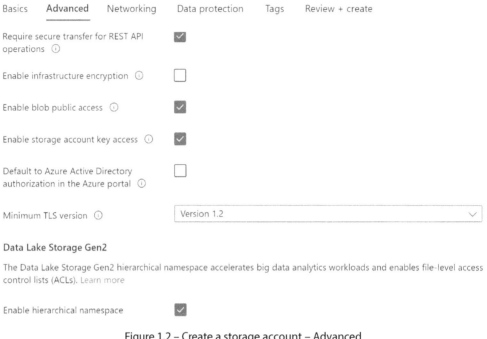

Figure 1.2 – Create a storage account – Advanced

5. In the **Networking** tab, we need to provide the connectivity method:

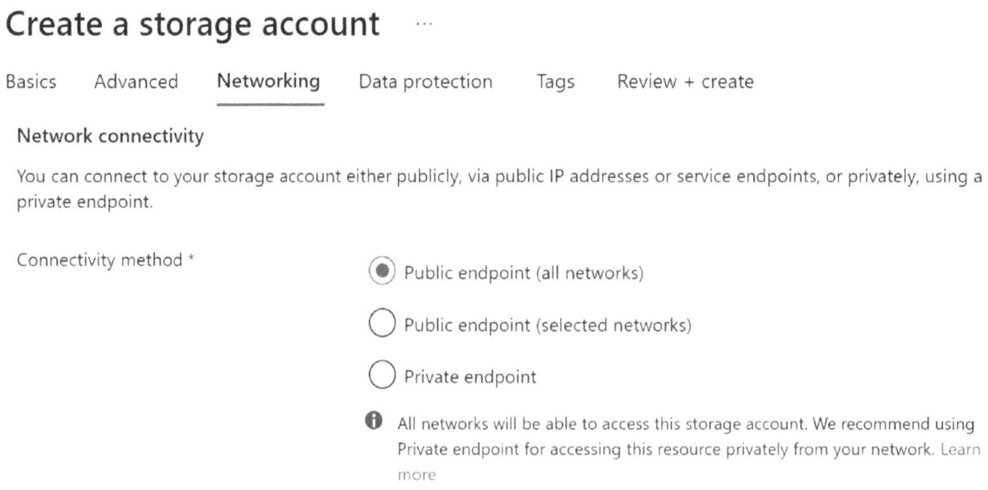

Create a storage account
Basics
Advanced
Networking
Data protection
Tags
Review + create
Network connectivity
You can connect to your storage account either publicly, via public IP addresses or service endpoints, or privately, using a private endpoint.
Connectivity method *
Public endpoint (all networks)
Public endpoint (selected networks)
Private endpoint
All networks will be able to access this storage account. We recommend using Private endpoint for accessing this resource privately from your network. Learn more

Figure 1.3 – Create a storage account – Networking

6.   In the **Review + create** tab, review the configuration settings and select **Create** to provision the Azure storage account:

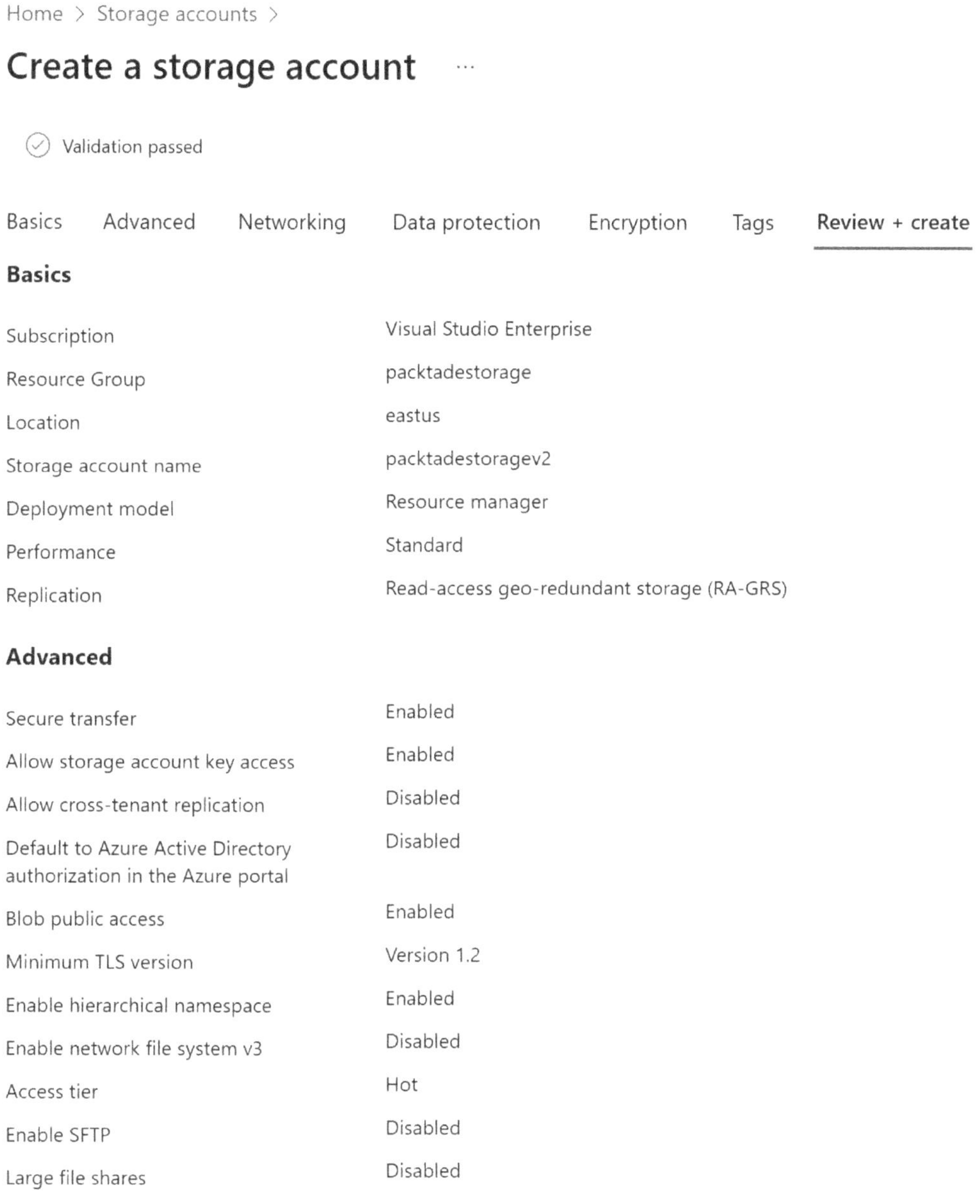

Home > Storage accounts >

Create a storage account   ...

⊘ Validation passed

| Basics | Advanced | Networking | Data protection | Encryption | Tags | **Review + create** |

Basics

Subscription	Visual Studio Enterprise
Resource Group	packtadestorage
Location	eastus
Storage account name	packtadestoragev2
Deployment model	Resource manager
Performance	Standard
Replication	Read-access geo-redundant storage (RA-GRS)

Advanced

Secure transfer	Enabled
Allow storage account key access	Enabled
Allow cross-tenant replication	Disabled
Default to Azure Active Directory authorization in the Azure portal	Disabled
Blob public access	Enabled
Minimum TLS version	Version 1.2
Enable hierarchical namespace	Enabled
Enable network file system v3	Disabled
Access tier	Hot
Enable SFTP	Disabled
Large file shares	Disabled

Figure 1.4 – Create a storage account – Review + create

How it works...

The Azure storage account is deployed in the selected subscription, resource group, and location. The **Performance** tier can be either **Standard** or **Premium**. A **Standard** performance tier is a low-cost magnetic drive-backed storage. It's suitable for applications such as static websites and bulk storing flat files. The **Premium** tier is a high-cost SSD-backed storage service. The **Premium** tier can only be used with Azure virtual machine disks for I/O-intensive applications.

The **Replication** options available are **Locally-redundant storage (LRS)**, **Zone-redundant storage (ZRS)**, **Geo-redundant storage (GRS)**, and **Geo-zone-redundant storage (GZRS)**. Local redundancy stores three local copies within the data center and provides fault tolerance for failures within it. Zone-redundant storage provides fault tolerance by copying data to additional data centers within the same region, while geo-redundant storage maintains copies across regions. Geo-zone-redundant storage combines geo- and zone-redundant features and offers the highest fault tolerance. The default **Geo-redundant storage (GRS)** option was selected, as it provides fault tolerance across regions.

Azure storage accounts can be accessed publicly over the internet, through selected networks (selected IPs and IP ranges), and from private endpoints.

Provisioning an Azure storage account using PowerShell

PowerShell is a scripting language used to programmatically manage various tasks. In this recipe, we will learn how to provision an Azure storage account using PowerShell.

Getting ready

Before you start, you need to log in to the Azure subscription from the PowerShell console. To do this, execute the following command in a new PowerShell window:

```
Connect-AzAccount
```

Then, follow the instructions to log in to the Azure account.

How to do it...

The steps for this recipe are as follows:

1. Execute the following command in a PowerShell window to create a new resource group. If you want to create the Azure storage account in an existing resource group, this step isn't required:

    ```
    New-AzResourceGroup -Name Packtade-powershell -Location
    'East US'
    ```

    You should get the following output:

    ```
    ResourceGroupName : Packtade-powershell
    Location          : eastus
    ProvisioningState : Succeeded
    Tags              :
    ResourceId        : /subscriptions/b85b0984-a391-4f22-a832-fb6e46c39f38/resourceGroups/Packtade-powershell
    ```

    Figure 1.5 – Creating a new resource group

2. Execute the following command to create a new Azure storage account in the Packtade-powershell resource group:

    ```
    New-AzStorageAccount -ResourceGroupName Packtade-
    powershell -Name packtstoragepowershellv2 -SkuName
    Standard_LRS -Location 'East US' -Kind StorageV2
    ```

    You should get the following output:

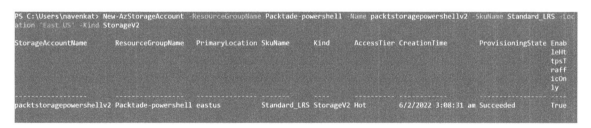

Figure 1.6 – Creating a new storage account

How it works...

There is a single command to create an Azure storage account using PowerShell – New-AzStorageAccount. The SkuName parameter specifies the performance tier, and the Kind parameter specifies the account kind.

In the later recipes, we will look at how to assign public/private endpoints to an Azure storage account using PowerShell.

Creating containers and uploading files to Azure Blob storage using PowerShell

In this recipe, we will create a new container and upload files to Azure Blob storage using PowerShell.

Getting ready

Before you start, perform the following steps:

1. Make sure you have an existing Azure storage account. If not, create one by following the *Provisioning an Azure storage account using the Azure portal* recipe.

2. Log in to your Azure subscription in PowerShell. To log in, run the `Connect-AzAccount` command in a new PowerShell window and follow the instructions.

How to do it...

The steps for this recipe are as follows:

1. Execute the following commands to create the container in an Azure storage account:

```
$storageaccountname="packtadestoragev2"
$containername="logfiles"
$resourcegroup="packtadestorage"
#Get the Azure Storage account context
$storagecontext = (Get-AzStorageAccount -ResourceGroupName
$resourcegroup -Name $storageaccountname).Context;
#Create a new container
New-AzStorageContainer -Name $containername -Context
$storagecontext
```

Container creation is usually very quick. You should get the following output:

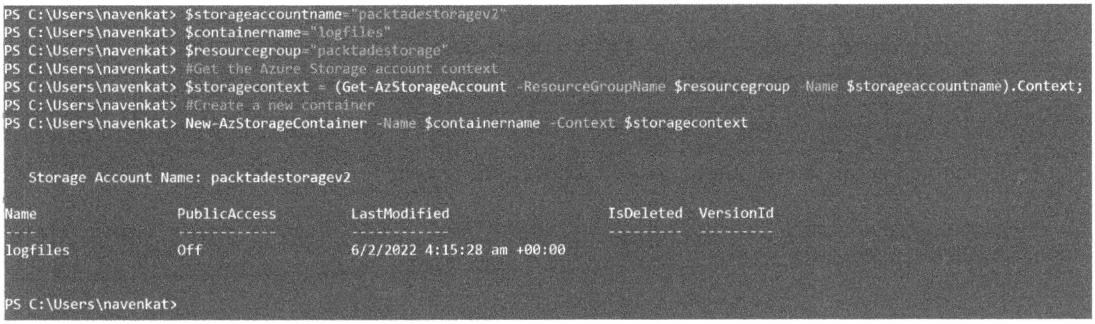

```
PS C:\Users\navenkat> $storageaccountname="packtadestoragev2"
PS C:\Users\navenkat> $containername="logfiles"
PS C:\Users\navenkat> $resourcegroup="packtadestorage"
PS C:\Users\navenkat> #Get the Azure Storage account context
PS C:\Users\navenkat> $storagecontext = (Get-AzStorageAccount -ResourceGroupName $resourcegroup  Name $storageaccountname).Context;
PS C:\Users\navenkat> #Create a new container
PS C:\Users\navenkat> New-AzStorageContainer -Name $containername -Context $storagecontext


    Storage Account Name: packtadestoragev2

Name                  PublicAccess        LastModified                    IsDeleted   VersionId
----                  ------------        ------------                    ---------   ---------
logfiles              Off                 6/2/2022 4:15:28 am +00:00


PS C:\Users\navenkat>
```

Figure 1.7 – Creating a new storage container

2. Execute the following commands to upload a text file to an existing container. Ensure that you create a folder in c:\ADECookbook\Chapter1\Logfiles\. Create any file inside the folder as Logfile1.txt:

```
#upload single file to container
Set-AzStorageBlobContent -File "C:\ADECookbook\Chapter1\
Logfiles\Logfile1.txt" -Context $storagecontext -Blob
logfile1.txt -Container $containername
```

You should get an output similar to the following screenshot:

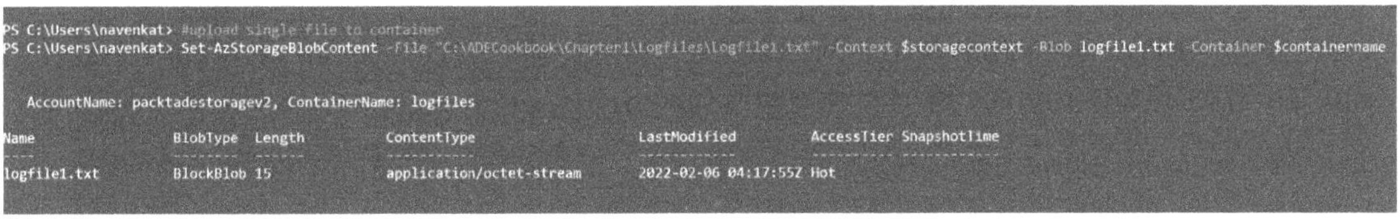

```
PS C:\Users\navenkat> #upload single file to container
PS C:\Users\navenkat> Set-AzStorageBlobContent -File "C:\ADECookbook\Chapter1\Logfiles\Logfile1.txt" -Context $storagecontext -Blob logfile1.txt -Container $containername

   AccountName: packtadestoragev2, ContainerName: logfiles

Name              BlobType   Length    ContentType                 LastModified              AccessTier  SnapshotTime
----              --------   ------    -----------                 ------------              ----------  ------------
logfile1.txt      BlockBlob  15        application/octet-stream    2022-02-06 04:17:55Z Hot
```

Figure 1.8 – Uploading a file to a storage container

3. Execute the following commands to upload all the files in a directory to an Azure container. Create additional copies of Logfile1.txt in the same folder for testing multiple file uploads:

```
#get files to be uploaded from the directory
$files = Get-ChildItem -Path "C:\ADECookbook\Chapter1\
Logfiles";
#iterate through each file int the folder and upload it
to the azure container
foreach($file in $files){
```

```
Set-AzStorageBlobContent -File $file.FullName -Context
$storagecontext -Blob $file.BaseName -Container
$containername -Force
}
```

You should get an output similar to the following screenshot:

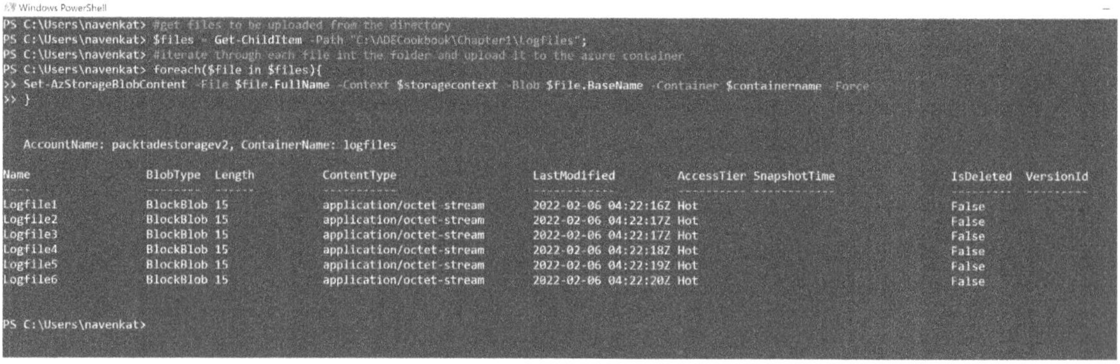

Figure 1.9 – Uploading multiple files to a storage container

How it works...

The storage container is created using the `New-AzStorageContainer` command. It takes two parameters – the container name and the storage context. The storage context can be set using the `Get-AzStorageAccount` command context property.

To upload files to the container, we used the `Set-AzStorageBlobContent` command. This command requires the storage context, a file path to be uploaded, and the container name. To upload multiple files, we can iterate through the folder and upload each file using the `Set-AzStorageBlobContent` command.

Managing blobs in Azure Storage using PowerShell

In this recipe, we will learn how to perform various management tasks on an Azure blob. We will perform operations such as copying, listing, modifying, deleting, and downloading files from Azure Blob storage.

Getting ready

Before you start, perform the following steps:

1. Make sure you have an existing Azure storage account. If not, create one by following the *Provisioning an Azure storage account using PowerShell* recipe.

2. Make sure you have an existing Azure storage container. If not, create one by following the *Creating containers and uploading files to Azure Blob storage using PowerShell* recipe.

3. Log in to your Azure subscription in PowerShell. To log in, run the `Connect- AzAccount` command in a new PowerShell window and follow the instructions.

How to do it...

Let's perform the following operations in this recipe:

1. Copy files/blobs between two blob storage containers.

2. List files from a blob container.

3. Modify the storage access tier of a blob from **Hot** to **Cool**.

4. Download a file/blob from a container.

5. Delete a file/blob from a container.

Let's look at each of them in detail. We'll begin by copying blobs between containers.

Copying blobs between containers

Perform the following steps:

1. Execute the following commands to create a new container in an Azure storage account:

```
#set the parameter values
$storageaccountname="packtadestoragev2"
$resourcegroup="packtadestorage"
$sourcecontainername="logfiles"
$destcontainername="textfiles"
#Get storage account context
$storagecontext = (Get-AzStorageAccount -ResourceGroupName
$resourcegroup -Name $storageaccountname).Context
# create the container
$destcontainer = New-AzStorageContainer -Name
$destcontainername -Context $storagecontext
$destcontainer
```

You should get an output similar to the following screenshot:

```
PS C:\Users\navenkat> $storageaccountname="packtadestoragev2"
PS C:\Users\navenkat> $resourcegroup="packtadestorage"
PS C:\Users\navenkat> $sourcecontainername="logfiles"
PS C:\Users\navenkat> $destcontainername="textfiles"
PS C:\Users\navenkat> #Get storage account context
PS C:\Users\navenkat> $storagecontext = (Get-AzStorageAccount -ResourceGroupName $resourcegroup -Name $storageaccountname).Context
PS C:\Users\navenkat> # create the container
PS C:\Users\navenkat> $destcontainer = New-AzStorageContainer -Name $destcontainername -Context $storagecontext
PS C:\Users\navenkat>
PS C:\Users\navenkat>
PS C:\Users\navenkat> $destcontainer


   Storage Account Name: packtadestoragev2

Name               PublicAccess       LastModified               IsDeleted  VersionId
----               ------------       ------------               ---------  ---------
textfiles          Off                6/2/2022 4:24:42 am +00:00
```

Figure 1.10 – Creating a new storage container

2. Execute the following command to copy the `Logfile1` blob from the source container to the destination container:

    ```
    #copy a single blob from one container to another

    Start-CopyAzureStorageBlob -SrcBlob "Logfile1"
    -SrcContainer $sourcecontainername -DestContainer
    $destcontainername -Context $storagecontext -DestContext
    $storagecontext
    ```

    You should get an output similar to the following screenshot:

Figure 1.11 – Copying a blob from one storage container to another

3. Execute the following command to copy all the blobs from the source container to the destination container:

    ```
    # copy all blobs in new container

    Get-AzStorageBlob -Container $sourcecontainername
    -Context $storagecontext | Start-CopyAzureStorageBlob
    -DestContainer $destcontainername -DestContext
    $storagecontext -force
    ```

You should get an output similar to the following screenshot:

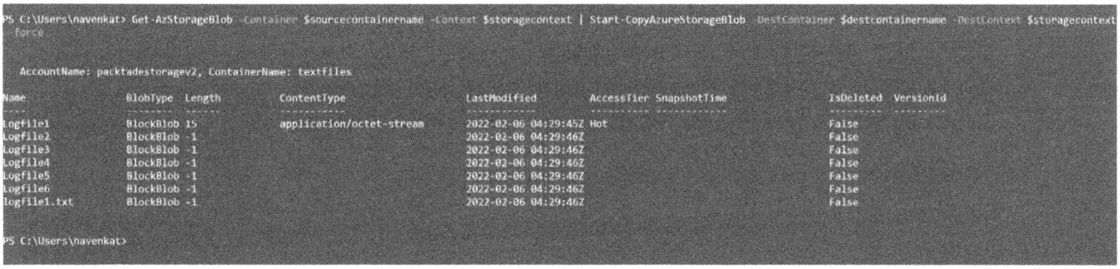

Figure 1.12 – Copying all blobs from one storage container to another

Listing blobs in an Azure storage container

Execute the following command to list the blobs from the destination container:

```
# list the blobs in the destination container


(Get-AzStorageContainer -Name $destcontainername -Context
$storagecontext).CloudBlobContainer.ListBlobs()
```

You should get an output similar to the following screenshot:

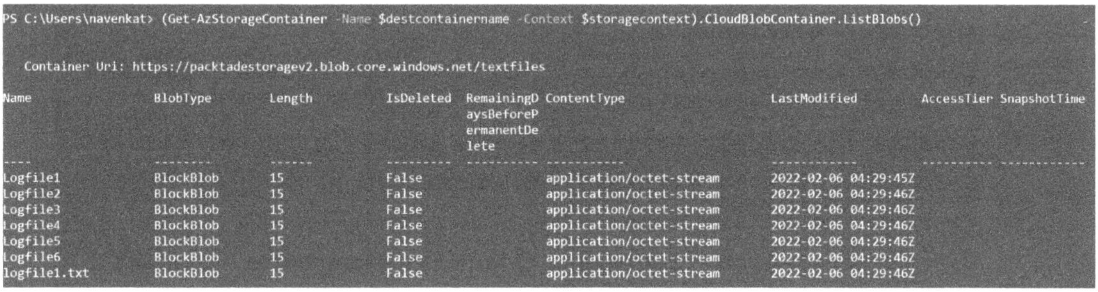

Figure 1.13 – Listing blobs in a storage container

Modifying a blob access tier

Perform the following steps:

1. Execute the following commands to change the access tier of a blob:

    ```
    # Get the blob reference
    $blob = Get-AzStorageBlob -Blob *Logfile2* -Container
    $sourcecontainername -Context $storagecontext
    #Get current access tier
    $blob
    ```

```
#change access tier to cool
$blob.ICloudBlob.SetStandardBlobTier("cool")
#Get the modified access tier
Get-AzStorageBlob -Blob *Logfile2* -Container
$sourcecontainername -Context $storagecontext
```

You should get an output similar to the following screenshot:

Figure 1.14 – Modifying the blob access tier

2. Execute the following commands to change the access tier of all the blobs in the container:

```
#get blob reference
$blobs = Get-AzStorageBlob -Container $destcontainername
-Context $storagecontext
#change the access tier of all the blobs in the container
$blobs.icloudblob.setstandardblobtier("Cool")
#verify the access tier
Get-AzStorageBlob -Container $destcontainername -Context
$storagecontext
```

You should get an output similar to the following screenshot:

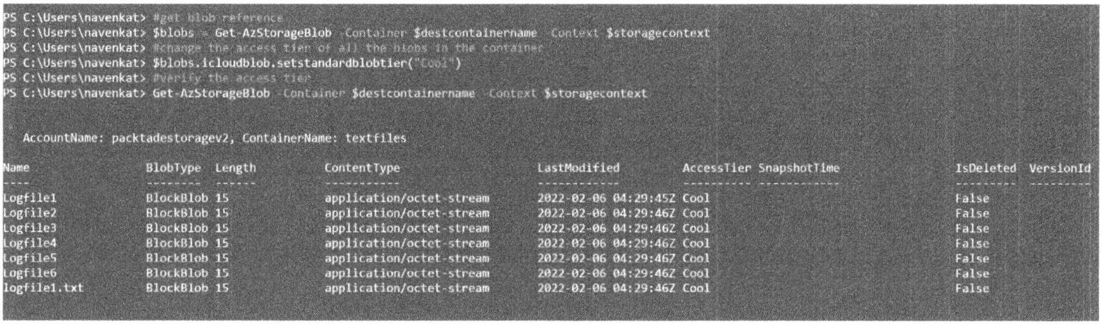

Figure 1.15 – Modifying the blob access tier of all the blobs in a storage container

Downloading a blob

Execute the following commands to download a blob from Azure Storage to your local computer:

```
#get the storage context
$storagecontext = (Get-AzStorageAccount -ResourceGroupName
$resourcegroup -Name $storageaccountname).Context
#download the blob
Get-AzStorageBlobContent -Blob "Logfile1" -Container
$sourcecontainername -Destination C:\ADECookbook\Chapter1\
Logfiles\ -Context $storagecontext -Force
```

Deleting a blob

Execute the following command to remove/delete a blob:

```
#get the storage context

$storagecontext = (Get-AzStorageAccount -ResourceGroupName
$resourcegroup -Name $storageaccountname).Context

Remove-AzStorageBlob -Blob "Logfile2" -Container
$sourcecontainername -Context $storagecontext
```

How it works...

Copying blobs across containers in the same storage account or a different storage account can be done easily by the PowerShell `Start-CopyAzureStorageBlob` command. The command takes the source and destination blobs, the source and destination containers, and the source and destination storage accounts as parameters. To copy all blobs in a container, we can run `Get-AzStorageBlob` to get all the blobs in the container and pipe the blobs to the `Start-CopyAzureStorageBlob` command.

A blob access tier can be modified by first getting the reference to the blob object using `Get-AzStorageBlob` and then modifying the access tier using the `setstandardblobtier` property. There are three access tiers – `Hot`, `Cool`, and `Archive`:

- The **Hot tier** is suitable for files that are accessed frequently. It has a higher storage cost and low access cost.

- The **Cool tier** is suitable for infrequently accessed files and has a lower access cost and a lower storage cost.

- The **Archive tier**, as the name suggests, is used for long-term archival and should be used for files that are seldom required. It has the highest access cost and the lowest storage cost.

To download a blob from Azure to a local system, we use `Get-AzStorageBlobContent`. The command accepts the blob name, the container name, the local file path, and the storage context.

To delete a blob, run `Remove-AzStorageBlob`. Provide the blob name, the container name, and the storage context.

Configuring blob lifecycle management for blob objects using the Azure portal

Azure Storage provides different blob access tiers such as **Hot**, **Cool**, **and Archive**. Each access tier has a different storage and data transfer cost. Applying a proper lifecycle rule to move a blob among different access tiers helps optimize the cost. In this recipe, we will learn how to apply a lifecycle rule to a blob using the Azure portal.

Getting ready

Before you start, perform the following steps:

1. Make sure you have an existing Azure storage account. If not, create one by following the *Provisioning an Azure storage account using PowerShell* recipe.

2. Make sure you have an existing Azure storage container. If not, create one by following the *Creating containers and uploading files to Azure Blob storage using PowerShell* recipe.

3. Make sure you have existing blobs/files in an Azure storage container. If not, you can upload blobs in accordance with the previous recipe. Then, log in to the Azure portal at `https://portal.azure.com`.

How to do it...

Follow the given steps to configure a blob lifecycle:

1. In the Azure portal, find and open the Azure Storage case. In our case, it is **packtadestoragev2**.

2. In the **packtadestoragev2** window, search for `Data management` and select **Lifecycle Management** under **Data management**, as shown in the following screenshot:

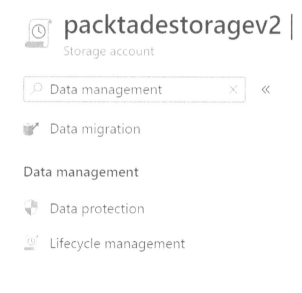

Figure 1.16 – Opening Lifecycle management

3. On the **Add a rule** page, create a rule to provide the lifecycle configuration. A lifecycle defines when to move a blob from a Hot to a Cool access tier, when to move a blob from a Cool to a Storage access tier, and when to delete the blob. Select **Limit blobs with filters** to create a lifecycle policy for a particular container. Click **Next**:

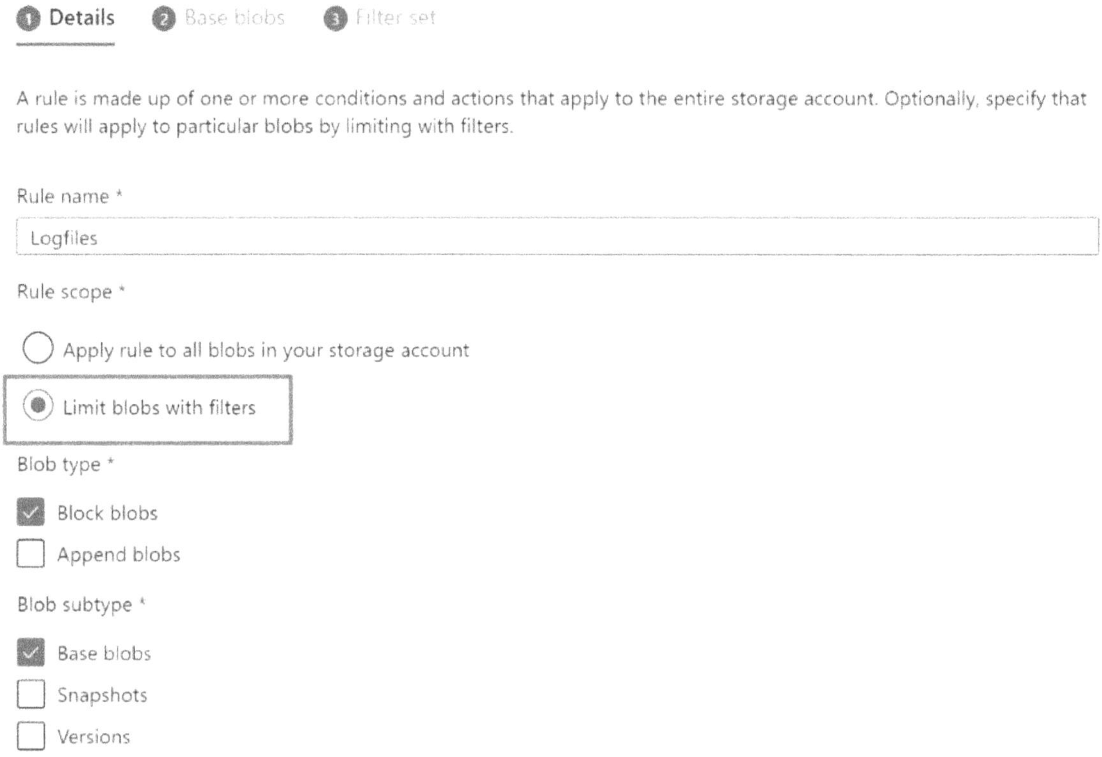

① **Details**   **②** Base blobs   **③** Filter set

A rule is made up of one or more conditions and actions that apply to the entire storage account. Optionally, specify that rules will apply to particular blobs by limiting with filters.

Rule name *

Logfiles

Rule scope *

◯ Apply rule to all blobs in your storage account

◉ Limit blobs with filters

Blob type *

☑ Block blobs

☐ Append blobs

Blob subtype *

☑ Base blobs

☐ Snapshots

☐ Versions

Figure 1.17 – Lifecycle management – the action set

4. Specify the condition you would like to use as a lifecycle policy. For example, the following rule moves the blobs that haven't been modified in the last 30 days to Cool storage and 60 days to Archive storage:

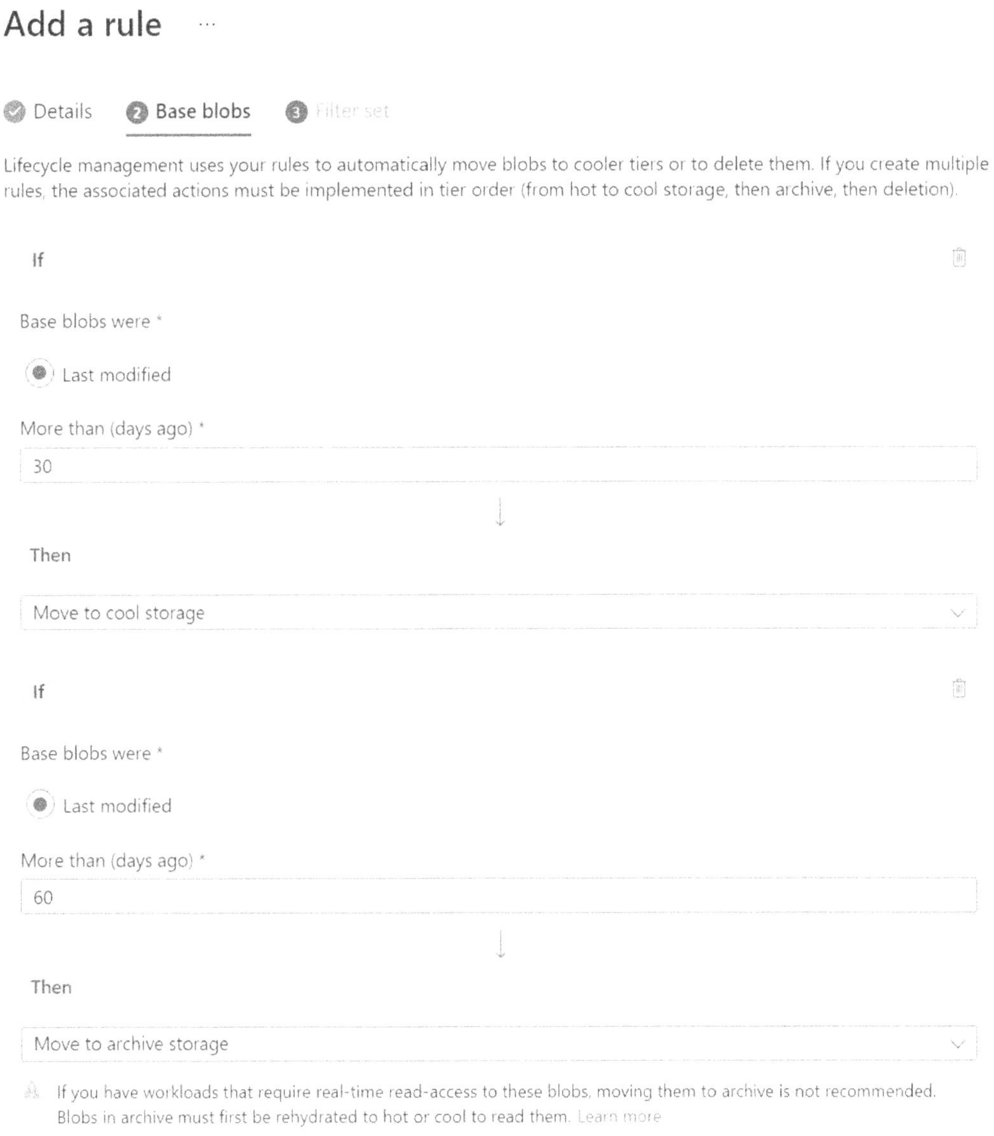

Add a rule   ...

✔ Details   ❷ **Base blobs**   ❸ Filter set

Lifecycle management uses your rules to automatically move blobs to cooler tiers or to delete them. If you create multiple rules, the associated actions must be implemented in tier order (from hot to cool storage, then archive, then deletion).

If   🗑

Base blobs were *

⦿ Last modified

More than (days ago) *

30

↓

Then

Move to cool storage	∨

If   🗑

Base blobs were *

⦿ Last modified

More than (days ago) *

60

↓

Then

Move to archive storage	∨

⚠ If you have workloads that require real-time read-access to these blobs, moving them to archive is not recommended. Blobs in archive must first be rehydrated to hot or cool to read them. Learn more

Figure 1.18 – Lifecycle management – the filter set

5. In **Filter set**, specify the specific container or filename for which you need to apply the rule:

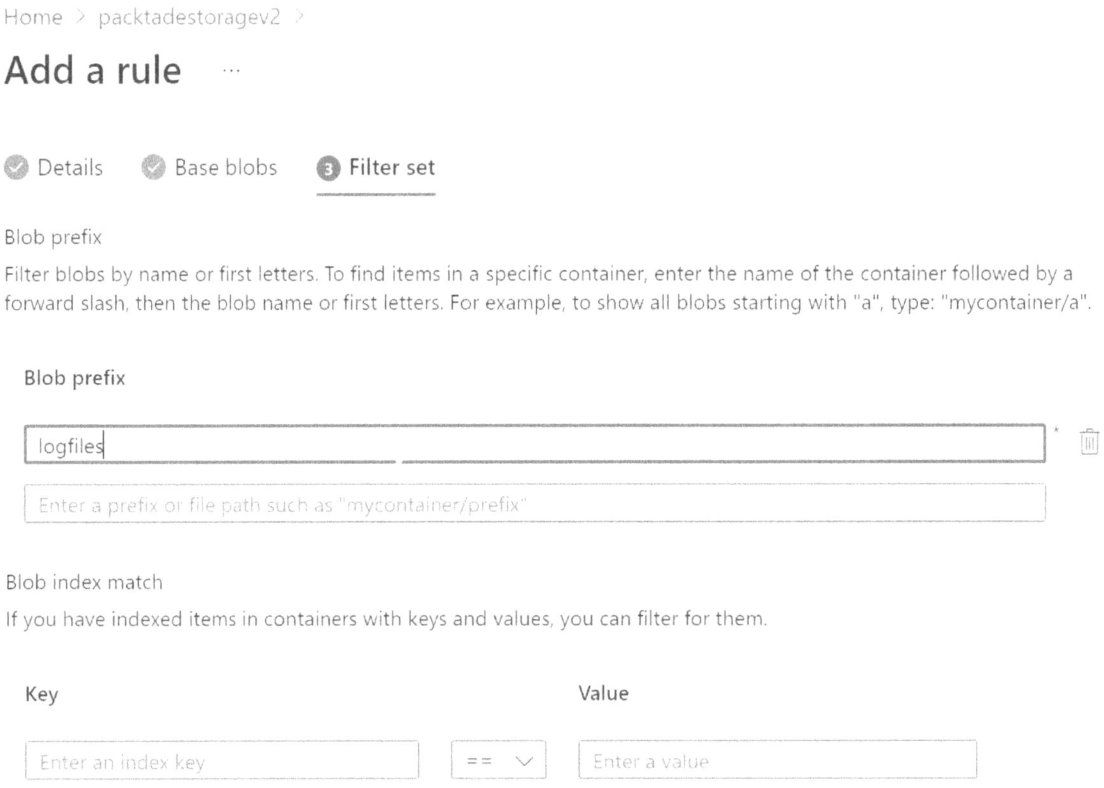

Home > packtadestoragev2 >

Add a rule ...

✅ Details   ✅ Base blobs   ③ **Filter set**

Blob prefix

Filter blobs by name or first letters. To find items in a specific container, enter the name of the container followed by a forward slash, then the blob name or first letters. For example, to show all blobs starting with "a", type: "mycontainer/a".

Blob prefix

logfiles

Enter a prefix or file path such as "mycontainer/prefix"

Blob index match

If you have indexed items in containers with keys and values, you can filter for them.

Key		Value
Enter an index key	== ∨	Enter a value

Figure 1.19 – Lifecycle management – reviewing and adding

How it works...

A blob lifecycle management rule helps in managing storage costs by modifying the access tier of blobs as per the specified rule. Consider a log processing application that reads the log file from Azure Storage, analyzes it, and saves the result in a database. As the log file is read and processed, it may not be needed any further. Therefore, moving it to a Cool access tier from a Hot access tier will save on storage costs.

Blob lifecycle management helps in automating the access tier modification as per the application requirement and is, therefore, a must-have for any storage-based application.

## 2

Securing and Monitoring Data in Azure Data Lake

Data Lake forms the key storage layer for data engineering pipelines. Security and the monitoring of Data Lake accounts are key aspects of Data Lake maintenance. This chapter will focus on configuring security controls such as firewalls, encryption, and creating private links to a Data Lake account. By the end of this chapter, you will have learned how to configure a firewall, virtual network, and private link to secure the Data Lake, encrypt Data Lake using Azure Key Vault, and monitor key user actions in Data Lake.

We will be covering the following recipes in this chapter:

- Configuring a firewall for an Azure Data Lake account using the Azure portal
- Configuring virtual networks for an Azure Data Lake account using the Azure portal
- Configuring private links for an Azure Data Lake account
- Configuring encryption using Azure Key Vault for Azure Data Lake
- Accessing Blob storage accounts using managed identities
- Creating an alert to monitor an Azure Data Lake account
- Securing an Azure Data Lake account with an SAS using PowerShell

Configuring a firewall for an Azure Data Lake account using the Azure portal

Data Lake account access can be restricted to an IP or a range of IPs by whitelisting the allowed IPs in the storage account firewall. In this recipe, we'll learn to restrict access to a Data Lake account using a firewall.

Getting ready

Before you start, perform the following steps:

1. Open a web browser and go to the Azure portal at `https://portal.azure.com`.

2. Make sure you have an existing storage account. If not, create one using the *Provisioning an Azure storage account using the Azure portal* recipe in *Chapter 1, Creating and Managing Data in Azure Data Lake*.

How to do it...

To provide access to an IP or range of IPs, follow these steps:

1. In the Azure portal, locate and open the Azure storage account. In our case, the storage account is **packtadestoragev2**, created in the *Provisioning an Azure storage account using the Azure portal* recipe of *Chapter 1, Creating and Managing Data in Azure Data Lake*.

2. On the storage account page, in the **Security + Networking** section, locate and select **Firewalls and virtual networks**.

    As the **packtadestoragev2** account was created with public access, it can be accessed from all networks.

3. To allow access from an IP or an IP range, click on the **Selected networks** option on the storage account on the **Firewalls and virtual networks** page:

Figure 2.1 – Azure Storage – Firewalls and virtual networks

4. In the **Selected networks** option, scroll down to the **Firewall** section. To give access to your machine only, select the **Add your client IP** address option. To give access to a different IP or range of IPs, type in the IPs in the **Address range** section:

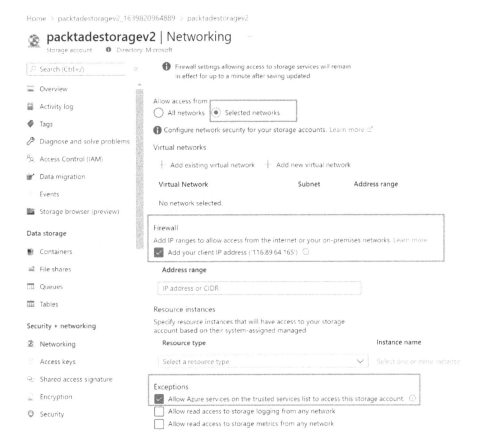

Figure 2.2 – The whitelist IPs in the Azure Storage Firewall section

5. To access storage accounts from Azure services such as Azure Data Factory and Azure Functions, check **Allow Azure services on the trusted services list to access this storage account** under the **Exceptions** heading.

6. Click **Save** to save the configuration changes.

How it works...

Firewall settings are used to restrict access to an Azure storage account to an IP or range of IPs. Even if a storage account is public, it will only be accessible to the whitelisted IPs defined in the firewall configuration.

Configuring virtual networks for an Azure Data Lake account using the Azure portal

A storage account can be public which is accessible to everyone, public with access to an IP or range of IPs, or private with access to selected virtual networks. In this recipe, we'll learn how to restrict access to an Azure storage account in a virtual network.

Getting ready

Before you start, perform the following steps:

1. Open a web browser and go to the Azure portal at `https://portal.azure.com`.

2. Make sure you have an existing storage account. If not, create one using the *Provisioning an Azure storage account using the Azure portal* recipe in *Chapter 1, Creating and Managing Data in Azure Data Lake*.

How to do it...

To restrict access to a virtual network, follow the given steps:

1. In the Azure portal, locate and open the storage account. In our case, it's **packtadestoragev2**. On the storage account page, in the **Security + Network** section, locate and select **Firewalls and virtual networks | Selected networks**:

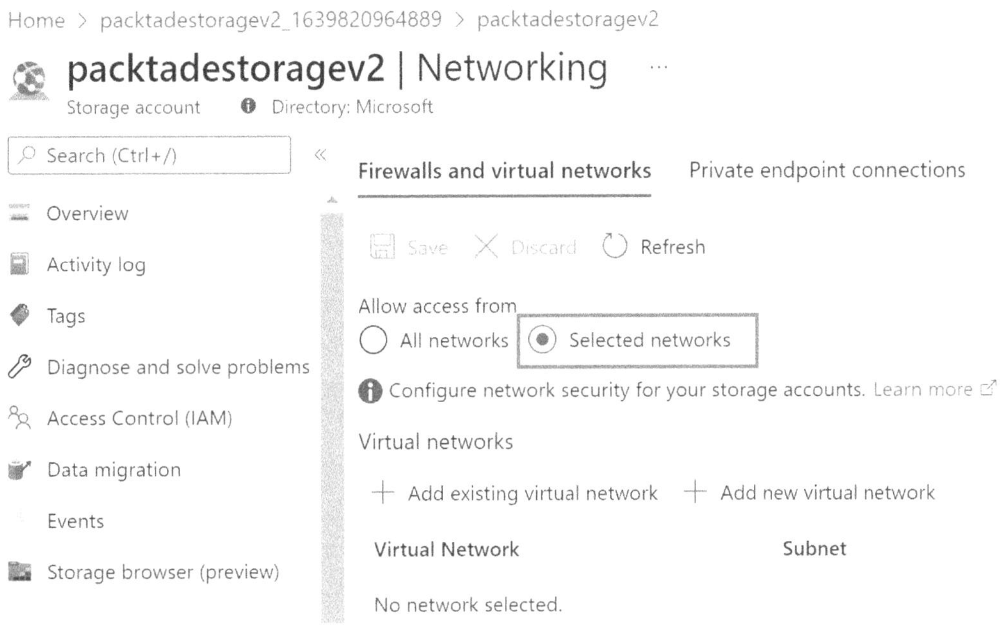

Home > packtadestoragev2_1639820964889 > packtadestoragev2
packtadestoragev2 | Networking
Storage account
Directory: Microsoft
Search (Ctrl+/)
Overview
Activity log
Tags
Diagnose and solve problems
Access Control (IAM)
Data migration
Events
Storage browser (preview)
Firewalls and virtual networks
Private endpoint connections
Save
Discard
Refresh
Allow access from
All networks
Selected networks
Configure network security for your storage accounts. Learn more
Virtual networks
Add existing virtual network
Add new virtual network
Virtual Network
Subnet
No network selected.

Figure 2.3 – Azure Storage – Selected networks

2. In the **Virtual networks** section, select **+ Add new virtual network**:

Home > packtadestoragev2_1639820964889 > packtadestoragev2
packtadestoragev2 | Networking
Storage account
Directory: Microsoft
Search (Ctrl+/)
Overview
Activity log
Tags
Diagnose and solve problems
Access Control (IAM)
Data migration
Events
Storage browser (preview)
Firewalls and virtual networks
Private endpoint connections
Save
Discard
Refresh
Allow access from
All networks
Selected networks
Configure network security for your storage accounts. Learn more
Virtual networks
Add existing virtual network
Add new virtual network
Virtual Network
Subnet

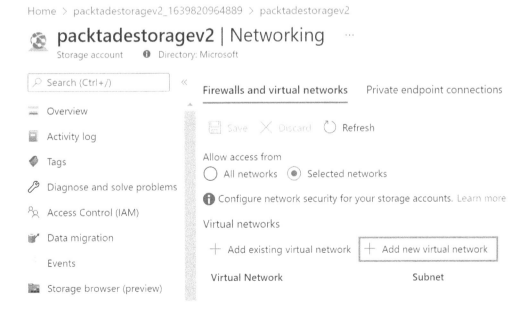

Figure 2.4 – Adding a virtual network

3. In the **Create virtual network** blade, provide the virtual network name, **Address space** details, and **Subnet** address range. The remainder of the configuration values are pre-filled, as shown in the following screenshot:

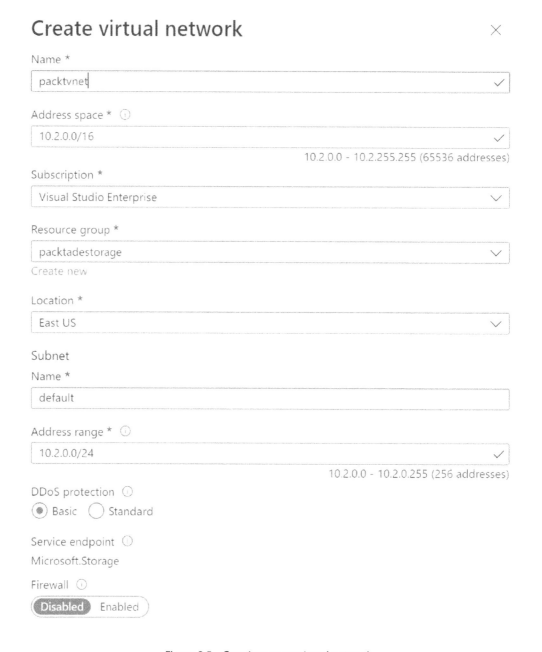

Figure 2.5 – Creating a new virtual network

4.   Click on **Create** to create the virtual network. This is created and listed in the **Virtual Network** section, as shown in the following screenshot:

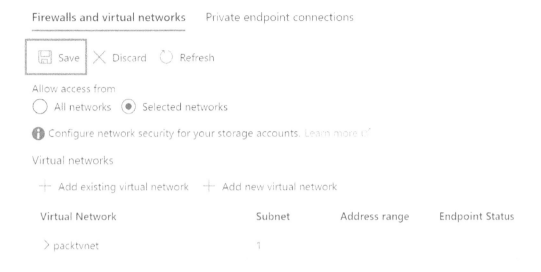

Figure 2.6 – Saving a virtual network configuration

5.   Click **Save** to save the configuration changes.

How it works...

We first created an Azure virtual network and then added it to the Azure storage account. Creating the Azure virtual network from the storage account page automatically fills in the resource group, location, and subscription information. The virtual network and the storage account should be in the same location.

The address space specifies the number of IP addresses in a given virtual network.

We also need to define the subnet within the virtual network that the storage account will belong to. We can also create a custom subnet. In our case, for the sake of simplicity, we have used the default subnet.

This allows the storage account to only be accessed by resources that belong to the given virtual network. The storage account is inaccessible to any network other than the specified virtual network.

Configuring private links for an Azure Data Lake account

In this recipe, we will be creating a private link to a storage account and using private endpoints to connect to it.

Private links and private endpoints ensure that all communication to the storage account goes through the Azure backbone network. Communications to the storage account don't use a public internet network, which makes them very secure.

Getting ready

Before you start, perform the following steps:

1. Open a web browser and go to the Azure portal at `https://portal.azure.com`.

2. Make sure you have an existing storage account. If not, create one using the *Provisioning an Azure storage account using the Azure Portal* recipe in *Chapter 1, Creating and Managing Data in Azure Data Lake.*

3. Make sure you have an existing virtual network configured to the storage account. If not, create one using the *Configuring virtual networks for an Azure Data Lake account using the Azure portal* recipe in this chapter.

How to do it...

Perform the following steps to configure private links to a Data Lake account:

1. Log in to the Azure portal and click on the storage account.

2. Click on **Networking** | the **Private Endpoints** tab.

3. Click on the + **Private endpoint** button, as shown here:

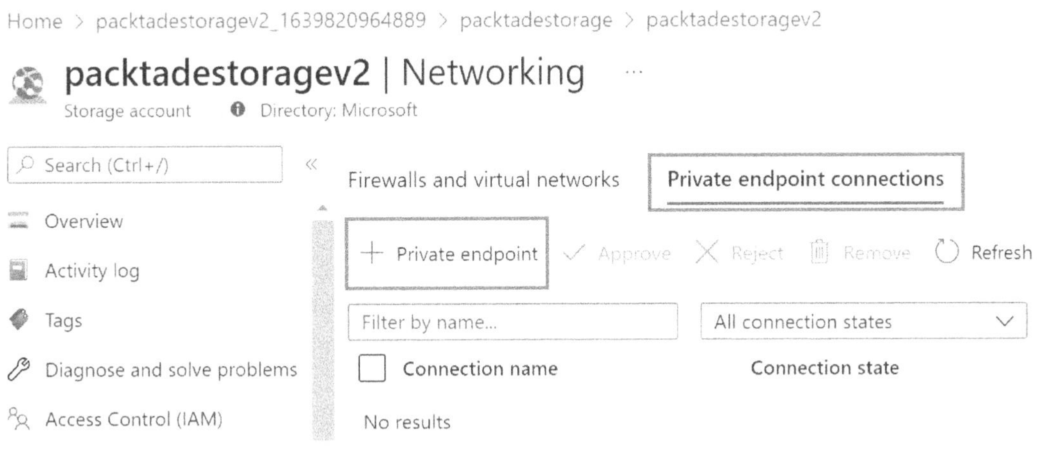

Home > packtadestoragev2_1639820964889 > packtadestorage > packtadestoragev2

⚙ packtadestoragev2 | Networking ...
Storage account   ⓘ Directory: Microsoft

🔍 Search (Ctrl+/) ≪	Firewalls and virtual networks	**Private endpoint connections**

- ▤ Overview
- 🖫 Activity log
- 🏷 Tags
- 🖉 Diagnose and solve problems
- 👥 Access Control (IAM)

＋ Private endpoint   ✓ Approve   ✗ Reject   🗑 Remove   ↻ Refresh

Filter by name...   All connection states   ∨

☐ Connection name   Connection state

No results

Figure 2.7 – Creating a private endpoint to a storage account

4. Provide an endpoint name, as shown in the following screenshot:

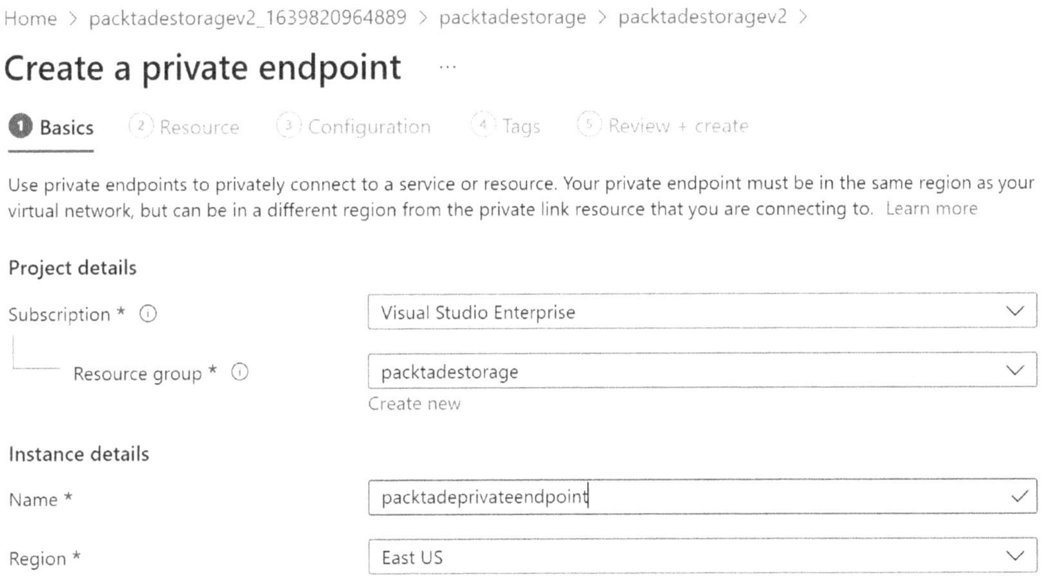

Home > packtadestoragev2_1639820964889 > packtadestorage > packtadestoragev2 >

Create a private endpoint   ...

① Basics   ② Resource   ③ Configuration   ④ Tags   ⑤ Review + create

Use private endpoints to privately connect to a service or resource. Your private endpoint must be in the same region as your virtual network, but can be in a different region from the private link resource that you are connecting to. Learn more

Project details

Subscription * ⓘ	Visual Studio Enterprise	∨

Resource group * ⓘ	packtadestorage	∨

Create new

Instance details

Name *	packtadeprivateendpoint	✓

Region *	East US	∨

Figure 2.8 – Providing an endpoint name

5. In the **Resource** tab, set **Target sub-resource** to **dfs**. **Distributed File Systems (DFS)** is sub-source if we are connecting to Data Lake Storage Gen2. The rest of the fields are auto-populated. Proceed to the **Configuration** section:

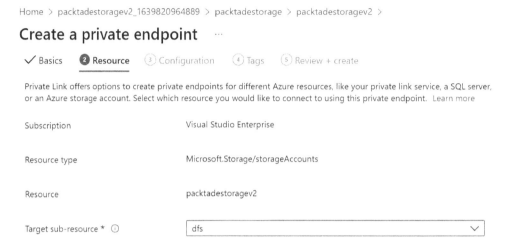

Figure 2.9 – Setting the target resource type to dfs

6. Create a private **Domain Name System (DNS)** zone by picking the same resource group where you created the storage account, as shown in the following screenshot:

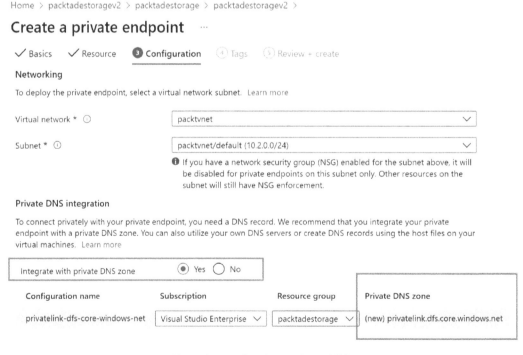

Figure 2.10 – Creating a private DNS

7. Hit the **Create** button to create the private DNS link.

8. After the private endpoint is created, open it in the Azure portal. Click on **DNS configuration**:

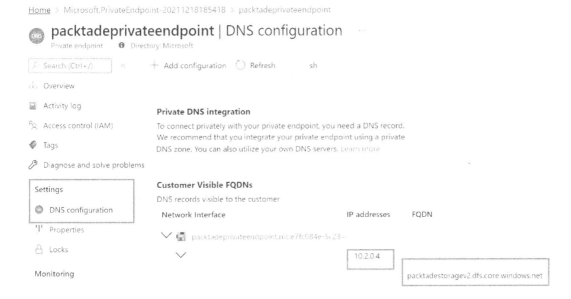

Figure 2.11 – Copy the FQD9

- Make a note of the **FQDN** and **IP addresses** details. The **FQDN** is the **Fully Qualified Domain Name**, which will resolve to the private IP address if, and only if, you are connected to the virtual network.

With the preceding steps, we have created a private endpoint that will use private links to connect to a storage account.

How it works...

We have created a private link to a storage account and ensured that traffic goes through the Microsoft backbone network (and not the public internet), as we will be accessing the storage account via a private endpoint. To show how it works, let's resolve the private URL link from the following locations. Let's perform the following:

- Use `nslookup` to look up a private URL link from your local machine.

- Use `nslookup` to look up a private URL link from a virtual machine inside the virtual network.

On your machine, open Command Prompt and type `nslookup <FQDN of private link>`, as shown in the following screenshot:

```
C:\Users\navenkat>nslookup packtadestoragev2.dfs.core.windows.net
Server:   UnKnown
Address:   2404:e801:2000:50d:3223:3ff:fec4:c8e2

Non-authoritative answer:
Name:     dfs.mnz22prdstr03a.store.core.windows.net
Address:   20.60.128.226
Aliases:  packtadestoragev2.dfs.core.windows.net
          packtadestoragev2.privatelink.dfs.core.windows.net
```

Figure 2.12 – Testing a private endpoint connection outside of the virtual network

`nslookup` resolves the private link to an incorrect IP address, as your machine is not part of the virtual network. To see it working, perform the following instructions:

1. Create a new virtual machine in the Azure portal. Ensure to allow a remote desktop connection to the virtual machine, as shown in the following screenshot:

Inbound port rules

Select which virtual machine network ports are accessible from the public internet. You can specify more limited or granular network access on the Networking tab.

Public inbound ports * ⓘ   ◯ None
                            ◉ Allow selected ports

Select inbound ports *   RDP (3389)   ⌄

⚠ **This will allow all IP addresses to access your virtual machine.** This is only recommended for testing. Use the Advanced controls in the Networking tab to create rules to limit inbound traffic to known IP addresses.

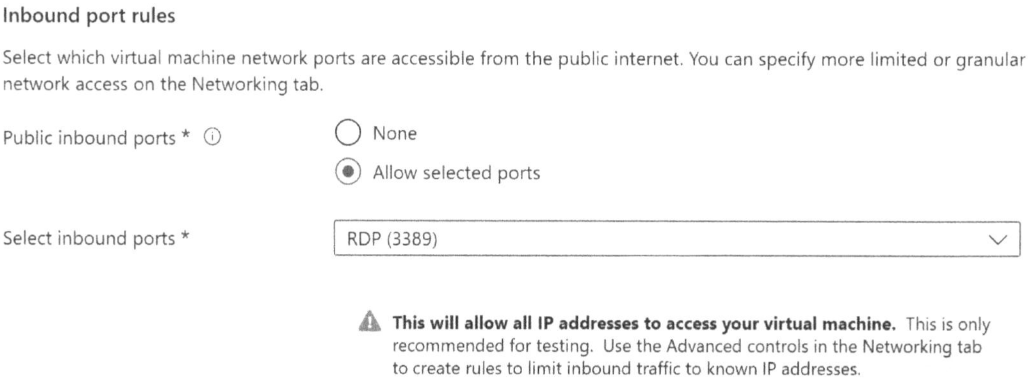

Figure 2.13 – Creating a new virtual machine and allowing a remote desktop

2. Under **Networking**, select the virtual network in which the storage account resides:

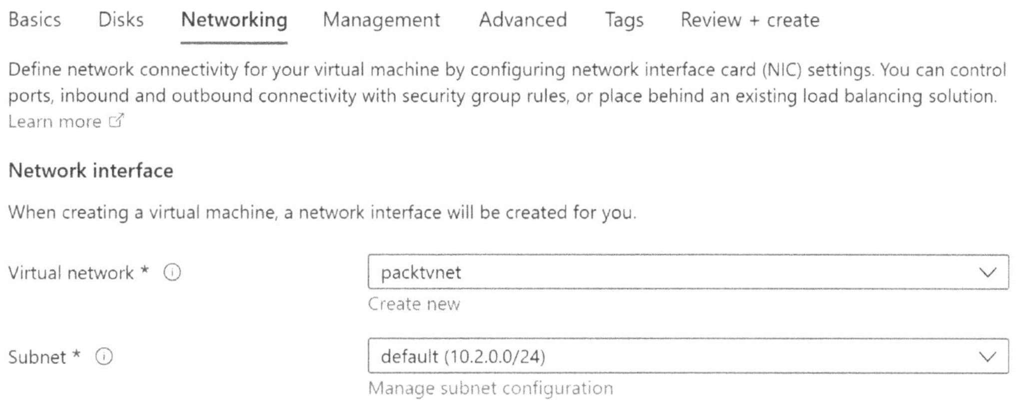

Basics   Disks   **Networking**   Management   Advanced   Tags   Review + create

Define network connectivity for your virtual machine by configuring network interface card (NIC) settings. You can control ports, inbound and outbound connectivity with security group rules, or place behind an existing load balancing solution. Learn more ⍿

Network interface

When creating a virtual machine, a network interface will be created for you.

Virtual network * ⓘ	packtvnet ∨
	Create new
Subnet * ⓘ	default (10.2.0.0/24) ∨
	Manage subnet configuration

Figure 2.14 – Configuring the virtual machine to use the virtual network

Once the virtual machine is created, log in to the virtual machine using a remote desktop and perform nslookup to look up the private link URL again to resolve its IP address. nslookup is a command that will resolve an URL to an IP address. We will use nslookup to verify whether the private link URL resolves to a private IP address (10.x.x.x) and not a public IP address.

nslookup from a virtual machine inside the virtual network resolves correctly to the private IP address of the private link, as shown in the following screenshot. This shows that the connection goes through a virtual network only and doesn't use public internet:

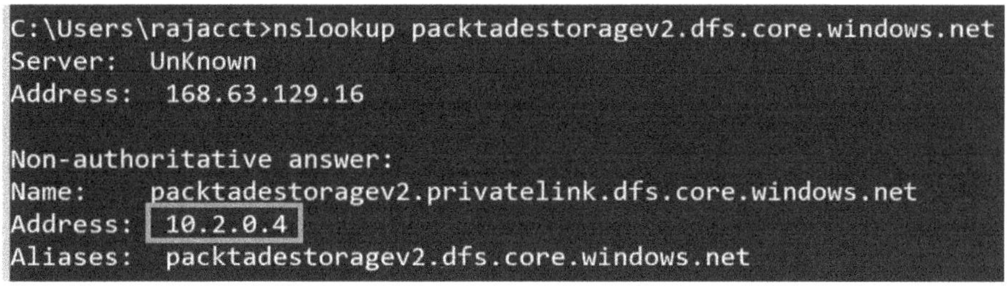

```
C:\Users\rajacct>nslookup packtadestoragev2.dfs.core.windows.net
Server:    UnKnown
Address:   168.63.129.16

Non-authoritative answer:
Name:      packtadestoragev2.privatelink.dfs.core.windows.net
Address:   10.2.0.4
Aliases:   packtadestoragev2.dfs.core.windows.net
```

Figure 2.15 – nslookup from the virtual network

With the previous recipe, we have successfully created a private link to a storage account, configured a private endpoint connection, and accessed it via a virtual machine to verify the connectivity. This recipe covers how you can securely connect to a storage account through virtual networks only by passing a public network.

Configuring encryption using Azure Key Vault for Azure Data Lake

In this recipe, we will create a key vault and use it to encrypt an Azure Data Lake account.

Azure Data Lake accounts are encrypted at rest by default using Azure managed keys. However, you have the option of bringing your own key to encrypt an Azure Data Lake account. Using your own key gives better control over encryption.

Getting ready

Before you start, perform the following steps:

1. Open a web browser and go to the Azure portal at `https://portal.azure.com`.

2. Make sure that you have an existing storage account. If not, create one using the *Provisioning an Azure storage account using the Azure portal* recipe in *Chapter 1, Creating and Managing Data in Azure Data Lake*.

How to do it...

Perform the following steps to add encryption to a Data Lake account using Azure Key Vault:

1. Log in to `portal.azure.com`, click on **Create a resource**, search for `Key Vault`, and click on **Create**. Provide the key vault details, as shown in the following screenshot. Click on **Review + Create**:

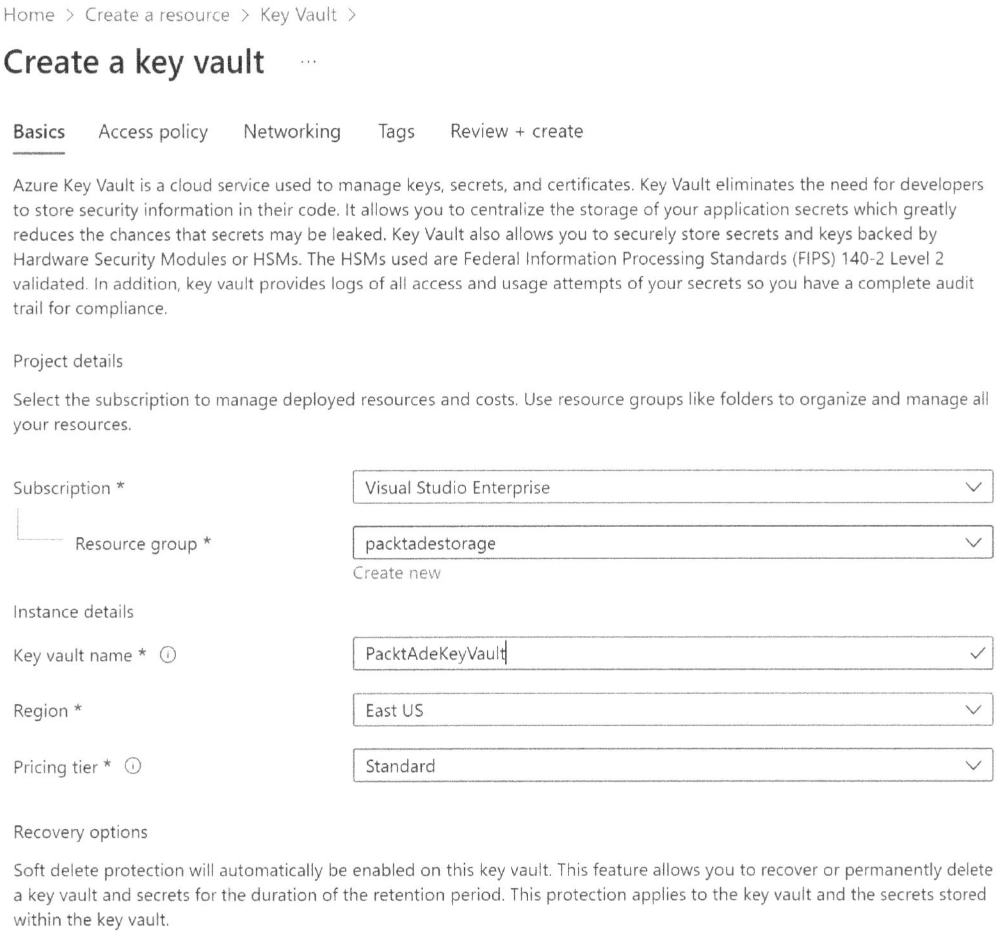

Home > Create a resource > Key Vault >

Create a key vault   ...

Basics   Access policy   Networking   Tags   Review + create

Azure Key Vault is a cloud service used to manage keys, secrets, and certificates. Key Vault eliminates the need for developers to store security information in their code. It allows you to centralize the storage of your application secrets which greatly reduces the chances that secrets may be leaked. Key Vault also allows you to securely store secrets and keys backed by Hardware Security Modules or HSMs. The HSMs used are Federal Information Processing Standards (FIPS) 140-2 Level 2 validated. In addition, key vault provides logs of all access and usage attempts of your secrets so you have a complete audit trail for compliance.

Project details

Select the subscription to manage deployed resources and costs. Use resource groups like folders to organize and manage all your resources.

Subscription *	Visual Studio Enterprise
Resource group *	packtadestorage
	Create new

Instance details

Key vault name * ⓘ	PacktAdeKeyVault
Region *	East US
Pricing tier * ⓘ	Standard

Recovery options

Soft delete protection will automatically be enabled on this key vault. This feature allows you to recover or permanently delete a key vault and secrets for the duration of the retention period. This protection applies to the key vault and the secrets stored within the key vault.

Figure 2.16 – Creating an Azure key vault

2.   Go to the storage account to be encrypted. Search for Encryption on the left. Click on
     Encryption and select **Customer-managed keys** as the **Encryption type**. Click on **Select a
     key vault and key** at the bottom:

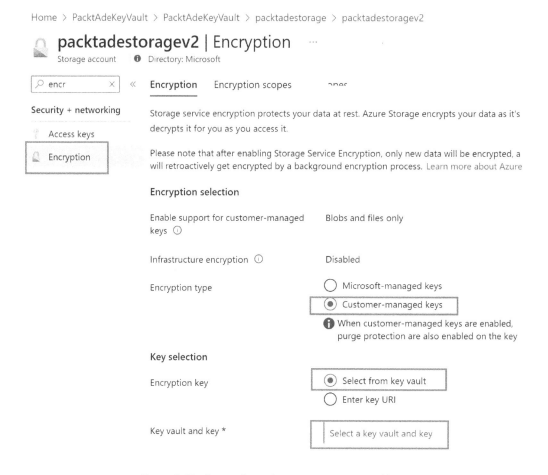

Figure 2.17 – Encrypting using customer-managed keys

3.   On the new screen, **Select a key**, select **Key vault** as **Key store type** and select the newly created
     PacktAdeKeyVault as **Key vault**. Click on **Create new key**, as shown in the following screenshot:

Home > PacktAdeKeyVault > PacktAdeKeyVault > packtadestorage > packtadestoragev2 >

Select a key   ...

Subscription *	Visual Studio Enterprise ∨
Key store type ⓘ	⦿ Key vault
	◯ Managed HSM
Key vault *	PacktAdeKeyVault ∨
	Create new key vault
Key *	∨
	Create new key

Figure 2.18 – Selecting Key Vault

4. Provide a name for the key to be used for encryption of the storage account. The default option, **Generate**, ensures that the key is generated automatically. Click on **Create**:

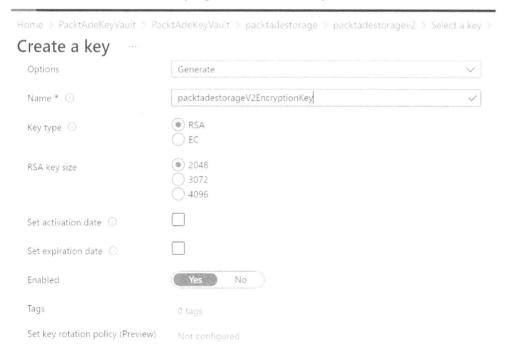

Home > PacktAdeKeyVault > PacktAdeKeyVault > packtadestorage > packtadestoragev2 > Select a key >

Create a key   ...

Options	Generate ∨
Name * ⓘ	packtadestorageV2EncryptionKey ✓
Key type ⓘ	⦿ RSA
	◯ EC
RSA key size	⦿ 2048
	◯ 3072
	◯ 4096
Set activation date ⓘ	☐
Set expiration date ⓘ	☐
Enabled	**Yes** No
Tags	0 tags
Set key rotation policy (Preview)	Not configured

Figure 2.19 – Creating a key

5. Once the key is created, the screen automatically moves to the key vault selection page in the Blob storage, and the newly created key is selected as the key. Click on **Select**:

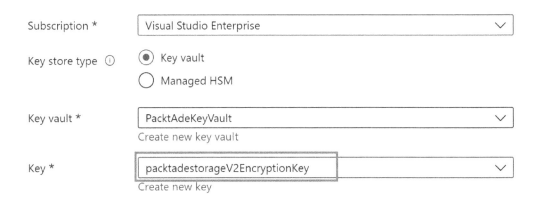

Home > PacktAdeKeyVault > PacktAdeKeyVault > packtadestorage > packtadestoragev2 >

Select a key   ...

ⓘ The key 'packtadestorageV2EncryptionKey' has been successfully created.

Subscription * | Visual Studio Enterprise   ⌄

Key store type ⓘ   ◉ Key vault
                     ◯ Managed HSM

Key vault * | PacktAdeKeyVault   ⌄
               Create new key vault

Key * | packtadestorageV2EncryptionKey   ⌄
         Create new key

Figure 2.20 – Selecting the key

6. The screen moves to the encryption page on the Blob storage page. Click on **Save** to complete the encryption configuration.

How it works...

As the newly created key vault has been set for encryption on an Azure Data Lake account, all Data Lake operations (`read`, `write`, and `metadata`) will use the key from Key Vault to encrypt and decrypt the data in Data Lake. The encryption and decryption operations are fully transparent and have no impact on users' operations.

The Data Lake account automatically gets permissions on the key vault to extract the key and perform encryption on data. You can verify this by *opening the key vault* in the Azure portal and clicking on **Access Policies**. Note that the storage account has been granted **Get**, **wrap**, and **unwrap** permissions on the keys, as shown in the next screenshot:

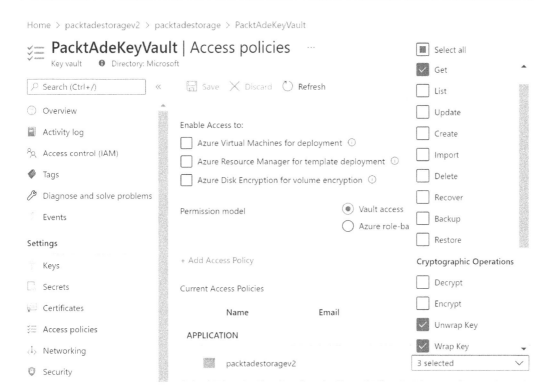

Figure 2.21 – Storage account permissions in Key Vault

Accessing Blob storage accounts using managed identities

In this recipe, we will grant permissions to managed identities on a storage account and showcase how you can use managed identities to connect to Azure Data Lake.

Managed identities are password-less service accounts used by Azure services such as Data Factory and Azure VMs to access other Azure services, such as Blob storage. In this recipe, we will show you how Azure Data Factory's managed identity can be granted permission on an Azure Blob storage account.

Getting ready

Before you start, perform the following steps:

1. Open a web browser and go to the Azure portal at `https://portal.azure.com`.

2. Make sure you have an existing storage account. If not, create one using the *Provisioning an Azure storage account using the Azure portal* recipe in *Chapter 1*, *Creating and Managing Data in Azure Data Lake*.

How to do it...

We will be testing accessing a Data Lake account using managed identities. To achieve this, we will create a Data Factory account and use Data Factory's managed identity to access the Data Lake account. Perform the following steps to test this:

1. Create an Azure Data Factory by using the following PowerShell command:

    ```
    $resourceGroupName = " packtadestorage";
    $location = 'east us'
    $dataFactoryName = "ADFPacktADE2";
    $DataFactory = Set-AzDataFactoryV2 -ResourceGroupName
    $resourceGroupName -Location $location -Name
    $dataFactoryName
    ```

2. Go to the storage account in the Azure portal. Click on **Access Control (IAM)** and then **Add**, as shown in the following screenshot:

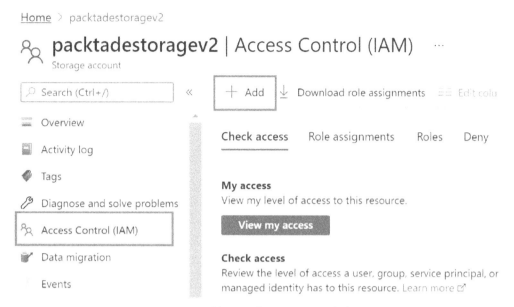

Figure 2.22 – Adding a role to a managed identity

3. Select **Add role assignment** and search for the `Storage Blob Data Contributor` role. Select the role and click **Next**. Select **Managed identity** in **Assign access to** and click on + **Select members**, as shown in the following screenshot:

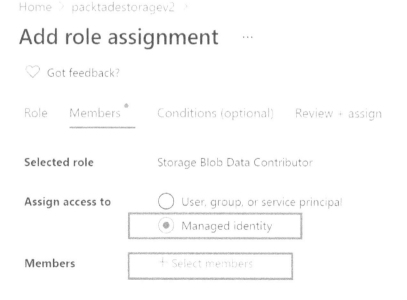

Figure 2.23 – Selecting the Data Factory managed identity

4. Your subscription should be selected by default. From the **Managed identity** dropdown, select **Data Factory (V2) (1)**. Select the recently created **ADFPacktADE2** Data Factory and click on the **Select** button:

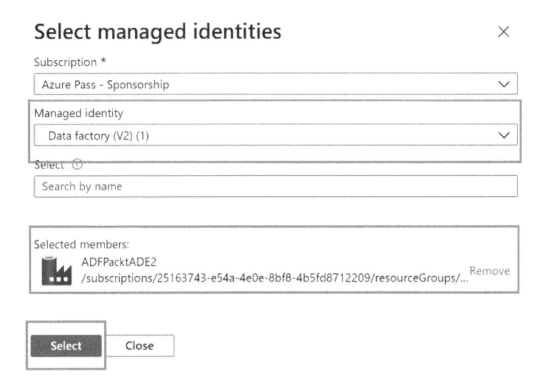

Figure 2.24 – Assigning a role to a managed identity

5. Click on **Review + Assign** to complete the assignment. To test whether it's working, open the **ADFPacktADE2** Data Factory that was created in *step 1*. Click on **Open Azure Data Factory Studio**, as shown in the next screenshot:

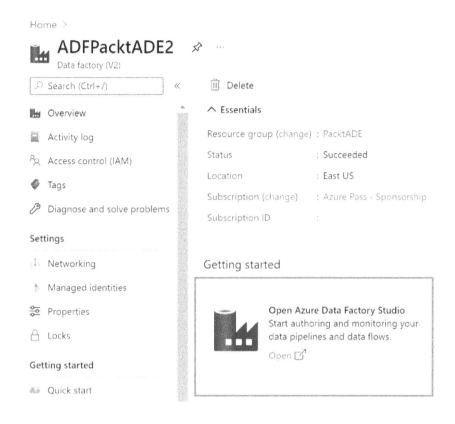

Figure 2.25 – Opening Azure Data Factory Studio

6. Click on the **Manage** button on the left and then **Linked services**. Click on + **New**, as shown in the following screenshot:

Figure 2.26 – Creating a linked service in Data Factory

7. Search for Data Lake and select **Azure Data Lake Storage Gen 2** as the data store. Select **Managed Identity** for **Authentication method**. Select the storage account (**packadestoragev2**) for **Storage account name**. Click on **Test connection**:

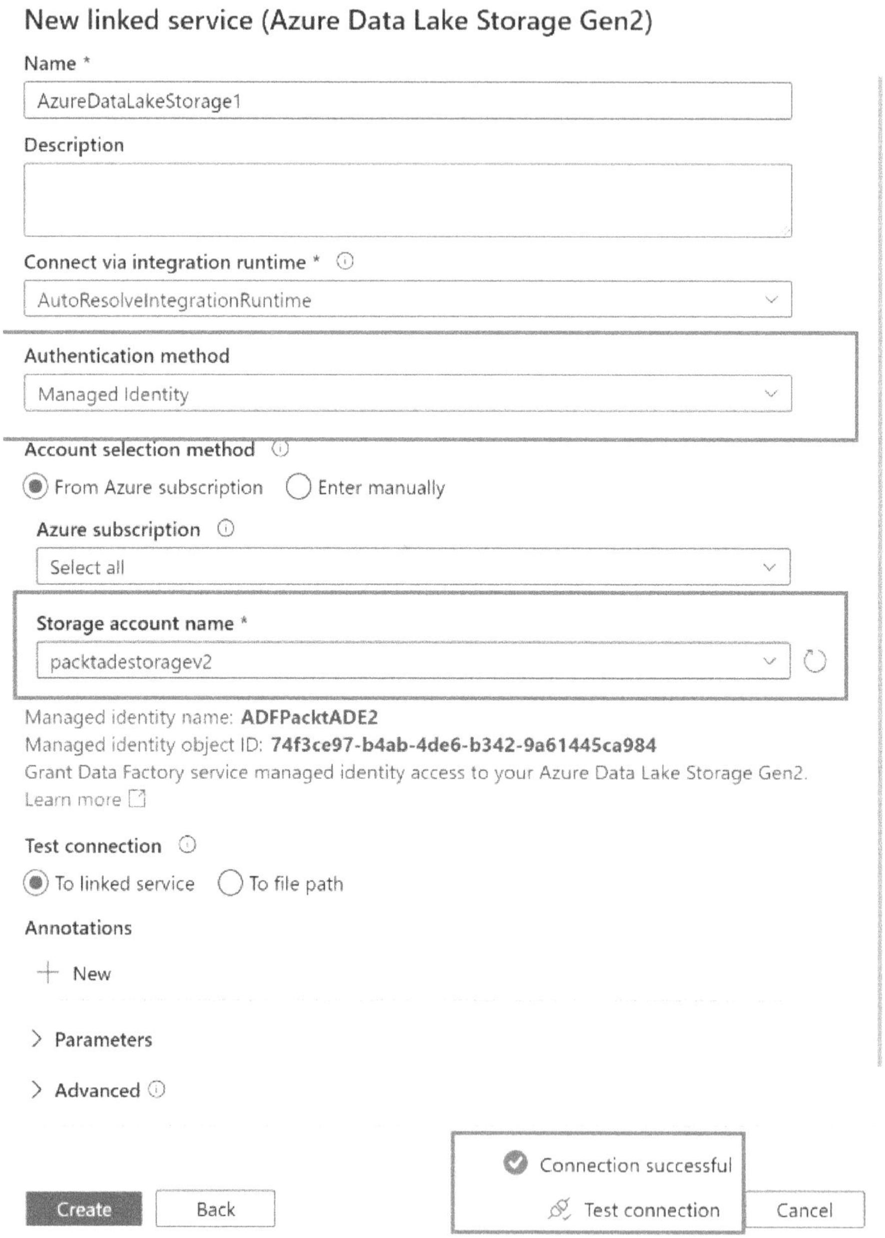

New linked service (Azure Data Lake Storage Gen2)

Name *

AzureDataLakeStorage1

Description

Connect via integration runtime * ⓘ

AutoResolveIntegrationRuntime   ∨

Authentication method

Managed Identity   ∨

Account selection method ⓘ

◉ From Azure subscription   ◯ Enter manually

Azure subscription ⓘ

Select all   ∨

Storage account name *

packtadestoragev2   ∨   ↻

Managed identity name: **ADFPacktADE2**
Managed identity object ID: **74f3ce97-b4ab-4de6-b342-9a61445ca984**
Grant Data Factory service managed identity access to your Azure Data Lake Storage Gen2.
Learn more ☐

Test connection ⓘ

◉ To linked service   ◯ To file path

Annotations

+ New

> Parameters

> Advanced ⓘ

```
                                        ✓ Connection successful
  Create        Back                    ✎  Test connection       Cancel
```

Figure 2.27 – Testing a managed identity connection in Data Factory

A successful test connection indicates that we can successfully connect to a storage account using a managed identity.

How it works...

A managed identity for the data factory was automatically created when the Data Factory account was created. We provided the **Storage Blob Data Contributor** permission on the Azure Data Lake storage account to the managed identity of Data Factory. Hence, Data Factory was successfully able to connect to the storage account in a secure way without using a key/password.

Creating an alert to monitor an Azure storage account

We can create an alert on multiple available metrics to monitor an Azure storage account. To create an alert, we need to define the trigger condition and the action to be performed when the alert is triggered. In this recipe, we'll create an alert to send an email if the used capacity metrics for an Azure storage account exceed 5 MB. The used capacity threshold of 5 MB is not a standard and is deliberately kept low to explain the alert functionality.

Getting ready

Before you start, perform the following steps:

1. Open a web browser and log in to the Azure portal at `https://portal.azure.com`.

2. Make sure you have an existing storage account. If not, create one using the *Provisioning an Azure storage account using the Azure portal* recipe in *Chapter 1, Creating and Managing Data in Azure Data Lake*.

How to do it...

Follow these steps to create an alert:

1. In the Azure portal, locate and open the storage account. In our case, the storage account is **packtadestoragev2**. On the storage account page, search for `alert` and open **Alerts** in the **Monitoring** section:

Figure 2.28 – Selecting Alerts

2. On the **Alerts** page, click on + **New alert rule**:

Figure 2.29 – Adding a new alert

3. On the **Alerts | Create alert rule** page, observe that the storage account is listed by default in the **Resource** section. You can add multiple storage accounts in the same alert. Under the **Condition** section, click **Add condition**:

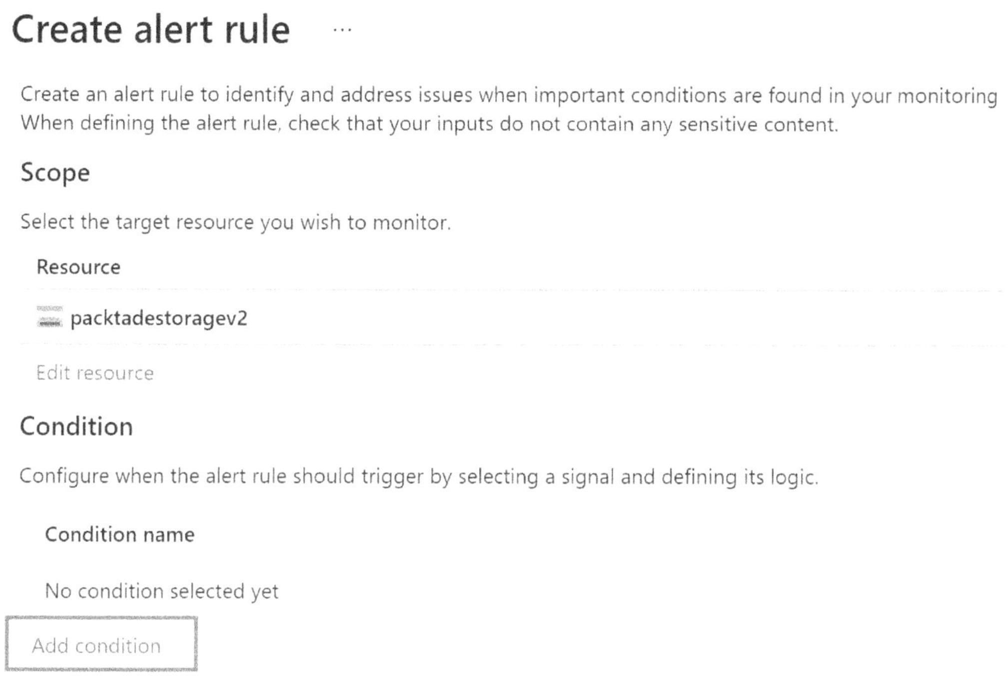

Create alert rule
Create an alert rule to identify and address issues when important conditions are found in your monitoring
When defining the alert rule, check that your inputs do not contain any sensitive content.
Scope
Select the target resource you wish to monitor.
Resource
packtadestoragev2
Edit resource
Condition
Configure when the alert rule should trigger by selecting a signal and defining its logic.
Condition name
No condition selected yet
Add condition

Figure 2.30 – Adding a new alert condition

4. On the **Configure signal logic** page, select **Used capacity** under **Signal name**:

Configure signal logic
Choose a signal below and configure the logic on the next screen
Signal type
All
Displaying 1 - 19 signals out of total 19 signals
Search by signal name
Signal name
Custon log search
Used capacity

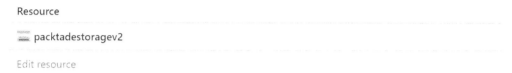

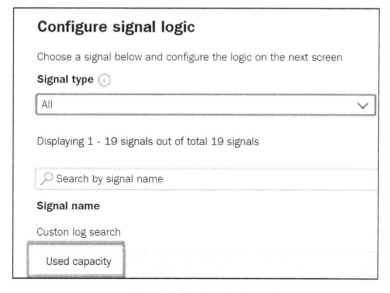

Figure 2.31 – Configuring the signal logic

5. On the **Configure signal logic** page, under **Alert logic**, set **Operator** to **Greater than**, **Aggregation type** to **Average**, and configure the threshold to **5 MiB**. We need to provide the value in bytes:

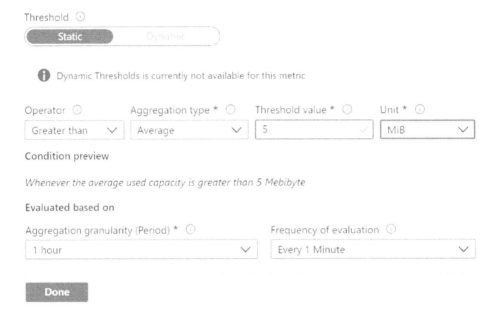

Figure 2.32 – Configuring alert logic

Click **Done** to configure the trigger. The condition is added, and we'll be taken back to the **Create alert rule** page:

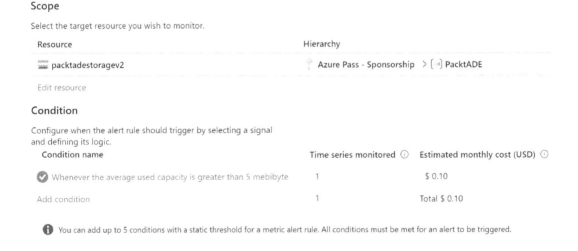

Figure 2.33 – Viewing a new alert condition

6. The next step is to add an action to perform when the alert condition is reached. On the **Create alert rule** page, in the **ACTIONS GROUPS** section, click **Create**:

Figure 2.34 – Creating a new alert action group

7. On the **Add action group** page, provide the **Action group name**, **Display name**, and **Resource group** details:

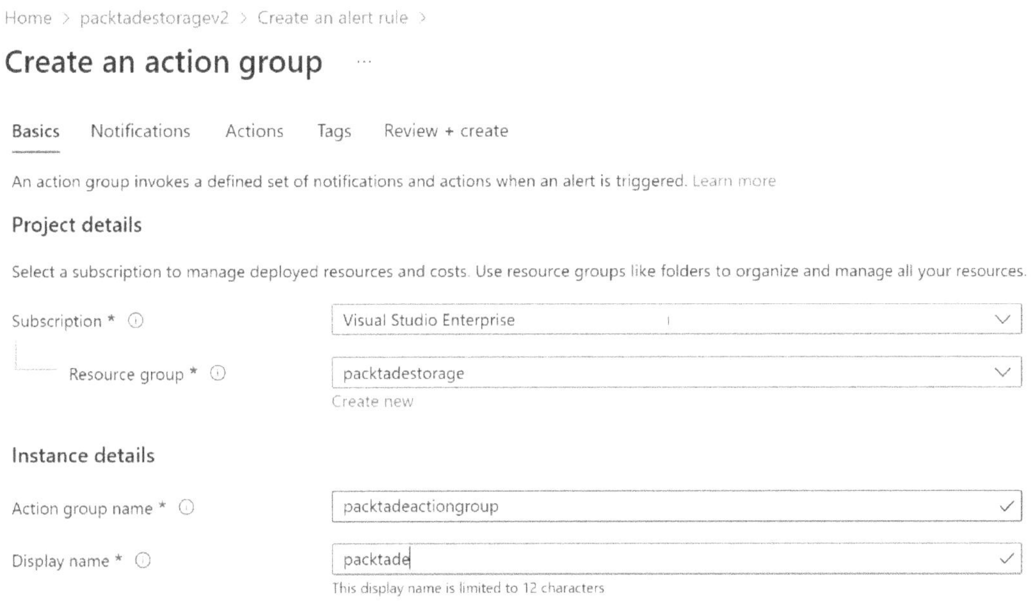

Home > packtadestoragev2 > Create an alert rule >

Create an action group   ...

Basics   Notifications   Actions   Tags   Review + create

An action group invokes a defined set of notifications and actions when an alert is triggered. Learn more

Project details

Select a subscription to manage deployed resources and costs. Use resource groups like folders to organize and manage all your resources.

Subscription * ⓘ	Visual Studio Enterprise	∨
Resource group * ⓘ	packtadestorage	∨
	Create new	

Instance details

Action group name * ⓘ	packtadeactiongroup	✓
Display name * ⓘ	packtade	✓
	This display name is limited to 12 characters	

Figure 2.35 – Adding a new alert action group

8. In **Notifications**, provide an email address. Click on **Review + Create**:

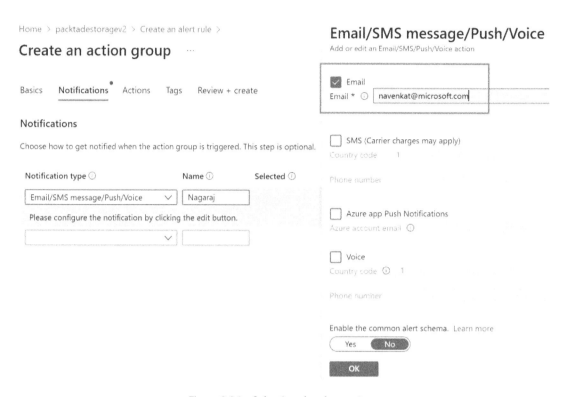

Figure 2.36 – Selecting the alert action

9. Click on **Create** to create the action group. We are then taken back to the **Create rule** page. The **Email** action is listed in the **Action Groups** section.

10. The next step is to define the **Severity**, **Alert rule name**, and **Alert rule description** details:

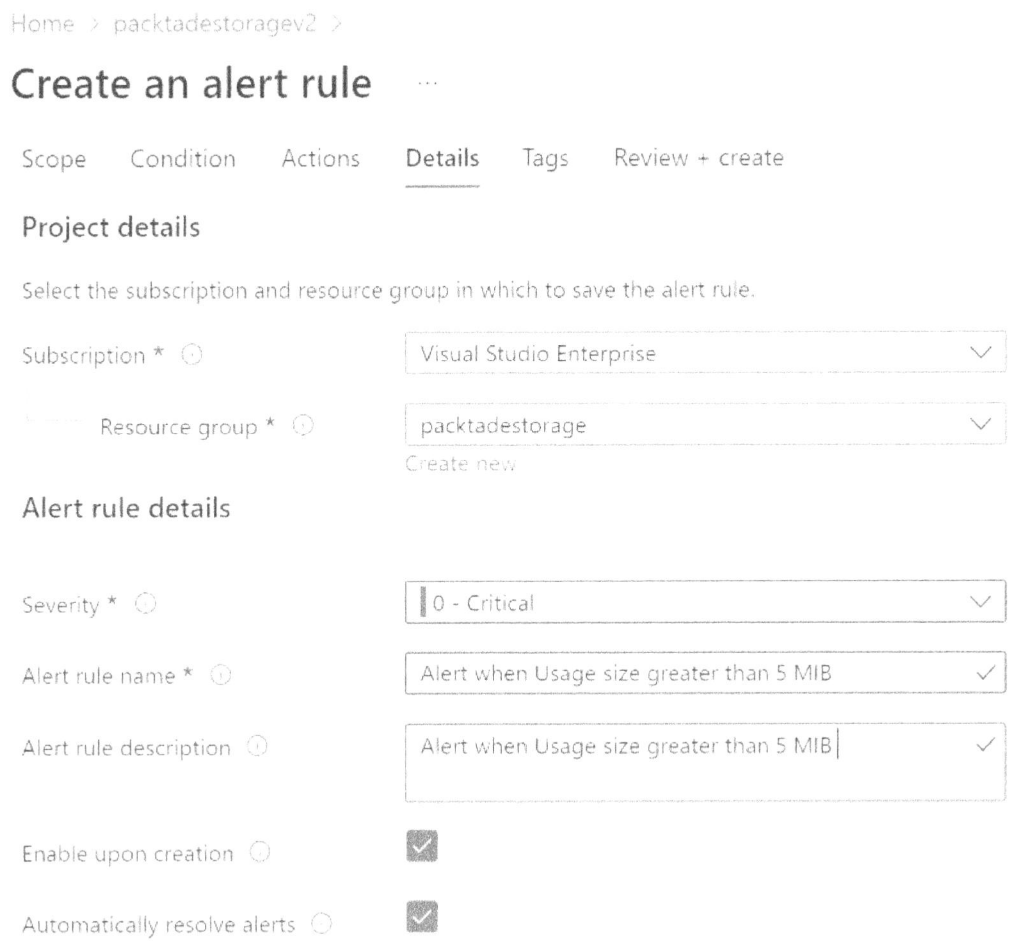

Home > packtadestoragev2 >

Create an alert rule   ...

Scope   Condition   Actions   **Details**   Tags   Review + create

Project details

Select the subscription and resource group in which to save the alert rule.

Subscription * ⓘ   Visual Studio Enterprise   ∨

Resource group * ⓘ   packtadestorage   ∨
                               Create new

Alert rule details

Severity * ⓘ   ❘ 0 - Critical   ∨

Alert rule name * ⓘ   Alert when Usage size greater than 5 MIB   ✓

Alert rule description ⓘ   Alert when Usage size greater than 5 MIB❘   ✓

Enable upon creation ⓘ   ☑

Automatically resolve alerts ⓘ ☑

Figure 2.37 – Creating an alert rule

11. Click the **Create alert rule** button to create the alert.

12. The next step is to trigger the alert. To do that, download `BigFile.csv` from the `https://github.com/PacktPublishing/Azure-Data-Engineering-Cookbook-2nd-edition/blob/main/Chapter2/BigFile.csv` file to the Azure storage account by following the steps mentioned in the *Creating containers and uploading files to Azure Blob storage using PowerShell* recipe of *Chapter 1, Creating and Managing Data in Azure Data Lake*. The triggered alerts are listed on the **Alerts** page, as shown in the following screenshot:

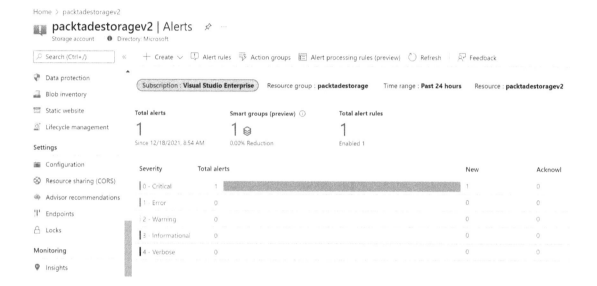

Figure 2.38 – Viewing alerts

13. An email is sent to the email ID specified in the email action group. The email appears as shown in the following screenshot:

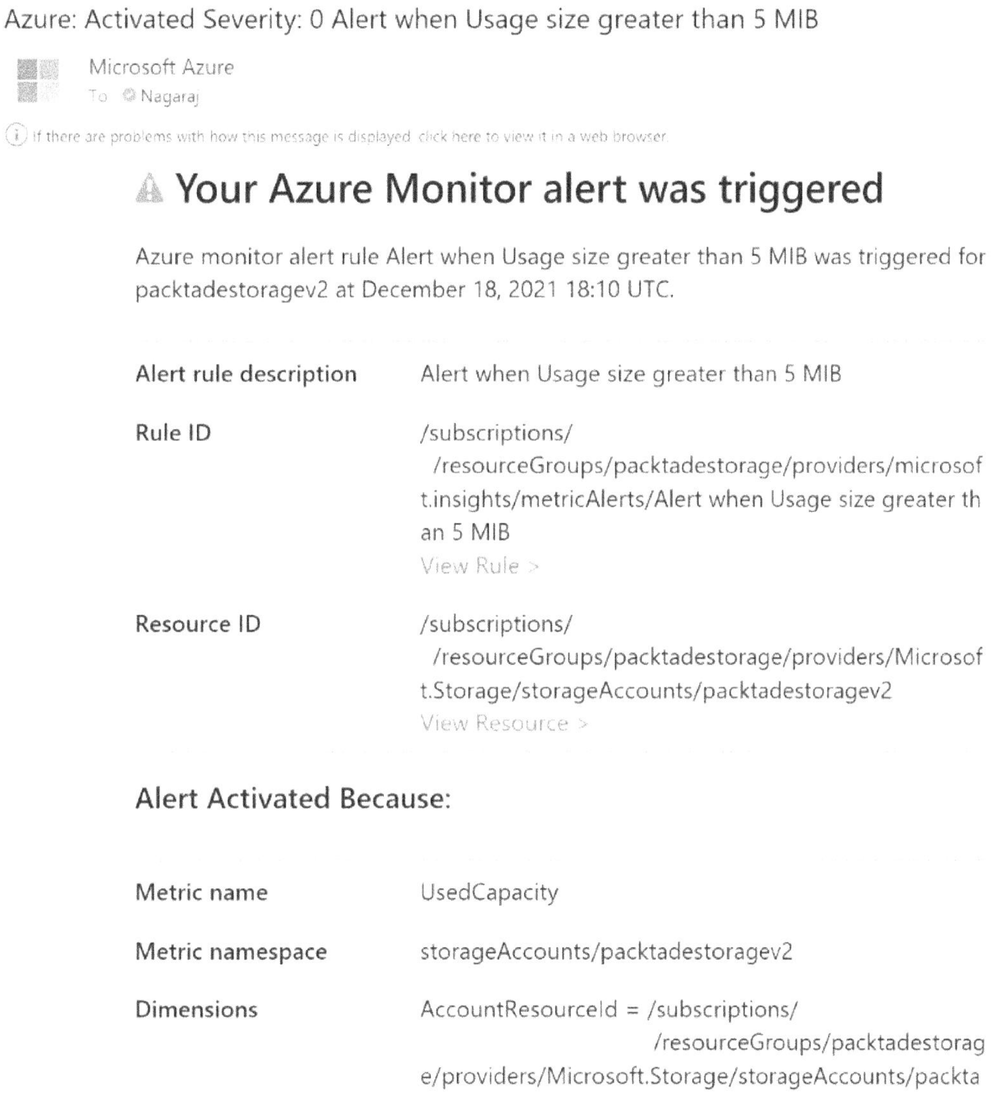

Azure: Activated Severity: 0 Alert when Usage size greater than 5 MIB

Microsoft Azure
To ☺ Nagaraj

ⓘ If there are problems with how this message is displayed, click here to view it in a web browser

⚠ Your Azure Monitor alert was triggered

Azure monitor alert rule Alert when Usage size greater than 5 MIB was triggered for packtadestoragev2 at December 18, 2021 18:10 UTC.

Alert rule description	Alert when Usage size greater than 5 MIB
Rule ID	/subscriptions/ /resourceGroups/packtadestorage/providers/microsoft.insights/metricAlerts/Alert when Usage size greater than 5 MIB View Rule >
Resource ID	/subscriptions/ /resourceGroups/packtadestorage/providers/Microsoft.Storage/storageAccounts/packtadestoragev2 View Resource >

Alert Activated Because:

Metric name	UsedCapacity
Metric namespace	storageAccounts/packtadestoragev2
Dimensions	AccountResourceId = /subscriptions/ /resourceGroups/packtadestorage/providers/Microsoft.Storage/storageAccounts/packta

Figure 2.39 – The alert email

How it works...

Setting up an alert is easy. At first, we need to define the alert condition (a trigger or signal). An alert condition defines the metrics and threshold that, when breached, trigger the alert. We can define more than one condition on multiple metrics for one alert.

We then need to define the action to be performed when the alert condition is reached. We can define more than one action for an alert. In our example, in addition to sending an email when the used capacity is more than 5 MB, we can configure Azure Automation to delete the old blobs/files in order to maintain the Azure storage capacity within 5 MB.

There are other signals, such as transactions, ingress, egress, availability, Success Server Latency, and Success E2E Latency, on which alerts can be defined. Detailed information on monitoring Azure storage is available at `https://docs.microsoft.com/en-us/azure/storage/common/storage-monitoring-diagnosing-troubleshooting`.

Securing an Azure storage account with SAS using PowerShell

A **Shared Access Signature** (**SAS**) provides more granular access to blobs by specifying an expiry limit, specific permissions, and IPs.

Using an SAS, we can specify different permissions to users or applications on different blobs, based on the requirement. For example, if an application needs to read one file/blob from a container, instead of providing access to all the files in the container, we can use an SAS to provide read access on the required blob.

In this recipe, we'll learn to create and use an SAS to access blobs.

Getting ready

Before you start, go through the following steps:

1. Make sure you have an existing Azure storage account. If not, create one by following the *Provisioning an Azure storage account using PowerShell* recipe in *Chapter 1, Creating and Managing Data in Azure Data Lake*.

2. Make sure you have an existing Azure storage container. If not, create one by following the *Creating containers and uploading files to Azure Blob storage using PowerShell* recipe.

3. Make sure you have existing blobs/files in an Azure storage container. If not, you can upload blobs by following the previous recipe.

4. Log in to your Azure subscription in PowerShell. To log in, run the `Connect- AzAccount` command in a new PowerShell window and follow the instructions.

How to do it...

Let's begin by securing blobs using an SAS.

Securing blobs using an SAS

Perform the following steps:

1. Execute the following command in the PowerShell window to get the storage context:

    ```
    $resourcegroup = "packtadestorage"


    $storageaccount = "packtadestoragev2"


    #get storage context


    $storagecontext = (Get-AzStorageAccount
    -ResourceGroupName $resourcegroup -Name $storageaccount).
    Context
    ```

2. Execute the following commands to get the SAS token for the `logfile1.txt` blob in the `logfiles` container with list and read permissions:

    ```
    #set the token expiry time
    $starttime = Get-Date
    $endtime = $starttime.AddDays(1)
    # get the SAS token into a variable
    $sastoken = New-AzStorageBlobSASToken -Container
    "logfiles" -Blob "logfile1.txt" -Permission lr -StartTime
    $starttime -ExpiryTime $endtime -Context $storagecontext
    # view the SAS token.
      $sastoken
    ```

3. Execute the following commands to list the blob using the SAS token:

    ```
    #get storage account context using the SAS token
    $ctx = New-AzStorageContext -StorageAccountName
    $storageaccount -SasToken $sastoken
    ```

```
#list the blob details
Get-AzStorageBlob -blob "logfile1.txt" -Container
"logfiles" -Context $ctx
```

You should get output as shown in the following screenshot:

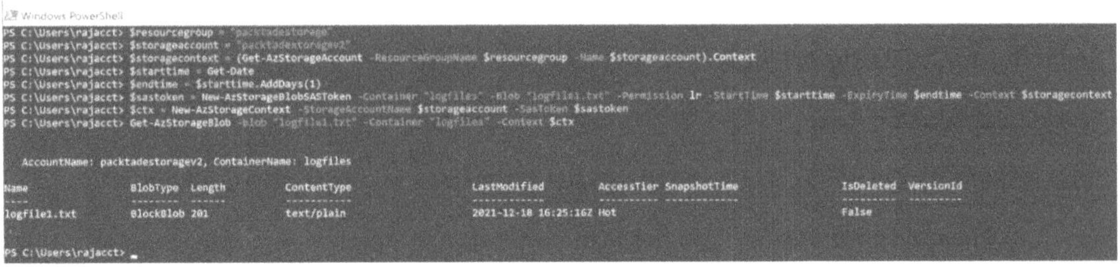

Figure 2.40 – Listing blobs using an SAS

4. Execute the following command to write data to logfile1.txt. Ensure you have the
    Logfile1.txt file in the C:\ADECookbook\Chapter1\ Logfiles\ folder in the
    machine you are running the script from:

```
Set-AzStorageBlobContent -File C:\ADECookbook\Chapter1\
Logfiles\Logfile1.txt -Container logfiles -Context $ctx
```

You should get output as shown in the following screenshot:

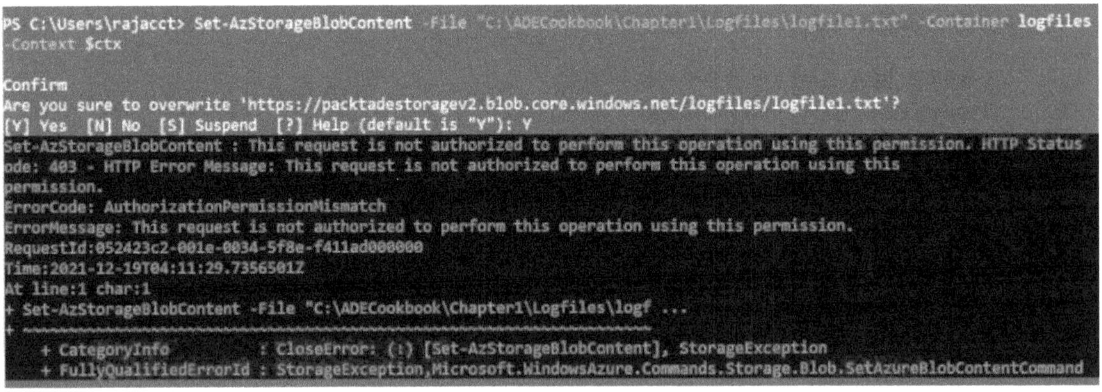

Figure 2.41 – Uploading a blob using an SAS

The write fails, as the SAS token was created with list and read access.

Securing a container with an SAS

Perform the following steps:

1. Execute the following command to create a container stored access policy:

    ```
    $resourcegroup = "packtadestorage"


    $storageaccount = "packtadestoragev2"


    #get storage context
    $storagecontext = (Get-AzStorageAccount
    -ResourceGroupName $resourcegroup -Name $storageaccount).
    Context
    $starttime = Get-Date
    $endtime = $starttime.AddDays(1)
    New-AzStorageContainerStoredAccessPolicy -Container
    logfiles -Policy writepolicy -Permission lw -StartTime
    $starttime -ExpiryTime $endtime -Context $storagecontext
    ```

2. Execute the following command to create the SAS token:

    ```
    #get the SAS token
    $sastoken = New-AzStorageContainerSASToken -Name logfiles
    -Policy writepolicy -Context
    ```

3. Execute the following commands to list all the blobs in the container using the SAS token:

    ```
    #get the storage context with SAS token
    $ctx = New-AzStorageContext -StorageAccountName
    $storageaccount -SasToken $sastoken
    #list blobs using SAS token
    Get-AzStorageBlob -Container logfiles -Context $ctx
    ```

How it works...

To generate a shared access token for a blob, use the `New-AzStorageBlobSASToken` command. We need to provide the blob name, container name, permission (`l` = list, `r` = read, and `w` = write), and storage context to generate an SAS token. We can additionally secure the token by providing IPs that can access the blob.

We then use the SAS token to get the storage context using the `New-AzStorageContext` command. We use the storage context to access the blobs using the `Get-AzStorageBlob` command. Note that we can only list and read blobs and can't write to them, as the SAS token doesn't have write permissions.

To generate a shared access token for a container, we first create an access policy for the container using the `New-AzStorageContainerStoredAccessPolicy` command. The access policy specifies the start and expiry time, permission, and IPs. We then generate the SAS token by passing the access policy name to the `New-AzStorageContainerSASToken` command.

We can now access the container and the blobs using the SAS token.

## 3

Building Data Ingestion Pipelines Using Azure Data Factory

Azure Data Factory is the data orchestration service in Azure. Using Azure Data Factory, you can build pipelines that are capable of reading data from multiple sources, transforming the data, and loading the data into data stores to be consumed by reporting applications such as Power BI. Azure Data Factory much like **SQL Server Integration Services (SSIS)** in an on-premises world, provides a code-free UI for developing, managing, and maintaining data engineering pipelines.

Azure Data Factory is the bread and butter for a data engineer and understanding its fundamentals is extremely essential in building efficient pipelines. By the end of the chapter, you will know how to provision a data factory account, copy data from an Azure SQL database to a data lake using copy activity, use control flow activities, move data from SQL Server to a data lake, and choose options to trigger a data factory pipeline.

In this chapter, we'll cover the following recipes:

- Provisioning Azure Data Factory

- Copying files to a database from a data lake using a control flow and copy activity

- Triggering a pipeline in Azure Data Factory

- Copying data from a SQL Server virtual machine to a data lake using the Copy data wizard

Technical requirements

For this chapter, you will need the following:

- A Microsoft Azure subscription
- PowerShell 7 and above
- Microsoft Azure PowerShell, and an additional PowerShell module that's required for managing Azure components

Provisioning Azure Data Factory

To get started with Azure Data Factory, you need to run an Azure Data Factory account. An Azure Data Factory account is comprised of the following key components:

- **Linked services**: A component that maintains the connection credentials to data sources. An example of this is a connection to a SQL database/text file.
- **Dataset**: The data that's obtained after connecting to the data source using a linked service. An example of this is a group of tables or files connected via a linked service.
- **Activity**: A task that will process the dataset. An example of this is a copy activity that moves the data from a flat file to a database.
- **Data flow**: These are specific tasks that perform data transformations on datasets. An example of this is pivoting or sorting a dataset while it is being moved from source to destination. This can be done by a data flow transformation task.
- **Integration runtime**: This is the Azure Data Factory engine that works behind the scenes and provides the compute and resources to run the activities or tasks.
- **Pipeline**: A single entity that combines all the aforementioned components to connect, process, transform, and ingest the data to the destination. A single pipeline may contain multiple linked services, datasets, activities, and data flows.

How to do it...

In this recipe, we will be provisioning an Azure Data Factory using the Azure portal. Follow these steps:

1. Log in to portal.azure.com, click on **Create a resource**, and search for Data Factory. Select **Data Factory** and click on **Create**. Provide the data factory name, the resource group name, and location as shown in the following screenshot:

Create Data Factory   ...

Basics   Git configuration   Networking   Advanced   Tags   Review + create

Project details

Select the subscription to manage deployed resources and costs. Use resource groups like folders to organize and manage all your resources.

Subscription * ⓘ

| Visual Studio Enterprise | ∨ |

Resource group * ⓘ

| (New) PacktADEADF | ∨ |
Create new

Instance details

Region * ⓘ

| East US | ∨ |

Name * ⓘ

| PacktADEADF | ∨ |

Version * ⓘ

| V2 (Recommended) | ∨ |

Figure 3.1 – Create Data Factory

2. Click on **Next: Git configuration**. Git configuration allows you to configure integration with Azure DevOps or GitHub. Git integration helps you save data factory pipelines as **Azure Resource Manager** (**ARM**) templates and lets you perform **continuous integration and continuous deployment** (**CI/CD**). For this recipe, we will choose **Configure Git later**:

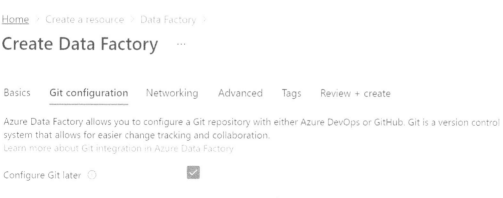

Home > Create a resource > Data Factory >

Create Data Factory   ...

Basics   Git configuration   Networking   Advanced   Tags   Review + create

Azure Data Factory allows you to configure a Git repository with either Azure DevOps or GitHub. Git is a version control system that allows for easier change tracking and collaboration.
Learn more about Git integration in Azure Data Factory

Configure Git later ⓘ   ✓

Figure 3.2 – Git configuration

3. As for the remaining tabs (**Networking/Advanced/Tags**), they can remain as is. Click **Review + create** to create the data factory.

How it works...

It is fairly straightforward to create a data factory instance. The **PacktADEADF** data factory that we created in this recipe will be used to hold several datasets, pipelines, and data sources that are to be created in the following recipes in this chapter.

Copying files to a database from a data lake using a control flow and copy activity

In this recipe, we will be building a pipeline that will copy a group of files in blob storage to Azure SQL Database, but only if the filenames contain today's date as a suffix. Follow these steps:

1. Get the list of files to be copied using the **Get Metadata** activity in the data factory.

2. Use the **Filter** activity to filter the file whose suffix is the current date.

3. Use the **ForEach** activity to loop through the files.

4. Use the **Copy** activity to load the file into Azure SQL Database.

Getting ready

Follow these steps:

1. Provision a data factory, as explained in the *Provisioning Azure Data Factory* recipe of this chapter.

2. Log in to Azure PowerShell using `Connect-AzAccount`.

3. Execute the following command to create a storage account and a container:

```
$storageaccountname="packtadeadfadl"
$resourcegroup="PacktADEADF"
$containername="dataloading"
New-AzStorageAccount -ResourceGroupName $resourcegroup
-Name $storageaccountname -SkuName Standard_LRS -Location
'East US' -Kind StorageV2
$storagecontext = (Get-AzStorageAccount -ResourceGroupName
$resourcegroup -Name $storageaccountname).Context;
New-AzStorageContainer -Name $containername -Context
$storagecontext
```

4. Execute the following script to create a SQL Server database in the same resource group:

```
$resourcegroup="PacktADEADF"
$serverName = "packadeadfsql"
```

```
$adminSqlLogin = "sqladmin"
$password = "SQLPwdW5!k"
$startIp = "0.0.0.0"
$endIp = "255.255.255.255"
$databasename = "sample"
$server = New-AzSqlServer -ResourceGroupName
$resourcegroup -ServerName $serverName -Location "Eastus"
-SqlAdministratorCredentials $(New-Object -TypeName
System.Management.Automation.PSCredential -ArgumentList
$adminSqlLogin, $(ConvertTo-SecureString -String
$password -AsPlainText -Force))
$serverFirewallRule = New-AzSqlServerFirewallRule
-ResourceGroupName $resourcegroup -ServerName $serverName
-FirewallRuleName "AllowedIPs" -StartIpAddress $startIp
-EndIpAddress $endIp
$database = New-AzSqlDatabase   -ResourceGroupName
$resourcegroup -ServerName $serverName -DatabaseName
$databaseName -RequestedServiceObjectiveName "S0"
```

5. Download the `orderdtls-20211118.csv`, `orderdtls-20211119.csv`, and `orderdtls-20211120.csv` files from `https://github.com/PacktPublishing/Azure-Data-Engineering-Cookbook-2nd-edition/tree/main/Chapter3` and upload them to the **dataloading** container in the **packtadeadfadl** storage account using the Azure portal, as shown in the following screenshot:

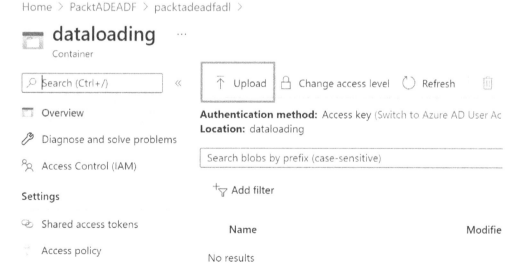

Figure 3.3 – Uploading the necessary files

6.   Select the files to be uploaded and hit the **Upload** button:

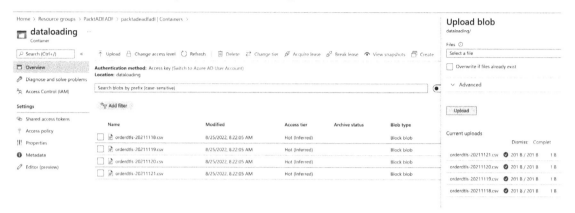

Figure 3.4 – The files that have been uploaded

How to do it...

To copy the files in blob storage that have the current date as the suffix into Azure SQL Database, we will do the following:

1.   Create linked services to connect the blob storage and Azure SQL Database.

2.   Add the **Get Metadata** activity to get a list of files.

3.   Add the **Filter** activity to filter the files with the current date as the suffix.

4.   Add the **Copy** activity to copy the files that have been filtered to Azure SQL Database.

Creating a linked service

First, let's create two connections (linked services) – one for Azure SQL Database and another for blob storage:

1.   In the Azure portal, open the data factory that we provisioned. Click on **Open Azure Data Factory Studio**. Once the **Azure Data Factory Studio** opens, click on the **Manage** button:

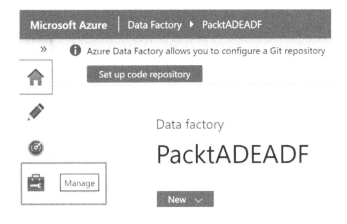

Figure 3.5 – Data Factory Studio – the Manage button

2. Click on **Linked Services**, then **+ New**, and search for data under **Data store**. Select **Azure Data Lake Storage Gen2** and click **Continue**:

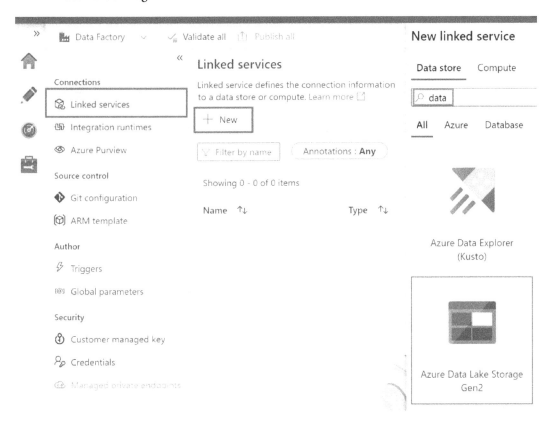

Figure 3.6 – Data Factory Studio – Data store

3.   Create a connection to the storage account, as shown in the following screenshot:

New linked service (Azure Data Lake Storage Gen2)

Name *

DataLoading

Description

Connect via integration runtime * ⓘ

AutoResolveIntegrationRuntime

Authentication method

Account key

Account selection method ⓘ

⦿ From Azure subscription   ◯ Enter manually

Azure subscription ⓘ

Visual Studio Enterprise

Storage account name *

packtadeadfadl

Test connection ⓘ

⦿ To linked service   ◯ To file path

Annotations

+ New

> Parameters

> Advanced ⓘ

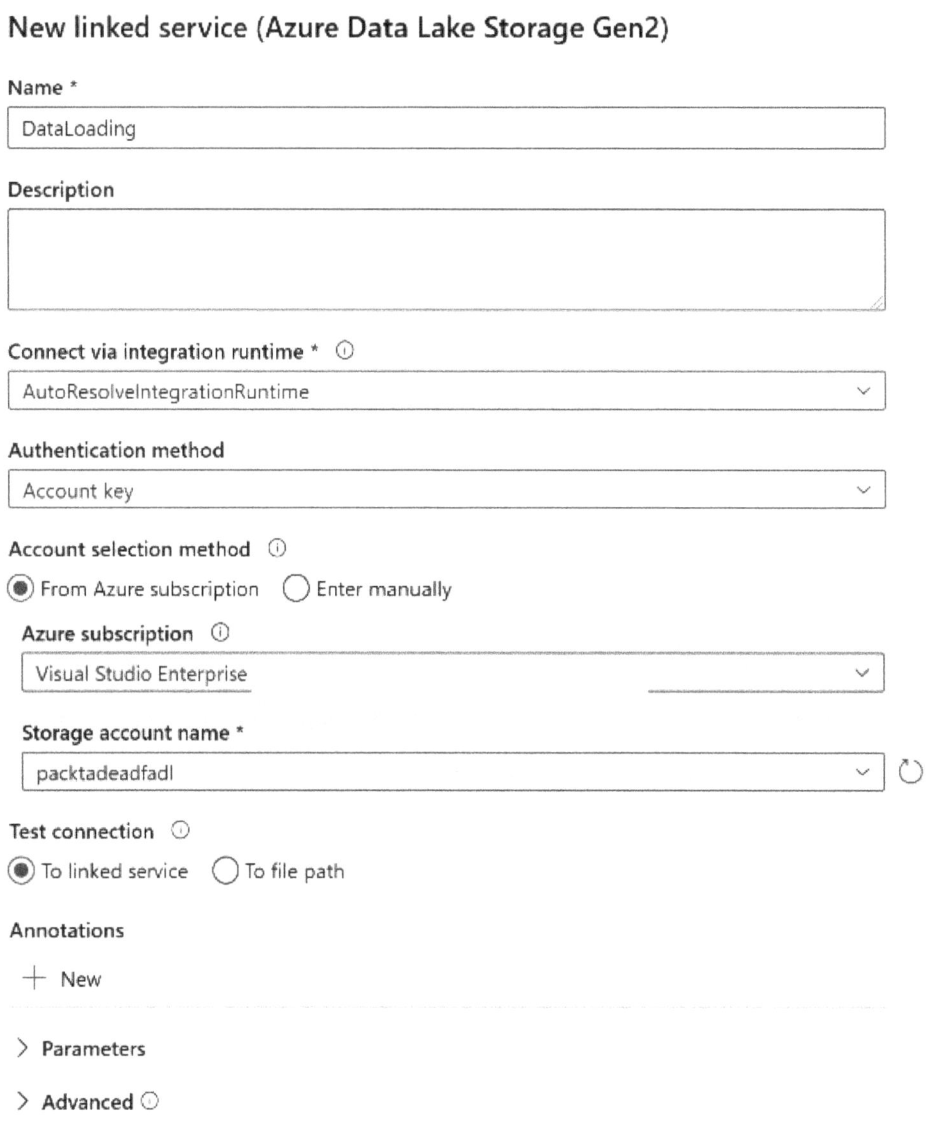

Figure 3.7 – New linked service (Azure Data Lake Storage Gen2)

4.   Similarly, create a linked service for **Azure SQL Database**, as shown in the following screenshot. Set **User name** to sqladmin and **Password** to SQLPwdW5!k:

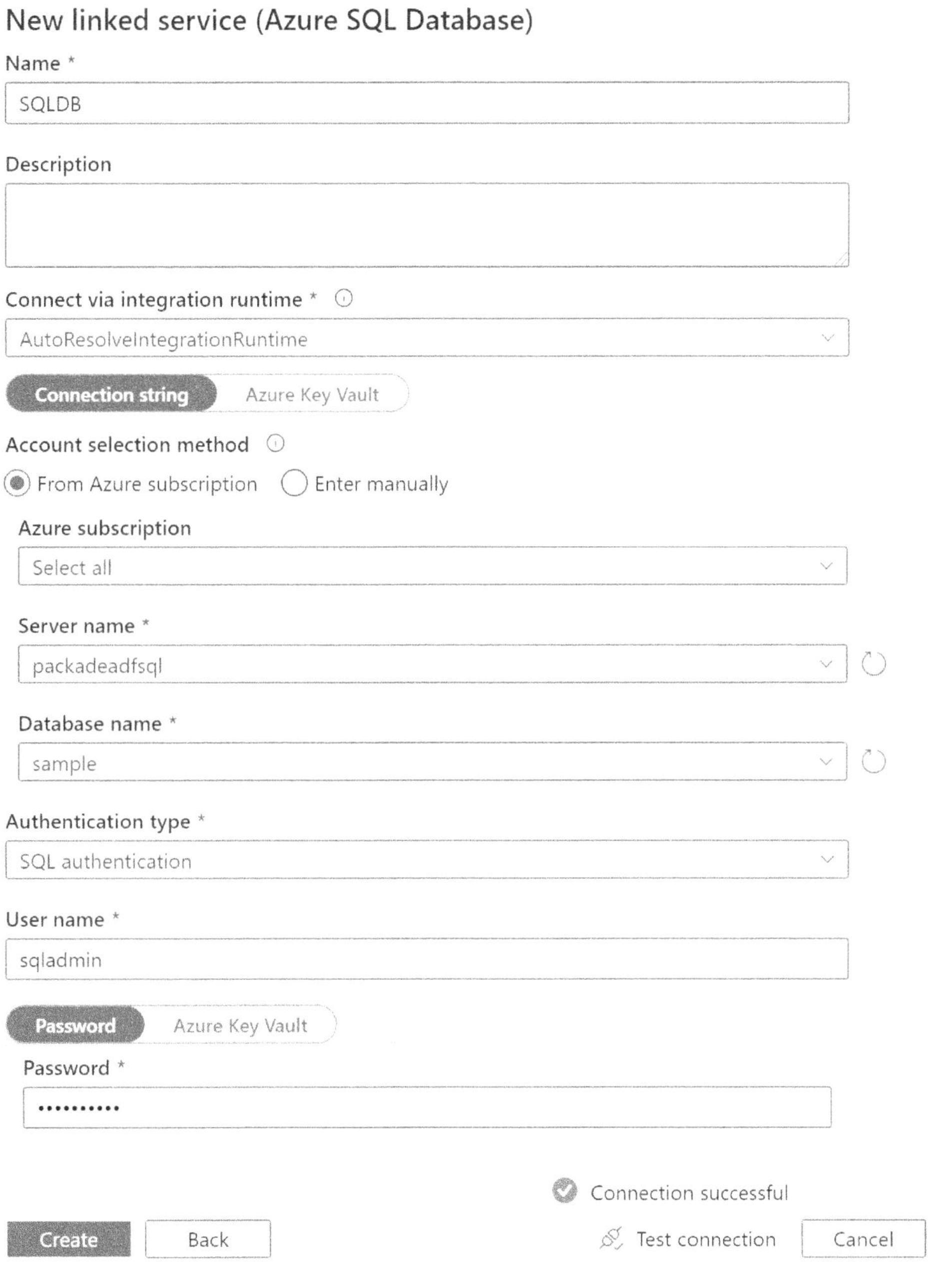

New linked service (Azure SQL Database)

Name *

```
SQLDB
```

Description

```


```

Connect via integration runtime * ⓘ

```
AutoResolveIntegrationRuntime                                    ∨
```

(**Connection string**   Azure Key Vault)

Account selection method ⓘ

⦿ From Azure subscription   ◯ Enter manually

Azure subscription

```
Select all                                                      ∨
```

Server name *

```
packadeadfsql                                             ∨    ◯
```

Database name *

```
sample                                                    ∨    ◯
```

Authentication type *

```
SQL authentication                                              ∨
```

User name *

```
sqladmin
```

(**Password**   Azure Key Vault)

Password *

```
··········
```

✓ Connection successful

[Create]   [Back]   ✎ Test connection   [Cancel]

Figure 3.8 – New linked service (Azure SQL Database)

Using the Get Metadata activity to get filenames

The first task is to get the list of files in the container. We'll do this using the **Get Metadata** activity. Follow these steps:

1. Create a new pipeline by clicking on the **Author** icon (the pencil-shaped icon on the left), then the + button, and then **Pipeline**:

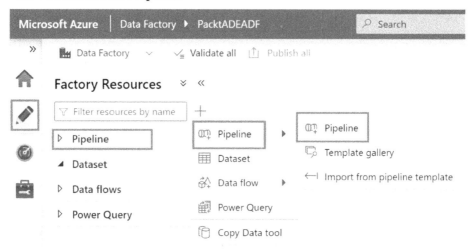

Figure 3.9 – Creating a pipeline

2. Under **Activities**, search for get and drag and drop the **Get Metadata** activity onto the pipeline:

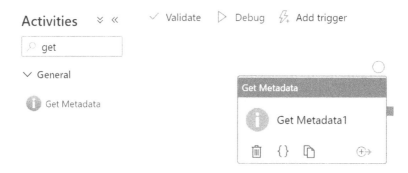

Figure 3.10 – Adding the Get Metadata activity

Set the name of the activity to GetFilename.

3. Under **Dataset**, click on the **+ New** button to add a new dataset:

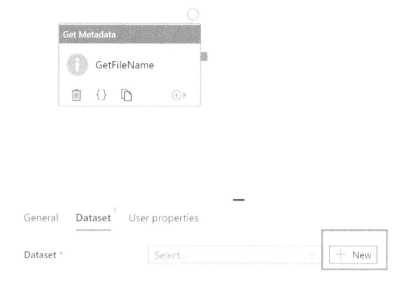

Figure 3.11 – Adding a dataset

4. Select **Azure Data Lake Storage Gen2** and select **CSV** as the file type. Name the dataset OrderdtlsCSV. Set **Linked service** to DataLoading, which we created earlier. Under **File path**, select the dataloading container. Then, check the **First row as header** box:

Set properties

Figure 3.12 – Set properties

5.   Under **Field list**, click on the **+ New** button and select **Child items**:

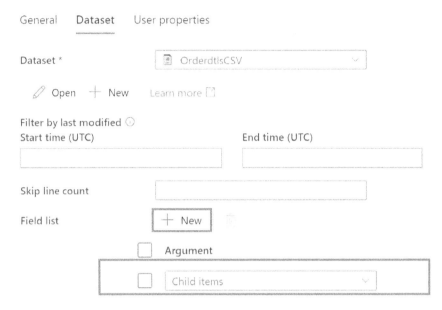

Figure 3.13 – Field list

6.   Hit the **Debug** button at the top and check the output. If no errors have been reported and the output shows the filenames, as shown in the following screenshot, then we have configured the **Get Metadata** task correctly and we can proceed to the next step:

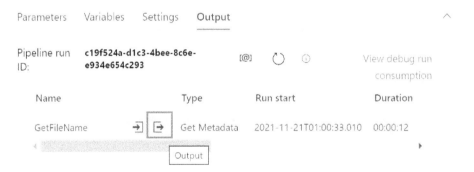

Figure 3.14 – Checking the output

The following screenshot displays the output details:

Figure 3.15 – Debug metadata

Filtering the current date files using the Filter activity

The next step is to filter the list of files that are returned by the **Get Metadata** activity for the files with the *current date* as the suffix. Let's add a **Filter** activity to the pipeline:

1. Search for filter in the **Activities** tab and drag and drop the activity onto the pipeline.

2. Connect the **Get Metadata** activity and the **Filter** activity.

3. Name the **Filter** activity FilterTodaysDate. Move to the **Settings** tab of the **Filter** activity.

4. For **Items,** click on the textbox and then click on **Add dynamic content (Alt + Shift + D)**:

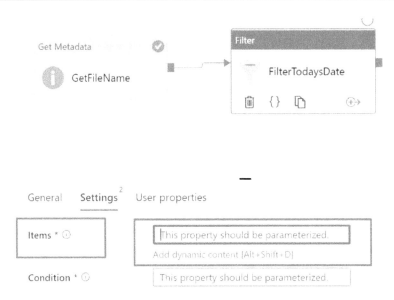

Figure 3.16 – Configuring the Filter activity

5.   Paste @activity('GetFileName').output.childitems into the **Items** field. This will retrieve the output array that was returned by the **Get Metadata** activity.

6.   Similarly, add @endswith(item().name,concat('-',formatDateTime(utcnow(),'yyyMMdd'),'.csv')) for **Condition**, as shown in the following screenshot:

Figure 3.17 – Adding a condition to the Filter activity

7. Hit the **Debug** button to test this. Ensure that the **FilterTodaysDate** activity has been completed successfully and shows the filename with the current date as the suffix:

Figure 3.18 – Adding a condition to the Filter activity

Adding a ForEach activity to loop through the files

Now, we will create a **ForEach** activity to iterate through the files that are returned by the **Filter** activity. Follow these steps:

1. Search for `ForEach` under **Activities** and add the activity to the pipeline. Link it to the **FilterTodaysDate** activity.

2. Go to the **Settings** tab of the **ForEach** activity. For **Items**, click on **Add dynamic content (Alt + Shift + D)**.

3. Under **Activity Outputs**, click on the **FilterTodaysDate** activity's output. This will automatically add `@activity('FilterTodaysDate').output`.

4. Append `.value` and set **Items** to `@activity('FilterTodaysDate').output.value`, as shown in the following screenshot. This will pass the filenames from the **FilterTodaysDate** activity to the **ForEach** activity:

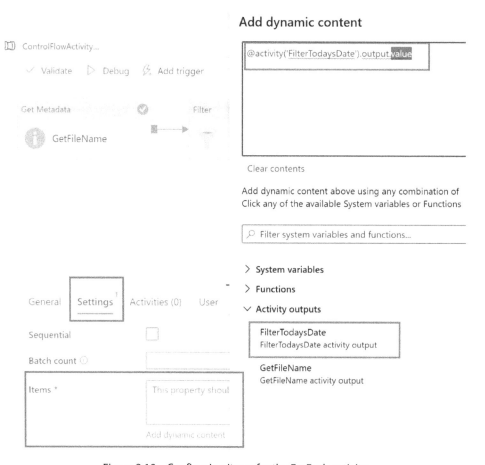

Figure 3.19 – Configuring items for the ForEach activity

5.   Go to the **Activities (0)** tab and click on the pencil button.

Adding the Copy data activity to ingest files to Azure SQL Database

Finally, we will ingest files from the data lake to Azure SQL Database. We can do this by passing the files listed by the **ForEach** activity to the **Copy** activity. Follow these steps:

1.   Search for Copy Data under **Activities** and add it to the pipeline.

2.   Name the **Copy data** activity CopyOrderDtltoSQL.

3.   Go to the **Source** tab. Select **OrderdtlsCSV** as the dataset since we need to copy CSV files to Azure SQL Database.

4.   Under **File path type**, select **Wildcard file path**. In the last textbox, type @item().name, as shown in the following screenshot:

Figure 3.20 – Configuring the Copy data activity

5.   Move to the **Sink** tab. Press **+ New** to add the dataset.

6.   Search for SQL and select **Azure SQL Database**.

7.   Select **SQLDB** as the linked service as we had created the connection to Azure SQL Database initially. Name the dataset as OrderdtlSQL. Click on the **Edit** checkbox below **Table name**. Provide table name as dbo.orderdtls as shown in the following screenshot. Select **None** option under **Import schema**. Press **OK**.:

Figure 3.21 – Adding a SQL dataset

8.   In the **Sink** tab, copy and paste the following script into the **Pre-copy script** field:

```
if not exists ( Select * from sys.objects where name like
'orderdtls')
Create table dbo.orderdtls(order_dt varchar(30),product
varchar(100),cost int, quantity int, location
varchar(100))
```

This will create a table called **orderdtls** for the first time and append rows to the same table in subsequent runs. All the other options can be left as is. Click on **pipeline1** to go back to the pipeline:

Figure 3.22 – Configuring Sink in the Copy data activity

9. Hit the **Debug** button to test all activities. The activities will complete, as shown in the following screenshot:

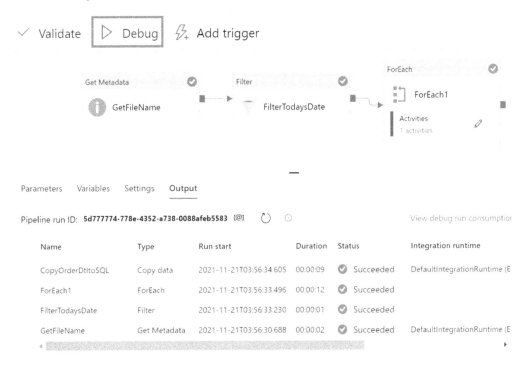

Figure 3.23 – Successful pipeline completion

10. Rename the pipeline `ControlFlowActivities` and hit the **Publish** button to save the pipeline. You can verify the result by querying Azure SQL Database from the Azure portal via **Query editor**, as shown in the following screenshot:

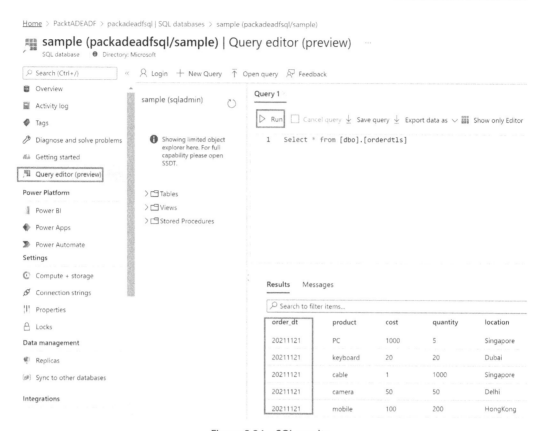

Figure 3.24 – SQL results

How it works...

In this recipe, we performed four key steps to move the data from blob storage to Azure SQL Database:

1. We got the list of files in the container using the **Get Metadata** task.

2. We filtered for files while using the current date as a prefix using the **Filter** task.

3. We iterated through all the current date files using the **ForEach** task.

4. We ingested the file's content in a SQL table using the **Copy** task.

The following steps explain how this works:

1. In the **Get Metadata** task, the dataset is set to **OrderdtlsCSV**, which is linked to the **dataloading** container.

2. In the **Get Metadata** task, we set **Field list** to **Child items**, which means that the output of the **Get Metadata** task will be the filenames from the **dataloading** container. This is because the task's dataset is linked to that container:

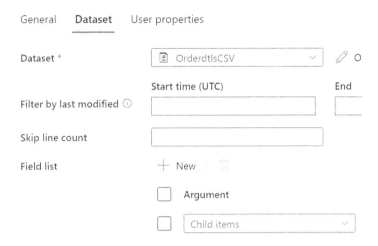

Figure 3.25 – Metadata child items

3. The **Filter** activity checks whether the filename of the **Get Metadata** activity has the current date as a suffix. In the **Filter** activity, two key settings are configured: **Items** and **Condition**. Let's look at how this works:

    A. The **Filter** activity's **Items** parameter, which can be found under the **Settings** tab, accepts an array as input.

    B. `@activity('GetFileName').output.childitems` is provided as input to **Items**, which means that all the filenames returned by the **Get Metadata** activity are passed as an array to the **Filter** activity.

    C. The entries from the array that satisfy the condition specified in the **Condition** box will be returned as the output of the **Filter** activity.

4. For the **Condition** parameter, we provided the `@endswith(item().name,concat('-',formatDateTime(utcnow(),'yyyMMdd'),'.csv'))` condition. Let's look at this in more detail:

    - `item().name`: This extracts each item from the array input, which will give us the filename

    - `formatDateTime(utcnow(),'yyyMMdd'`: This converts the current date into yyyymmdd format

    - `concat('-',formatDateTime(utcnow(),'yyyMMdd'),'.csv'))`: This concatenates – , the current date in yyyymmdd format, and the `.csv` string

    - `endswith`: This checks for files (`item().name`) whose names end with the current date and with a `.csv` string as an extension:

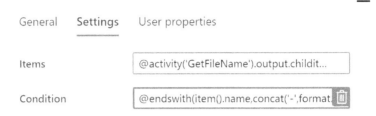

Figure 3.26 – The filename array and the date filter

5. The **ForEach** activity will perform a task for each item it receives from the **Filter** activity. In the **ForEach** activity, we received the filtered filenames as an array, which we then passed to the **Copy** activity.

6. We passed `@activity('FilterTodaysDate').output.value` for the **ForEach** activity's **Item** parameter. This ensures that the output of the **Filter** activity is passed as input to the **ForEach** activity:

Figure 3.27 – The ForEach activity's array input

7. Under the **Activities** section of the **ForEach** activity, we added the **Copy** activity.

8. The **Copy** activity copies the files it received to Azure SQL Database. In the **Copy** activity, first, we set the source dataset to **OrderdtlsCSV** since it's linked to **dataloading**. We were unable to provide the filename as that would be dynamic. We had to set **File path type** to **Wildcard file path** since we need to provide the name dynamically. `@item().name` is passed as input in **Wildcard paths**, where `@item()` references the array item coming from the **ForEach** activity, and the `name` field, which is the name inside the array:

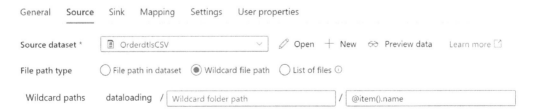

Figure 3.28 – The dynamic filename to copy

9. For the **Sink** tab, we created the dataset for Azure SQL Database. We set the table name to `dbo.orderdtls` using the **Edit** option, though the table doesn't exist in the database. The table will be created automatically in the first run. **Pre-copy script** has an `if exists` clause, which ensures a table is only created if it doesn't exist. Since the column names in the table and file match, no mapping needs to be performed during the transfer:

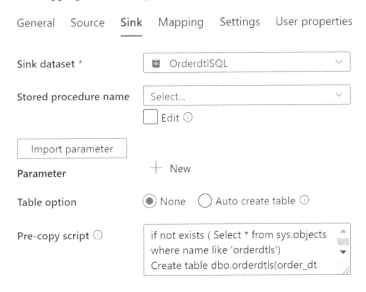

Figure 3.29 – Creating a table using Pre-copy script

By using various control activities, we can successfully transfer data from files to a database. Transferring files to a database is a common scenario in ETL workloads; you can use the preceding framework and customize the data transfer based on your requirements.

Triggering a pipeline in Azure Data Factory

An Azure Data Factory pipeline can be triggered manually, scheduled, or triggered by an event. In this recipe, we'll configure an event-based trigger to run the pipeline that we created in the previous recipe whenever a new file is uploaded to the Data Lake Store.

Getting ready

Before you start, perform the following steps:

- Log in to the Azure portal via PowerShell. To do this, execute the following command and follow the instructions to log in to Azure:

  ```
  Connect-AzAccount
  ```

- Go to `https://portal.azure.com` and log in using your Azure credentials.

- Create the **ControlFlowActivities** pipeline, as specified in the previous recipe, if you haven't created it already.

How to do it...

To create the trigger, follow these steps:

1. The event trigger requires the `eventgrid` resource to be registered in the subscription. To do that, execute the following PowerShell command:

   ```
   Register-AzResourceProvider -ProviderNamespace Microsoft.
   EventGrid
   ```

2. In the Azure portal, under **All resources**, open the **PacktADEADF** data factory that you created in the *Provisioning Azure Data Factory* recipe. On the **PacktADEADF** data factory overview page, select **Open Azure Data Factory Studio**. Click on the **Author** button on the left. Expand **Pipeline** and click on **ControlFlowActivities**, which was created in the previous recipe:

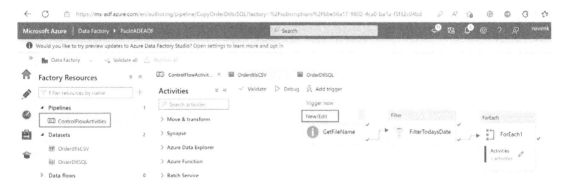

Figure 3.30 – Adding a trigger to a pipeline

3. Select **Add trigger** and then select **New/Edit**.

> **Note**
> To create **ControlFlowActivities**, please refer to the previous recipe.

4. In the **Add triggers** window, select **Choose trigger** and select **New**.

5. In the **New trigger** window, set **Name** as `NewFileTrigger`. Set the event's **Type** to **Storage event**. Use the **Storage account name** and **Container name** properties you created earlier. Under **Event**, select **blob Created**. This will ensure that any time a file is uploaded to the **dataloading** container, the pipeline will be triggered:

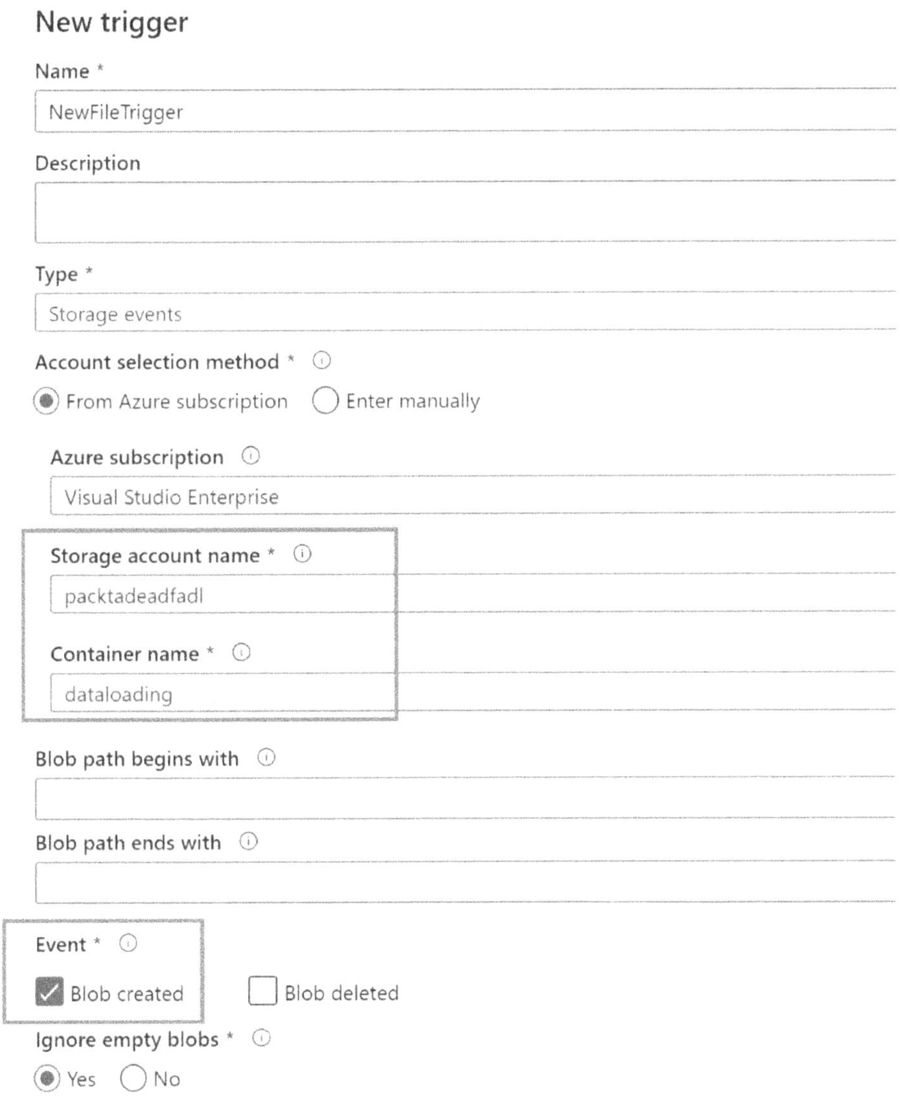

New trigger

Name *

NewFileTrigger

Description

Type *

Storage events

Account selection method * ⓘ

◉ From Azure subscription   ◯ Enter manually

Azure subscription ⓘ

Visual Studio Enterprise

Storage account name * ⓘ

packtadeadfadl

Container name * ⓘ

dataloading

Blob path begins with ⓘ

Blob path ends with ⓘ

Event * ⓘ

☑ Blob created   ☐ Blob deleted

Ignore empty blobs * ⓘ

◉ Yes   ◯ No

Figure 3.31 – Creating a trigger

6. Click **Continue** to create the trigger.

7. In the **Data preview** window, all the files in the **dataloading** container will be listed:

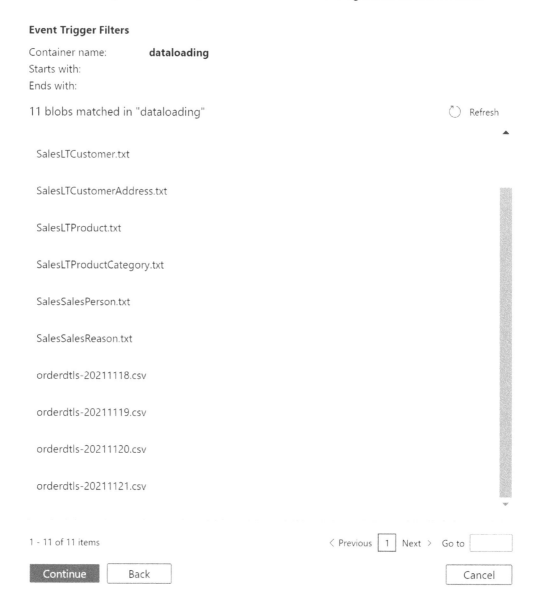

Figure 3.32 – The Data preview window

8. Click **Continue**.

9. In the **New trigger** window, we can specify the parameter values, if any, that are required by the pipeline to run:

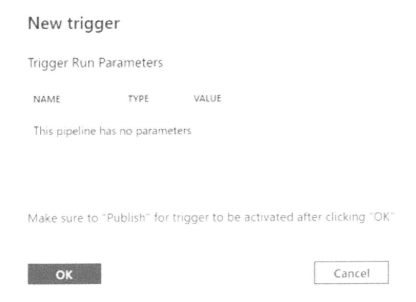

New trigger

Trigger Run Parameters

NAME   TYPE   VALUE

This pipeline has no parameters

Make sure to "Publish" for trigger to be activated after clicking "OK"

OK   Cancel

Figure 3.33 – Creating a new trigger

10. Click **OK** to create the trigger.

11. The trigger will be created. Click **Publish all** to save and apply these changes.

12. To see the trigger in action, do the following:

    I.   Download `orderdtls-Trigger.csv` from `https://github.com/ PacktPublishing/Azure-Data-Engineering-Cookbook-2nd-edition/ tree/main/Chapter3` to a local folder.

    II.   Log in to Azure PowerShell using `Connect-AzAccount`.

    III.   Execute the following command to upload the file to the **dataloading** container:

```
$storageaccountname="packtadeadfadl"
$containername="dataloading"
$resourcegroup="PacktADEADF"
#Get the Azure Storage account context
$storagecontext = (Get-AzStorageAccount
-ResourceGroupName
$resourcegroup -Name $storageaccountname).Context;
```

```
Set-AzStorageBlobContent -File "C:\temp\orderdtls-
Trigger.csv" -Context $storagecontext -Blob orderdtls-
Trigger.csv -Container $containername
```

You will receive the following output:

Figure 3.34 – File upload output

13. Once the file has been uploaded, **NewFileTrigger** will trigger the **ControlFlowActivities** pipeline.

14. To check the trigger and pipeline execution, open the **Monitor** window:

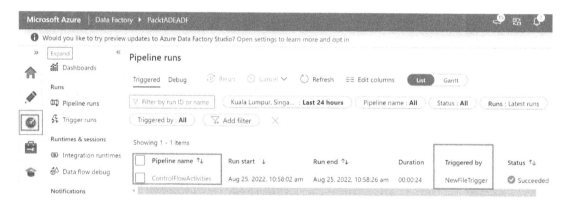

Figure 3.35 – Viewing the trigger's status

You will see that the **ControlFlowActivities** pipeline was executed and that it was triggered by **NewFileTrigger**. This proves that the execution was triggered by the file being uploaded.

How it works...

Azure Event Grid, which we registered at the subscription level using the `Register-AzureResourceProvider` PowerShell command, helps track and trigger the pipeline when a file is uploaded to the data lake container. The storage event-based trigger makes it a powerful feature in data engineering projects, since pipelines need to be triggered when a file is uploaded by another batch process or other similar scenarios. You can add conditions such as blob prefix/suffix filters or parameters to trigger the pipeline, but only when specific files are loaded/deleted or when granular conditions must be met.

Copying data from a SQL Server virtual machine to a data lake using the Copy data wizard

A common scenario in data engineering projects is where you need to ingest data from a relational database engine such as SQL Server, Oracle, or MySQL to a data lake. This recipe will show you how to ingest data from SQL Server, which has been installed in an Azure VM, to an Azure Data Lake. This method will work in **on-premises** SQL Server to Azure Data Lake instances too, but you will need to install an integration runtime. This will be covered in the next chapter. In this recipe, we will focus on copying data from SQL Server in an Azure VM to a data lake. We will be using the user-friendly Copy data wizard to transfer the data.

Getting ready

Follow these steps:

1. Provision Azure Data Factory, as explained in the *Provisioning Azure Data Factory* recipe.

2. Log in to Azure PowerShell using the `Connect-AzAccount` command and provision the blob storage account and container, as shown in the following code block (only do this if you didn't do this in the previous recipe). Execute the following command to create a storage account and a container:

```
$storageaccountname="packtadeadfadl"
$resourcegroup="PacktADEADF"
$containername="dataloading"
New-AzStorageAccount -ResourceGroupName $resourcegroup
-Name $storageaccountname -SkuName Standard_LRS -Location
'East US' -Kind StorageV2
$storagecontext = (Get-AzStorageAccount -ResourceGroupName
$resourcegroup -Name $storageaccountname).Context;
New-AzStorageContainer -Name $containername -Context
$storagecontext
```

Provision the SQL Server VM by doing the following:

1. Log in to `portal.azure.com`.

2. Click on **Create a Resource**.

3. Search for **SQL Server**.

4. Select **SQL Server 2019 on Windows Server 2019**. Pick the **Free SQL Server License: SQL 2019 Developer on Windows Server 2019** option, as shown in the following screenshot:

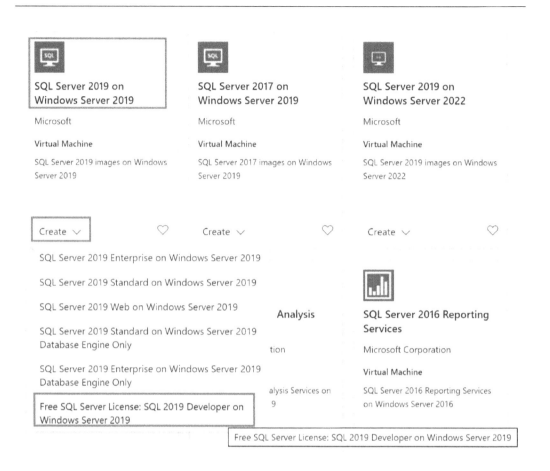

Figure 3.36 – Selecting the SQL Server 2019 image

5. Set **Resource group** to PacktADEADF and **Virtual machine name** to SQLVM. Then, set **Availability options** to **No infrastructure redundancy required**. After that, ensure that **Username** is set to sqladmin and that **Password** is set to SQLvmPwdW5!k. Leave **Select inbound ports** as is to allow the (RDP) 3389 port since it is allowed by default:

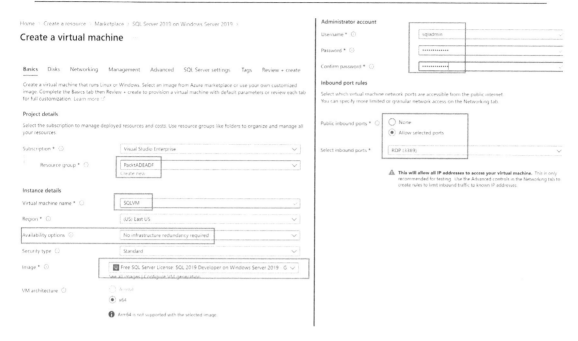

Figure 3.37 – Creating a SQL Server 2019 image

6. Click on the **SQL Server settings** tab and set **SQL connectivity** to **Public (Internet)**. For **SQL Authentication**, choose **Enable**:

Figure 3.38 – Configuring the SQL Server 2019 network

7. Click on **Review + create**. It will take around 15 minutes to create the VM. Once the VM has been created, go to the VM's overview page in the Azure portal, get the **Public IP Address** information, and perform a remote desktop connection to the VM. Log in using your user ID and password – that is, `sqladmin/SQLvmPwdW5!k`.

8. Open **Windows PowerShell** in the SQL VM and run the following commands. These commands will create a folder and download the `adventureworks` backup file into the folder:

```
New-Item -Path c:\temp -ItemType directory
cd c:\temp
Invoke-WebRequest "https://github.com/
Microsoft/sql-server-samples/releases/download/
adventureworks/AdventureWorksLT2019.bak"  -OutFile
"AdventureWorksLT2019.bak"
```

9. Open Command Prompt in the SQL VM and type the following:

```
Sqlcmd -e
```

10. Paste the following command in Command Prompt to restore the database to SQL Server:

```
RESTORE DATABASE [AdventureWorksLT2019] FROM  DISK
= N'c:\temp\AdventureWorksLT2019.bak' WITH  FILE =
1,  MOVE N'AdventureWorksLT2012_Data' TO N'F:\data\
AdventureWorksLT2012.mdf',  MOVE N'AdventureWorksLT2012_
Log' TO N'F:\log\AdventureWorksLT2012_log.
ldf',  NOUNLOAD,  STATS = 5
GO
```

11. Hit *Enter* on Command Prompt. This will restore the database, as shown in the following screenshot:

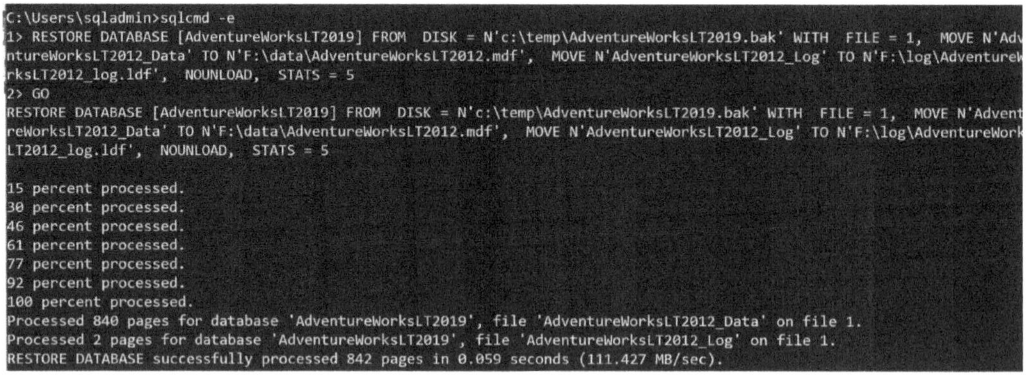

Figure 3.39 – Restoring the database

How to do it...

Now that the database has been restored, let's copy the data from the database into our data lake using the **Copy data** wizard in the data factory. Follow these steps:

1. Log in to `portal.azure.com`. Go to the data factory that you created earlier. Open **Azure Data Factory Studio**. Then, click on the **Ingest** button on the home page:

Figure 3.40 – Ingesting data

2. Select **Built-in copy task** and choose the **Run once now** option, as shown in the following screenshot:

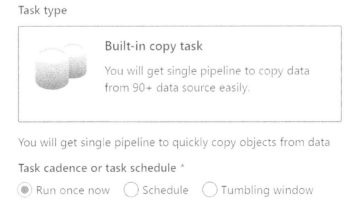

Figure 3.41 – Built-in copy task

3.   Set **Source type** to **SQL server** and click **+ New connection**, as shown in the following screenshot:

Figure 3.42 – New connection

4.   Provide the SQL VM's public IP address under **Server name**. Set **Database name** as
     `AdventureWorksLT2019` and pick **SQL authentication** under **Authentication type**.
     Finally, set **User name** as `sqladmin` and **Password** as `SQLvmPwdW5!k`. Then, click **Create**:

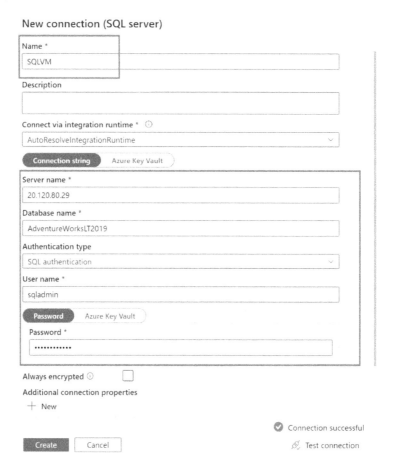

Figure 3.43 – Configuring the source dataset

5. Select a few tables you want to copy to the data lake:

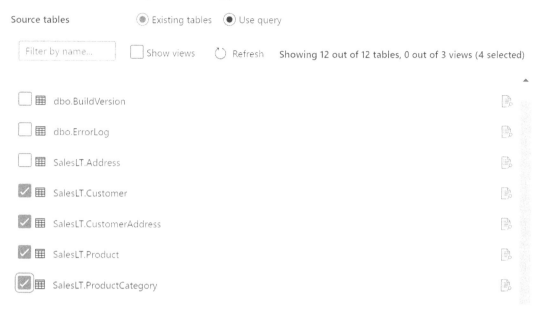

Figure 3.44 – Selecting tables

6. Hit **Next** twice to go to the **Destination data store** configuration screen. Select **Azure Data Lake Storage Gen2** under **Target type** and click + **New connection**:

Destination data store

Specify the destination data store for the copy task. You can use an existing data store connection or specify a new data store.

Figure 3.45 – Destination connection

7. Pick the storage account you created earlier and click **Create** to create the connection:

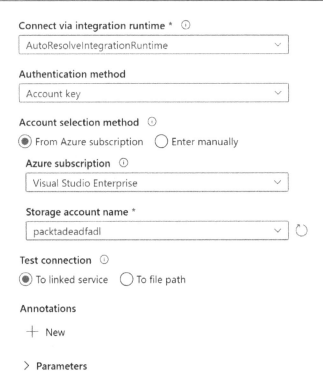

Figure 3.46 – Configuring the destination connection

8. Under **Folder path**, click on **Browse**. Select the **dataloading** container you created earlier. Then, click **OK**:

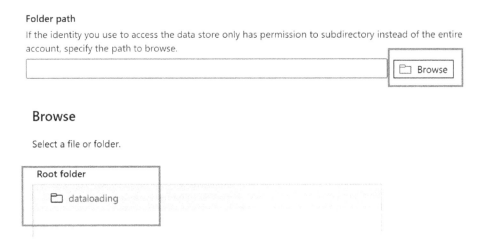

Figure 3.47 – Browsing to the Root folder

9. Pick a **File format**. Check the **Add header to file** box to ensure that the column names are shown. Then, click **Next**:

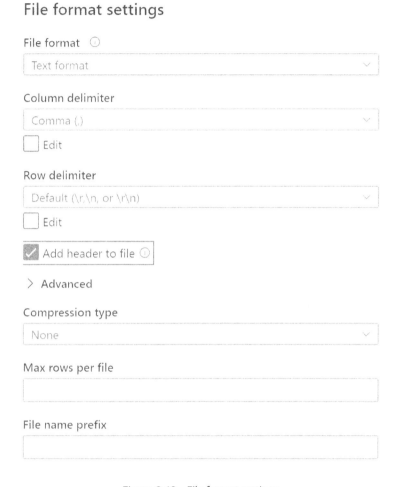

Figure 3.48 – File format settings

10. Set **Task name** as CopySQLVMtoADL and click **Next**:

Settings

Enter name and description for the copy data task, more options for data moveme

Task name *   CopySQLVMtoADL

Figure 3.49 – Using CopySQLVMtoADL as the task's name

11. Review your configuration on the **Summary** page and click **Next**. This will create the pipelines automatically and execute them. Click **Finish**:

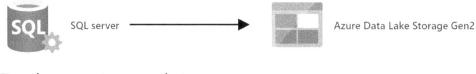

Figure 3.50 – The Summary page

12. Verify that the files have been created by checking the blob storage container in the Azure portal.

How it works...

Go to **Azure Data Factory Studio** and click on the **Author** button on the left. Expand **Pipeline** and notice that a new pipeline, **CopySQLVMtoADL**, has been created. Click on it. You will see that there's a **ForEach** activity in the pipeline. The **ForEach** activity is used to iterate through each table that needs to be copied:

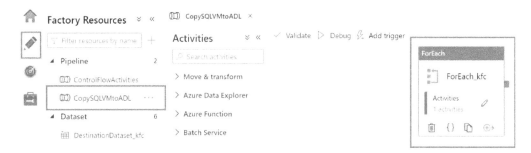

Figure 3.51 – The Copy data wizard pipeline

Click on **Activities**. You will notice a **Copy data** task whose source is **SQL Server** and the destination is **Azure Data Lake Storage Gen2**. The **Copy data** task will copy one table at a time from SQL Server to the data lake.

You will also notice that the source table name and destination filename come from the **ForEach** activity (item().source.table /item().destination.fileName). The Copy data wizard has automatically created the pipeline with the relevant activities to move the data:

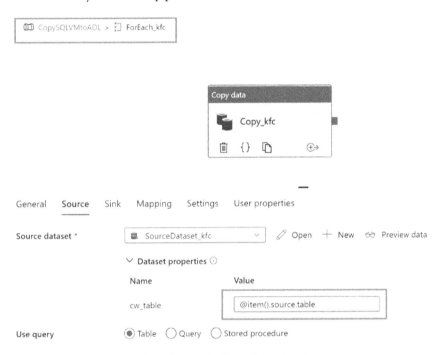

Figure 3.52 – The Copy data wizard

You can easily customize the pipeline or schedule it based on your needs to transfer data periodically. Copy data wizard saves so much time when it comes to configuring the datasets and the **ForEach** activity, which makes it easy to get started with data movement tasks.

Ensure that you delete the **PacktADEADF** resource group once you have finished since you will incur Azure consumption costs otherwise.

4

Azure Data Factory Integration Runtime

The Azure Data Factory **Integration Runtime (IR)** is the compute infrastructure that is responsible for executing data flows, pipeline activities, data movement, and **SQL Server Integration Services (SSIS)** packages. There are three types of IR: Azure, self-hosted, and Azure SSIS.

Azure IR is used to process data flows, data movement, and activities involving cloud-based data sources. By default, a type of Azure IR called **AutoResolveIntegrationRuntime** is created whenever a new data factory is created. AutoResolveIntegrationRuntime automatically determines the best location to run the IR. In addition to AutoResolveIntegrationRuntime, you can provision additional instances of Azure IR as well. In provisioned Azure IR instances, you may specify the preferred location to run the IR.

A self-hosted IR can be installed on-premises or on a virtual machine that uses the Windows OS and can be used to work with data on-premises or in the cloud. It can be used for data movement and activities.

The Azure SSIS IR is used to lift and shift existing SQL SSIS.

In this chapter, we'll learn how to use a self-hosted IR and Azure SSIS IR through the following recipes:

- Configuring a self-hosted IR
- Configuring a shared self-hosted IR
- Configuring high availability for a self-hosted IR
- Patching a self-hosted IR
- Migrating an SSIS package to Azure Data Factory

By the end of the chapter, you will know how to configure a self-hosted IR, share an IR across data factories, configure high availability for a self-hosted IR, upgrade a self-hosted IR, and migrate SSIS packages to Azure.

Technical requirements

For this chapter, the following are required:

- A Microsoft Azure subscription
- PowerShell 7
- Microsoft Azure PowerShell

Configuring a self-hosted IR

In this recipe, we'll learn how to configure a self-hosted IR and then use the IR to copy files from on-premises to Azure Storage using the **Copy data** activity.

Getting ready

To get started, do the following:

1. Log in to `https://portal.azure.com` using your Azure credentials.

2. Open a new PowerShell prompt and execute the following command to log into your Azure account from PowerShell:

   ```
   Connect-AzAccount
   ```

3. You will need an existing Data Factory account. If you don't have one, create one by executing the following PowerShell script:

   ```
   $resourceGroupName = "PacktADE";
   $location = 'east us'
   $dataFactoryName = "ADFPacktADE2";
   $DataFactory = Set-AzDataFactoryV2 -ResourceGroupName
   $resourceGroupName -Location $location -Name
   $dataFactoryName
   ```

How to do it...

To configure a self-hosted IR, follow these steps:

1. In the Azure portal, open **Data Factory**, and then open the **Manage** tab:

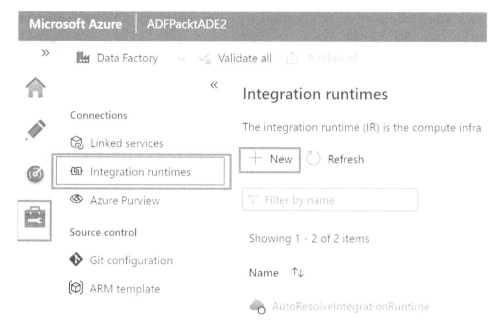

Figure 4.1 – Opening the Manage tab under Data Factory

2. Select **New** and then select **Azure, Self-Hosted**:

Integration runtime setup

Integration Runtime is the native compute used to execute or dispatch activities. Choose what integration runtime to create based on required capabilities. Learn more

Azure, Self-Hosted

Perform data flows, data movement and dispatch activities to external compute.

Azure-SSIS

Lift-and-shift existing SSIS packages to execute in Azure.

Figure 4.2 – Selecting the Azure, Self-Hosted IR

3. Select **Azure, Self-Hosted** and click **Continue** to go to the next step. In the **Network environment** section, select **Self-Hosted**:

Integration runtime setup

Network environment:

Choose the network environment of the data source / destination or external compute to which the integration runtime will connect to for data flows, data movement or dispatch activities:

Azure

Use this for running data flows, data movement, external and pipeline activities in a fully managed, serverless compute in Azure.

Self-Hosted

Use this for running activities in an on-premises / private network
View more ∨

External Resources:

You can use an existing self-hosted integration runtime that exists in another resource. This way you can reuse your existing infrastructure where self-hosted integration runtime is setup.

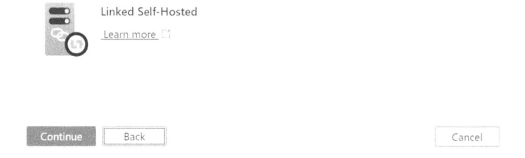

Linked Self-Hosted

Learn more ☐

Continue   Back   Cancel

Figure 4.3 – Selecting a network environment

4. Click **Continue** to go to the next step. In the next window, name the IR `selfhosted-onpremise` and click **Create**:

Figure 4.4 – Creating the IR

5. The next step is to download and install the IR on the on-premises machine:

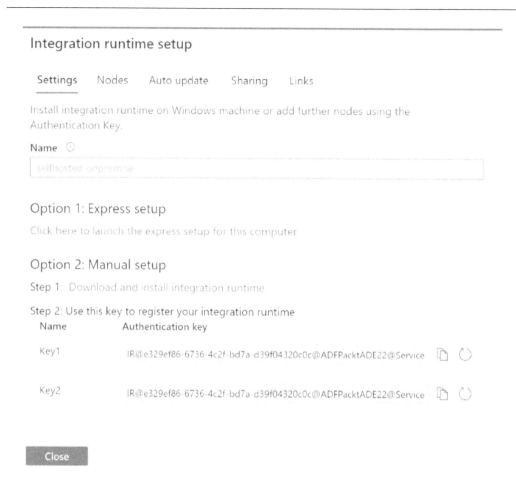

Figure 4.5 – Installing the IR

Note down the **Key1** value, as it's required in a later step.

6. If you are logged into the Azure portal from the machine where you wish to install the IR, select **Option 1: Express setup**. If you want to install it onto another machine, select **Option 2: Manual setup**.

    We'll be using **Option 2: Manual setup**. Click on **Download and install integration runtime**. Download the latest version from the download center. Double-click the downloaded file and click **Run** to start the installation wizard. Follow the wizard to install the IR. When the installation is complete, copy and paste the authentication key, and click **Register**:

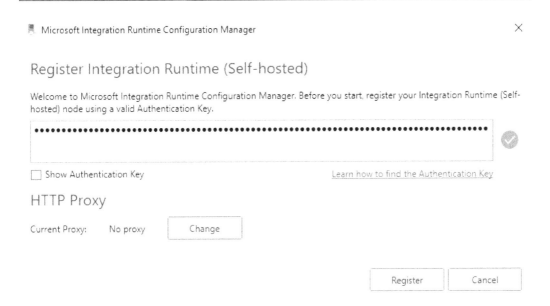

Figure 4.6 – Registering a self-hosted IR

7. After the successful verification of the key, the computer or the node has now been registered as part of the `selfhosted-onpremise` IR. Hit the **Finish** button:

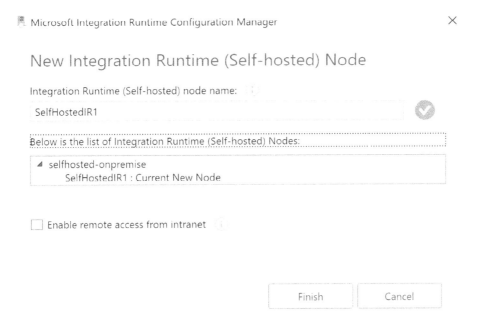

Figure 4.7 – Completing the self-hosted IR node configuration

8. Switch to the Azure portal and close the IR setup window. The new `selfhosted-onpremise` IR will be listed in the **Integration runtimes** window:

Figure 4.8 – Viewing the self-hosted IR status

9. Click on the **selfhosted-onpremise** IR to view the details:

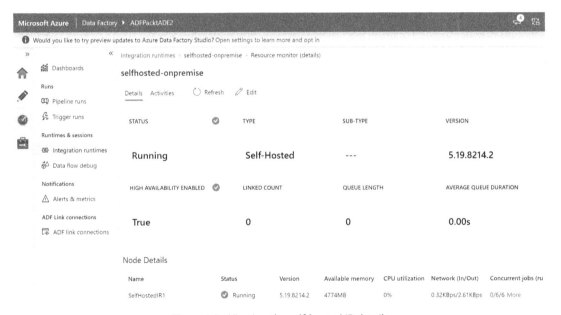

Figure 4.9 – Viewing the self-hosted IR details

The **SelfHostedIR1** node is the name of the computer where we installed the IR. We'll now use the self-hosted IR to upload data from the files stored on-premises to Azure SQL Database. For this example, we will store the files in the self-hosted runtime machine (**SelfHostedIR1**). If the file needs to reside in a different machine other than that of the self-hosted IR, you can facilitate this by creating a file share and making it accessible from the self-hosted IR machine.

10. As the next step, let's prepare the source and destination for our file copy task, which is the file and Azure SQL Database. Download the `orderdtls.csv` file from `https://github.com/PacktPublishing/Azure-Data-Engineering-Cookbook-2nd-edition/tree/main/Chapter04`.

11. Upload to the IR machine (**SelfHostedIR1**) on a `C:\Chapter04\Data` path. We will be copying `orderdtls.csv` to **AzureSQLDatabase**.

12. Please execute the following PowerShell script to create a SQL database if you don't have Azure SQL Database to use as a destination. The following script creates a database called `sample` in the `packadesql` server:

```
$resourcegroup="PacktADE"
$serverName = "packadesql"
$adminSqlLogin = "sqladmin"
$password = "SQLPwdW5!k"
$startIp = "0.0.0.0"
$endIp = "255.255.255.255"
$databasename = "sample"
$server = New-AzSqlServer -ResourceGroupName
$resourcegroup -ServerName $serverName -Location "Eastus"
-SqlAdministratorCredentials $(New-Object -TypeName
System.Management.Automation.PSCredential -ArgumentList
$adminSqlLogin, $(ConvertTo-SecureString -String
$password -AsPlainText -Force))
$serverFirewallRule = New-AzSqlServerFirewallRule
-ResourceGroupName $resourcegroup -ServerName $serverName
-FirewallRuleName "AllowedIPs" -StartIpAddress $startIp
-EndIpAddress $endIp
$database = New-AzSqlDatabase   -ResourceGroupName
$resourcegroup -ServerName $serverName -DatabaseName
$databaseName -RequestedServiceObjectiveName "S0"
```

13. Let's use the copy wizard to perform a quick data transfer:

    I.   Go to the **Home** page of Azure Data Factory Studio and click the **Ingest** button:

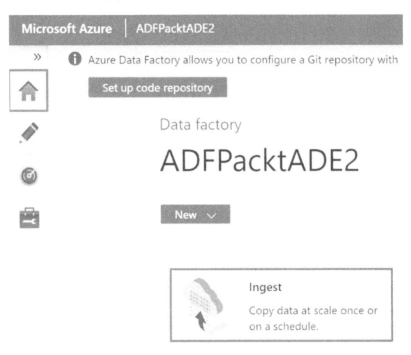

Figure 4.10 – Starting the copy wizard

    II.   On **New connection**, search for **File**, as shown in the following screenshot:

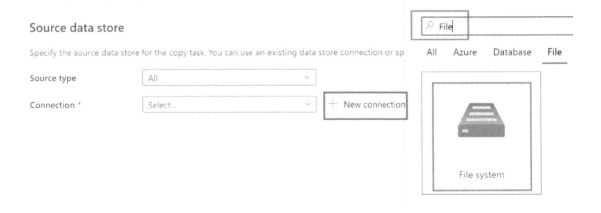

Figure 4.11 – Starting the copy wizard

III.   Provide the IR machine credentials and path to the file, as shown in the following screenshot:

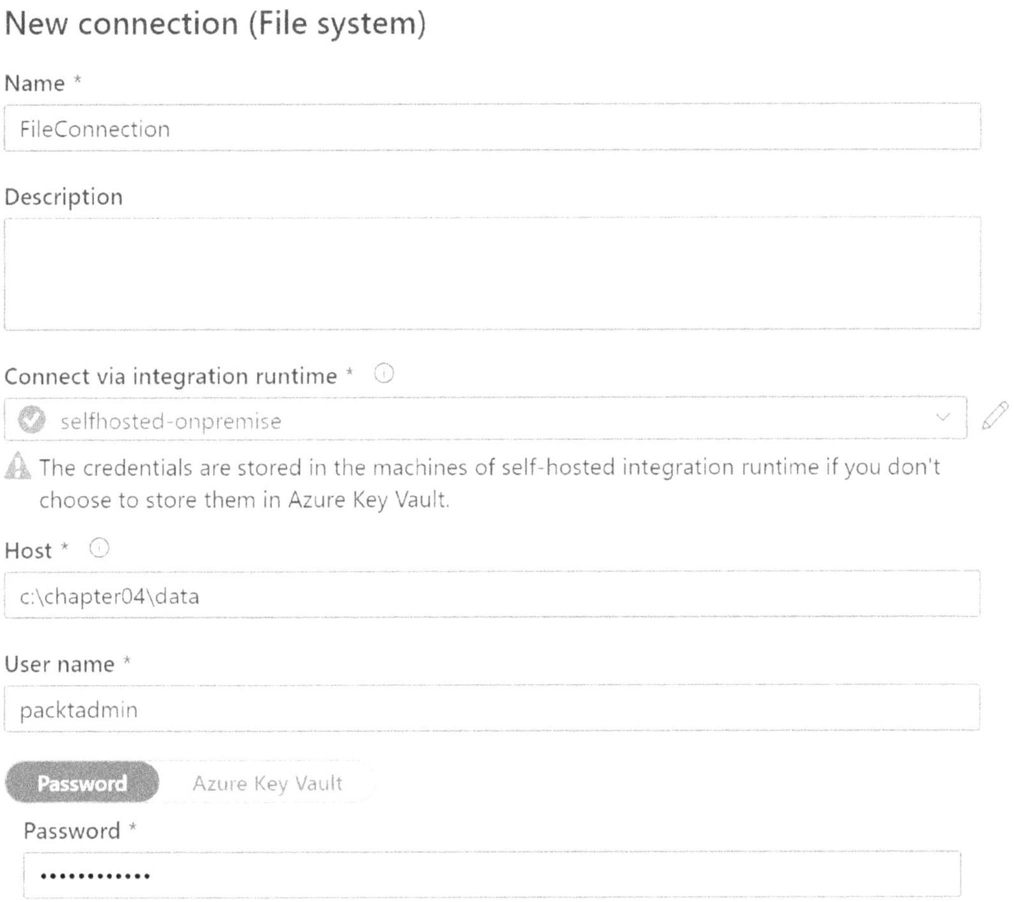

New connection (File system)

Name *

FileConnection

Description

Connect via integration runtime * ⓘ

✅ selfhosted-onpremise   ⌄   ✏️

⚠️ The credentials are stored in the machines of self-hosted integration runtime if you don't
choose to store them in Azure Key Vault.

Host * ⓘ

c:\chapter04\data

User name *

packtadmin

| Password | Azure Key Vault |

Password *

············

Figure 4.12 – Configure file connection

IV.   Once the connection is created, you will be taken back to the **Source data store** page
under **Copy Data tool**. Provide the file to be copied. Click on **Next**:

Copy Data tool

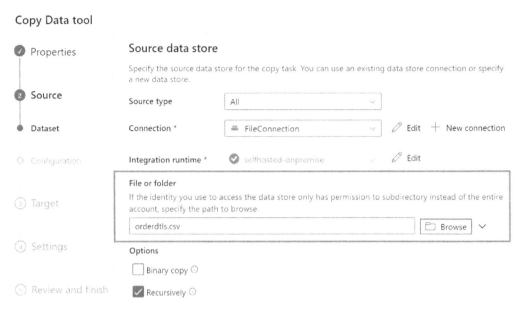

Figure 4.13 – File details

V.   The **Configuration** section in the **Source** section of **Copy Data tool** can remain with the defaults. On the **Target** page of **Copy Data tool**, click + **New connection** and pick **Azure SQL Database** as the destination data store, as shown in the following screenshot:

Copy Data tool

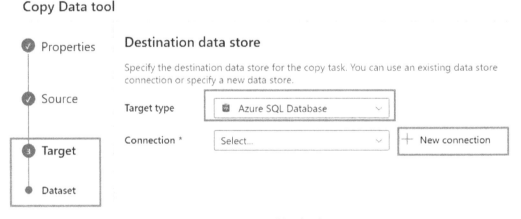

Figure 4.14 – Azure SQL database

VI.   Provide the database details and the credentials (`sqladmin` / `SQLPwdW5!k`), as shown in the following screenshot:

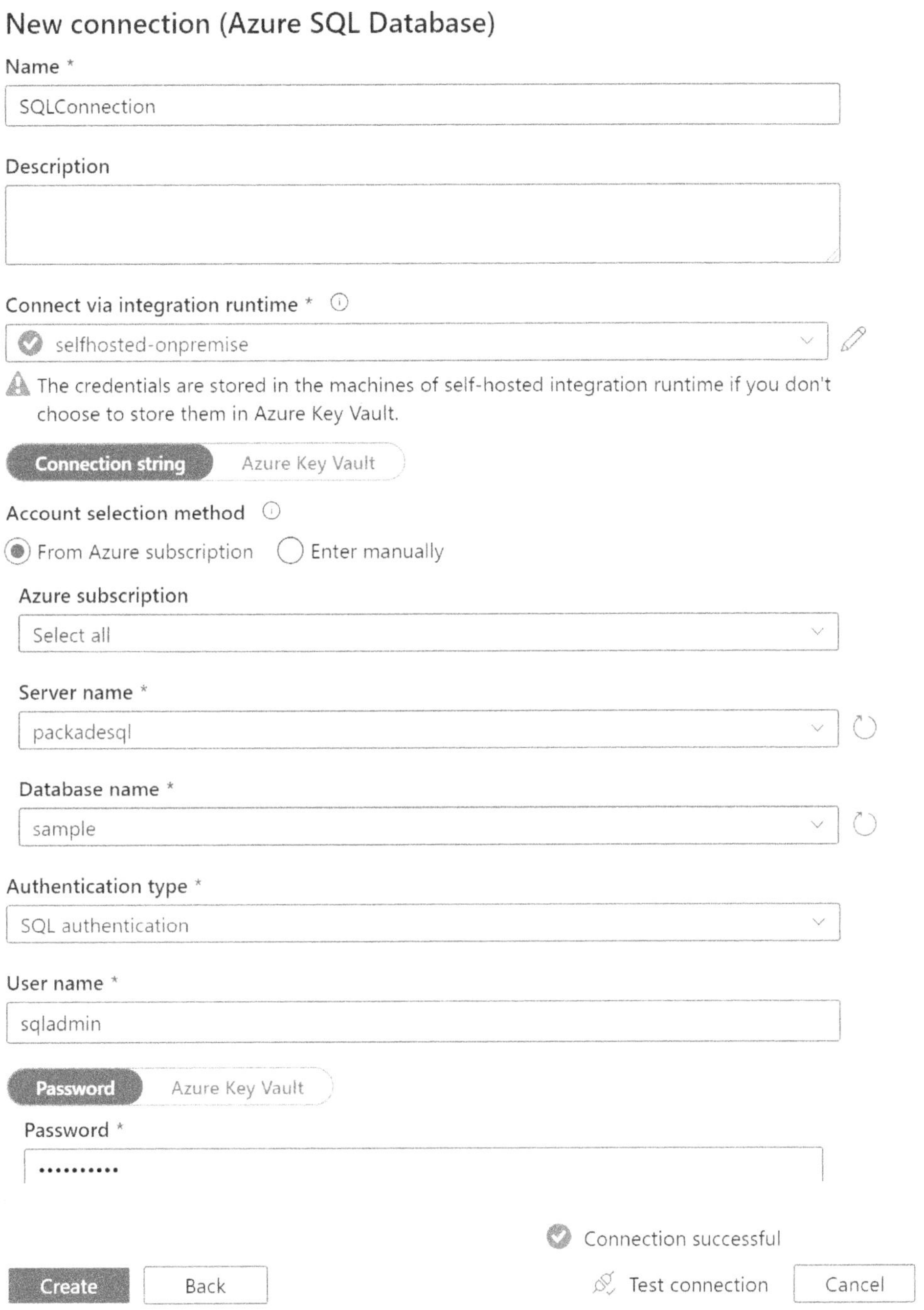

New connection (Azure SQL Database)

Name *

SQLConnection

Description

Connect via integration runtime * ⓘ

✓ selfhosted-onpremise   ✎

⚠ The credentials are stored in the machines of self-hosted integration runtime if you don't
   choose to store them in Azure Key Vault.

Connection string   Azure Key Vault

Account selection method ⓘ

◉ From Azure subscription   ◯ Enter manually

Azure subscription

Select all   ✓

Server name *

packadesql   ✓   ↻

Database name *

sample   ✓   ↻

Authentication type *

SQL authentication   ✓

User name *

sqladmin

Password   Azure Key Vault

Password *

••••••••••

                                               ✓ Connection successful

Create   Back   ✎ Test connection   Cancel

Figure 4.15 – Azure SQL database credentials

VII. Provide a destination table name. Check the **Skip column mapping for all tables** checkbox:

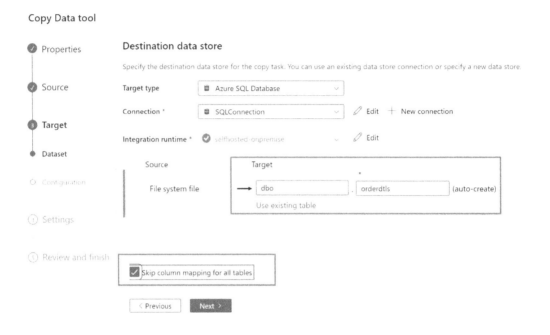

Figure 4.16 – Table details

VIII. Click through the **Configuration** and **Review and finish** screens to complete building the pipeline. Notice that the pipeline was created successfully and moved the data as well:

Figure 4.17 – Data transfer completion

Thus, we have successfully copied a file from an on-premises file system to Azure SQL Database using a self-hosted IR.

How it works...

Self-hosted IRs act as the gateway between Azure and the on-premise data source. Installing the **Microsoft IR** on the **SelfHostedIR1** machine ensured that the libraries and components required for moving the data from the on-premises file system to Azure are now available on **SelfHostedIR1**. Providing the self-hosted IR key during the installation registered the **SelfHostedIR1** machine in Azure as an IR. Once the self-hosted IR was configured, we were successfully able to move the file stored in **SelfHostedIR1** to Azure SQL Database (`packadesql.database.windows.net`) – `Sample` database.

The health of the IR machine plays an important role in the performance and stability of the data movement pipelines. It is recommended to monitor the self-hosted IR machine closely and upsize its configuration if required.

Configuring a shared self-hosted IR

A shared self-hosted IR, as the name suggests, can be shared among more than one data factory. This helps us use a single self-hosted IR to run multiple pipelines. In this activity, we'll learn how to share a self-hosted IR.

Getting ready

To get started, do the following:

1. Log in to `https://portal.azure.com` using your Azure credentials.

2. Open a new PowerShell prompt in your machine. Execute the following command to log in to your Azure account from PowerShell:

    ```
    Connect-AzAccount
    ```

3. You will need an existing Data Factory account. If you don't have one, create one by executing the following PowerShell script:

    ```
    $resourceGroupName = "PacktADE";
    $location = 'east us'
    $dataFactoryName = "ADFPacktADE2";
    $DataFactory = Set-AzDataFactoryV2 -ResourceGroupName
    $resourceGroupName -Location $location -Name
    $dataFactoryName
    ```

4. You need a self-hosted IR. If you don't have one, follow the *Configuring a self-hosted IR* recipe in this chapter to create one.

How to do it...

Perform the following steps to create a linked self-hosted IR:

1. Let's create a new data factory account called `ADFPacktSharedIR`. The `ADFPacktSharedIR` data factory will use the IR from `ADFPacktADE2`. Use the following PowerShell command to create the `ADFPacktSharedIR` data factory:

```
$resourceGroupName = "PacktADE";
$location = 'east us'
$dataFactoryName = "ADFPacktSharedIR";
$DataFactory = Set-AzDataFactoryV2 -ResourceGroupName
$resourceGroupName -Location $location -Name
$dataFactoryName
```

2. In the Azure portal, open `ADFPacktADE2`, and then open the **Manage** tab. On the **Manage** tab, select **Integration runtimes**.

3. Select the **selfhosted-onpremise** IR:

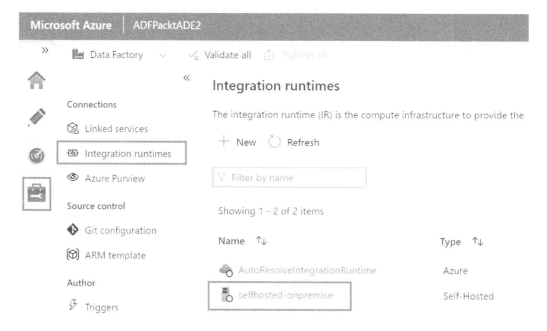

Figure 4.18 – Opening the packtdatafactory Manage tab

> **Note**
>
> This recipe requires an existing self-hosted IR. If you don't have one, follow the previous recipe to create one.

4. In the **Edit integration runtime** window, select the **Sharing** tab. Click **Grant permission to another Data Factory or user-assigned managed identity**:

Edit integration runtime

Settings   Nodes   Auto update   **Sharing**   Links

You can share your self-hosted integration runtime (IR) with another Data Fac

To enable sharing:
1. Grant permission to the Data Factory in which you would like to reference
2. Copy the below 'Resource ID' and use it while creating a new linked self-ho
   other Data Factory.

Resource ID ⓘ

/subscriptions/	/resourcegroups/PacktAD

> ╋ Grant permission to another Data Factory or user-assigned managed identity

Resource name   Resource type

\- - -

Apply   Cancel

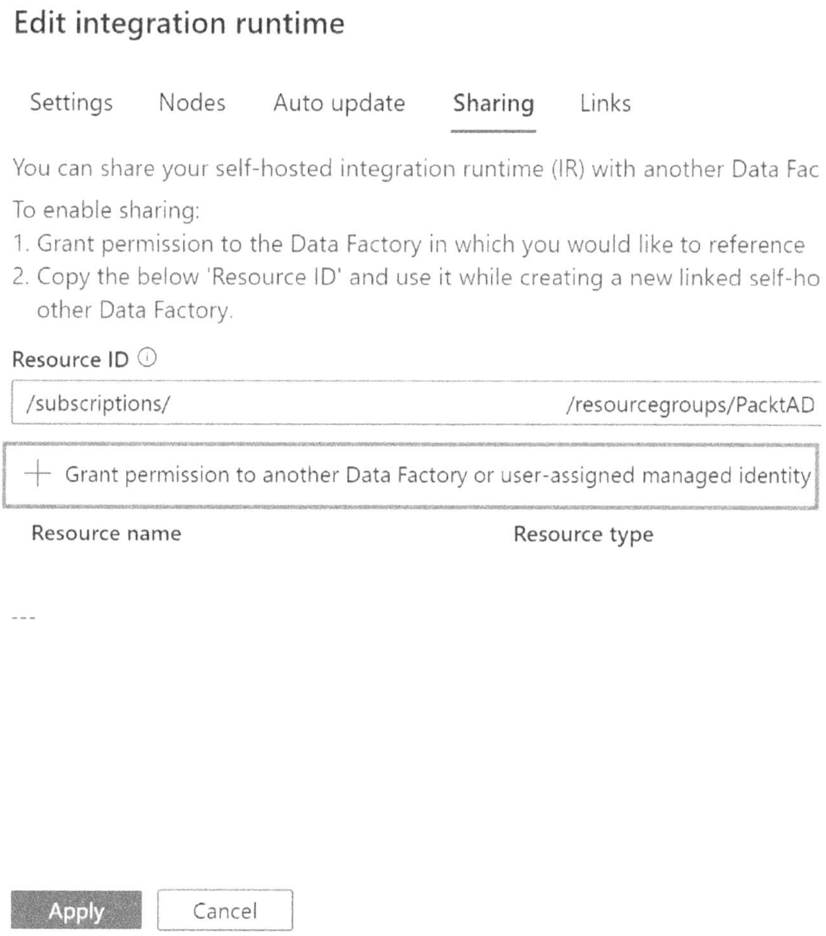

Figure 4.19 – Granting permission to another data factory

5. In the **Assign permissions** window, type `ADFPacktSharedIR` in the search box, and then select `ADFPacktSharedIR` from the list. Hit the **Apply** button:

Edit integration runtime

Settings   Nodes   Auto update   **Sharing**   Links

You can share your self-hosted integration runtime (IR) with another Data Factory.

To enable sharing:

1. Grant permission to the Data Factory in which you would like to reference this IR (shared).
2. Copy the below 'Resource ID' and use it while creating a new linked self-hosted IR in the other Data Factory.

Resource ID ⓘ

/subscriptions/(/resourcegroups/PacktADE/providers

＋ Grant permission to another Data Factory or user-assigned managed identity   ⟲ Refresh

Resource name	Resource type	Remove
ADFPacktSharedIR	Data Factory	✕

Figure 4.20 – The IR setup window

Note the resource ID for use in later steps. Click **Apply** to save the configuration settings.

6. In the Azure portal, open the **Manage** tab of ADFPacktSharedIR. On the **Connections** tab, select **Integration runtimes**. In the **Integration runtimes** panel, click + **New**:

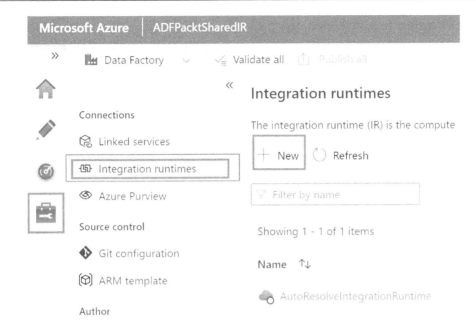

Figure 4.21 – Adding the IR to packtdatafactory2

7. In the **Integration runtime setup** window, select **Azure, Self-Hosted**:

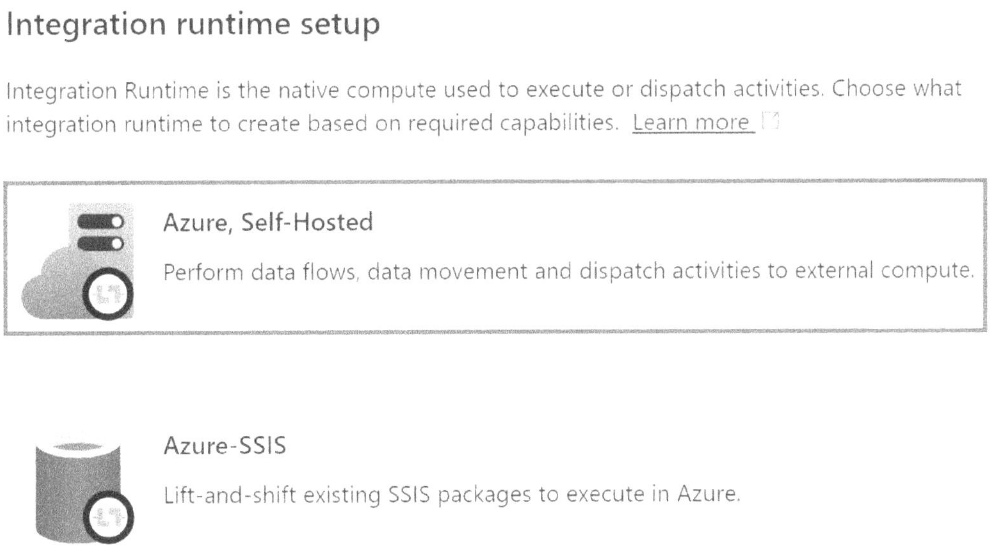

Integration runtime setup

Integration Runtime is the native compute used to execute or dispatch activities. Choose what integration runtime to create based on required capabilities. Learn more

Azure, Self-Hosted

Perform data flows, data movement and dispatch activities to external compute.

Azure-SSIS

Lift-and-shift existing SSIS packages to execute in Azure.

Figure 4.22 – Selecting the IR type

Click **Continue** to go to the next step.

8.   In the next window, under **External Resources**, select **Linked Self-Hosted**:

Integration runtime setup

Network environment:

Choose the network environment of the data source / destination or external compute to which the integration runtime will connect to for data flows, data movement or dispatch activities:

Azure

Use this for running data flows, data movement, external and pipeline activities in a fully managed, serverless compute in Azure.

Self-Hosted

Use this for running activities in an on-premises / private network
View more ∨

External Resources:

You can use an existing self-hosted integration runtime that exists in another resource. This way you can reuse your existing infrastructure where self-hosted integration runtime is setup.

Linked Self-Hosted

Learn more

Figure 4.23 – Selecting Linked Self-Hosted

Click **Continue** to go to the next step.

9. In the next window, name the IR `selfhosted-onpremise`. Under **Resource ID**, enter the **Resource ID** value from *step 5*:

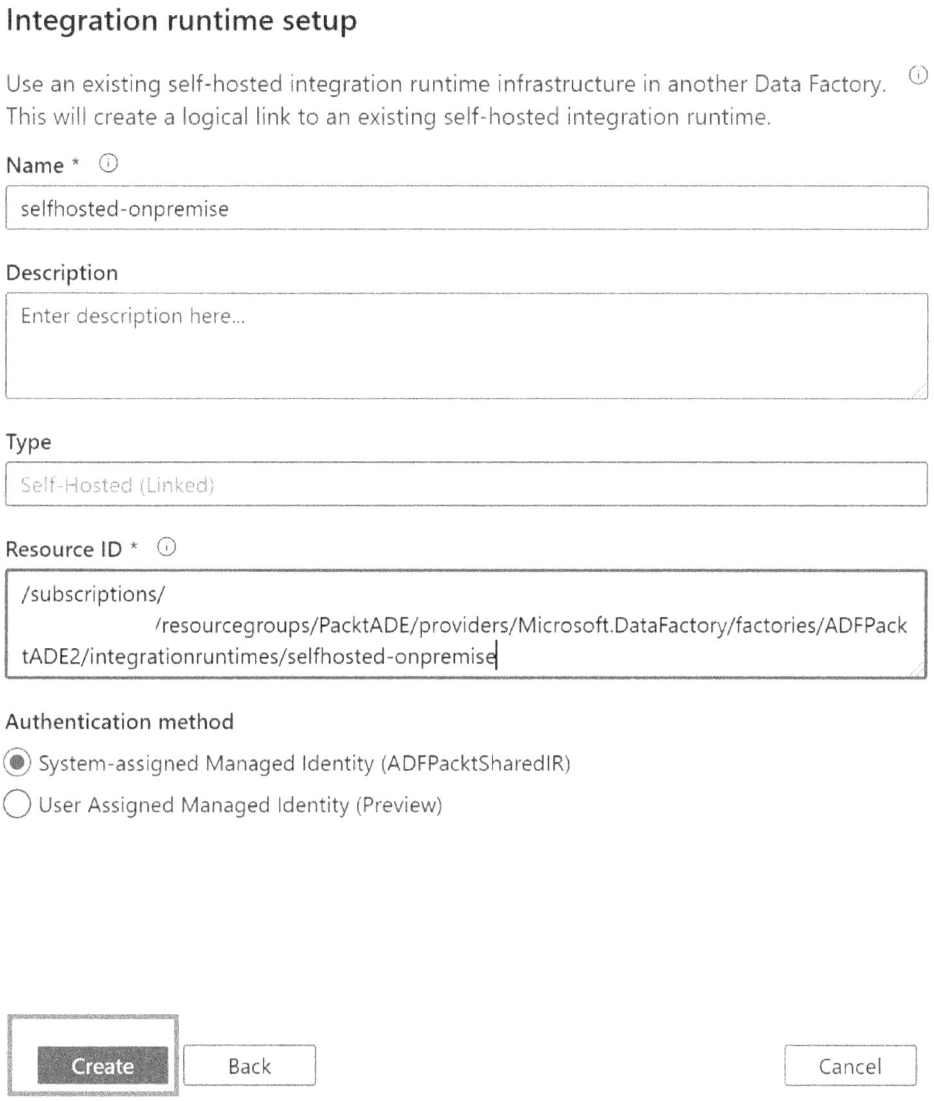

Integration runtime setup

Use an existing self-hosted integration runtime infrastructure in another Data Factory. ⓘ
This will create a logical link to an existing self-hosted integration runtime.

Name * ⓘ

selfhosted-onpremise

Description

Enter description here...

Type

Self-Hosted (Linked)

Resource ID * ⓘ

/subscriptions/   /resourcegroups/PacktADE/providers/Microsoft.DataFactory/factories/ADFPack tADE2/integrationruntimes/selfhosted-onpremise

Authentication method

⦿ System-assigned Managed Identity (ADFPacktSharedIR)

◯ User Assigned Managed Identity (Preview)

Create	Back		Cancel

Figure 4.24 – Providing a shared IR resource ID

The IR is added to `ADFPacktSharedIR`. The `selfhosted-onpremise` IR is now shared between `ADFPacktADE2` and `ADFPacktSharedIR`. The pipelines from the two data factories can benefit from one self-hosted IR.

Configuring high availability for a self-hosted IR

A self-hosted IR is a critical component for transferring data from on-premises data sources or data sources in virtual networks to Azure. Configuring additional nodes for a self-hosted IR ensures there is no single point of failure and provides high availability. Having additional nodes makes the self-hosted IR function as usual even when one of the nodes is under maintenance. Configuring additional nodes also allows load sharing with multiple data transfer jobs being executed in parallel on different nodes.

In the following recipe, we will install an IR on an additional machine to provide high availability and load-sharing capabilities.

Getting ready

Perform the following steps before working on this recipe:

1. Log in to `https://portal.azure.com` using your Azure credentials.

2. You will need an existing Data Factory account. If you don't have one, create one by executing the following PowerShell script. Open a new PowerShell prompt. Execute the following commands to log in to your Azure account from PowerShell and create a Data Factory account:

    ```
    Connect-AzAccount
    $resourceGroupName = "PacktADE";
    $location = 'east us'
    $dataFactoryName = "ADFPacktADE2";
    $DataFactory = Set-AzDataFactoryV2 -ResourceGroupName
    $resourceGroupName -Location $location -Name
    $dataFactoryName
    ```

3. Configure a self-hosted IR, as explained in the *Configuring a self-hosted IR* section of this chapter.

4. You will need an additional machine to be used as a secondary node for the IR.

How to do it...

Perform the following steps to configure high availability for a self-hosted IR:

1. Log in to `portal.azure.com`. Go to **Data Factory**. Click **Open Azure Data Factory Studio**.

2. Click on **Manage -> Integration runtimes**. Make note of the version number of the self-hosted IR. Click on **selfhosted-onpremise**, as shown in the following screenshot:

Figure 4.25 – The self-hosted IR

3. Copy the IR key and save it in Notepad for future reference.

Edit integration runtime

Settings   Nodes   Auto update   Sharing   Links

Install integration runtime on Windows machine or add further nodes using the
Authentication Key.

Name ⓘ

selfhosted-onpremise

Description

Option 1: Express setup

Click here to launch the express setup for this computer

Option 2: Manual setup

Step 1: Download and install integration runtime

Step 2: Use this key to register your integration runtime

Name	Authentication key
Key1	IR@42b666c7-2d9e-4e28-9fe3-6383f169383c@ADFPacktADE2@ServiceEr
Key2	IR@42b666c7-2d9e-4e28-9fe3-6383f169383c@ADFPacktADE2@ServiceEr

Figure 4.26 – Copy the key

4.   Click on **Download and install integration runtime** under **Option 2: Manual setup**.

Edit integration runtime

Settings   Nodes   Auto update   Sharing   Links

Install integration runtime on Windows machine or add further nodes using the Authentication Key.

Name ⓘ

selfhosted-onpremise

Description

Option 1: Express setup

Click here to launch the express setup for this computer

Option 2: Manual setup

Step 1: Download and install integration runtime

Step 2: Use this key to register your integration runtime

Name	Authentication key		
Key1	IR@42b666c7-2d9e-4e28-9fe3-6383f169383c@ADFPacktADE2@ServiceEr	🗋	↻
Key2	IR@42b666c7-2d9e-4e28-9fe3-6383f169383c@ADFPacktADE2@ServiceEr	🗋	↻

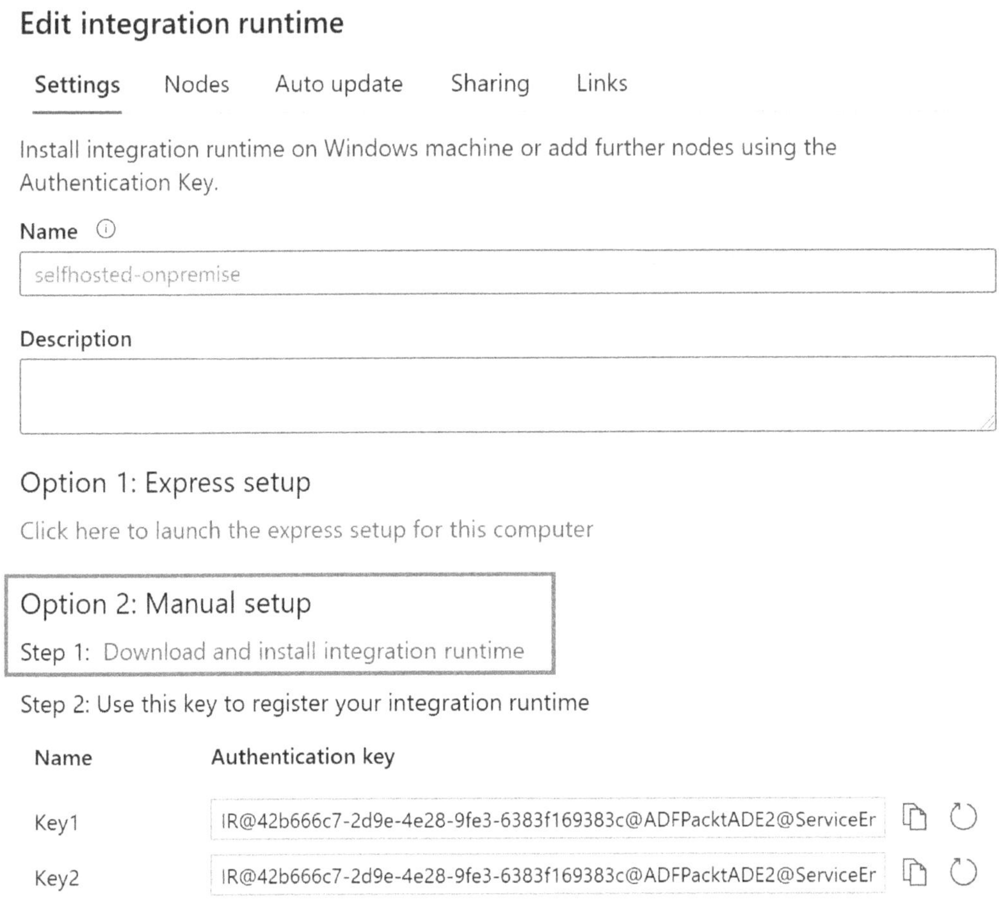

Figure 4.27 – Download the IR

5.   Download the version of the IR noted in *step 2*. In my case, it is **IntegrationRuntime_5.19.8214.2.msi**.

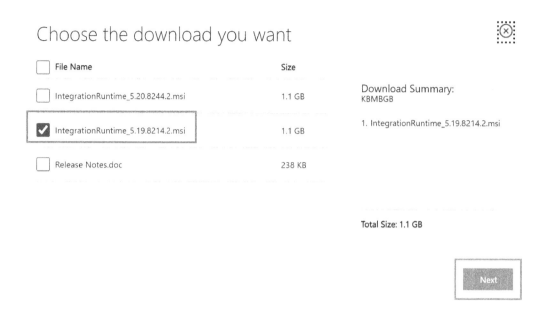

Figure 4.28 – Choose the IR version

6. To add a secondary node, the **Remote access from intranet** setting needs to be enabled on the active node of the self-hosted IR machine via the Microsoft Integration Runtime tool. Log in to the **SelfhostedIR1** machine. Go to the **Start** menu and find **Microsoft Integration Runtime**. On the **Settings** tab, hit the **Change** button to enable **Remote access from intranet**:

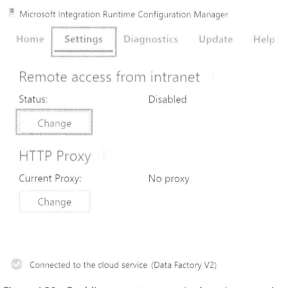

Figure 4.29 – Enabling remote access in the primary node

7. Select the **Enable without TLS/SSL certificate (Basic)** option. Make note of the TCP port 8060 used by the IR. Hit the **OK** button. The integration service on the primary node will restart for the change to take effect:

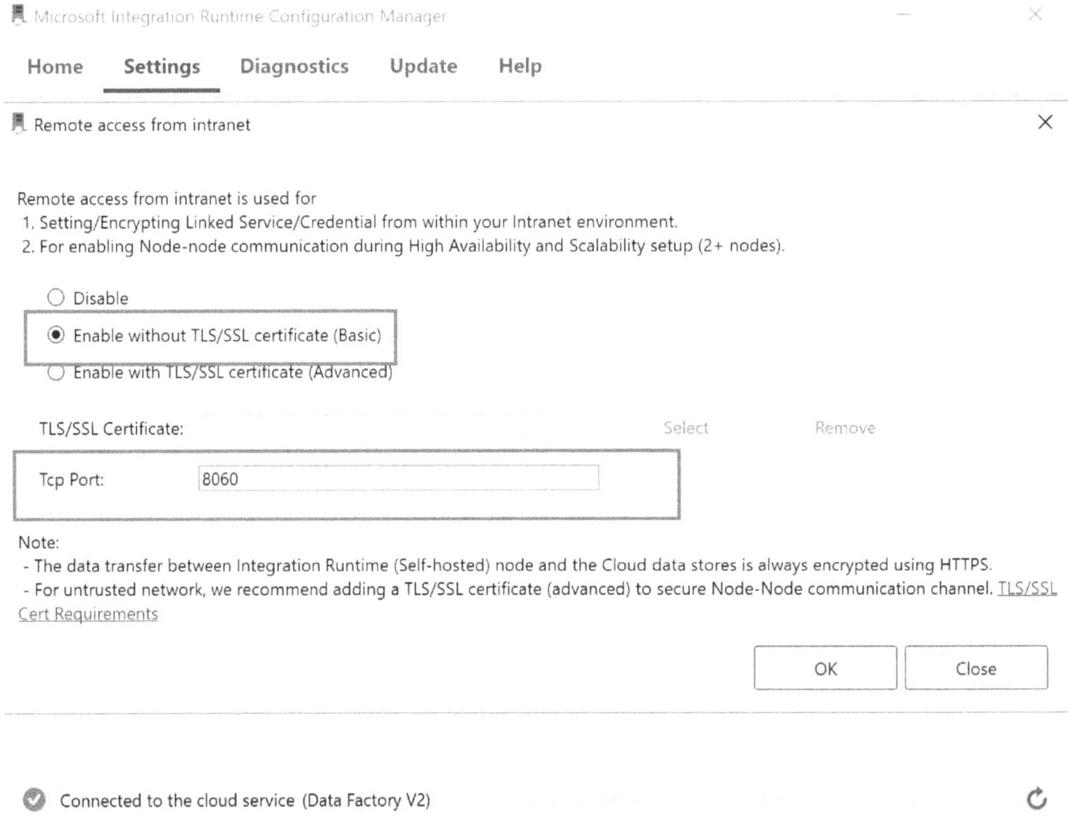

Figure 4.30 – Enabling the basic option

8. Log into the machine that'll be used as the additional node as secondary node for the IR. Copy the **IntegrationRuntime_5.19.8214.2.msi** file downloaded to the machine to be used as a secondary node for the IR. In this recipe, the machine name is **SelfHostedIR2**.

9. Install the **IntegrationRuntime_5.19.8214.2.msi** file in **SelfHostedIR2**. At the end of the installation, you will be prompted to register the key, which we copied in *step 3*. Paste the key and click on the **Register** button:

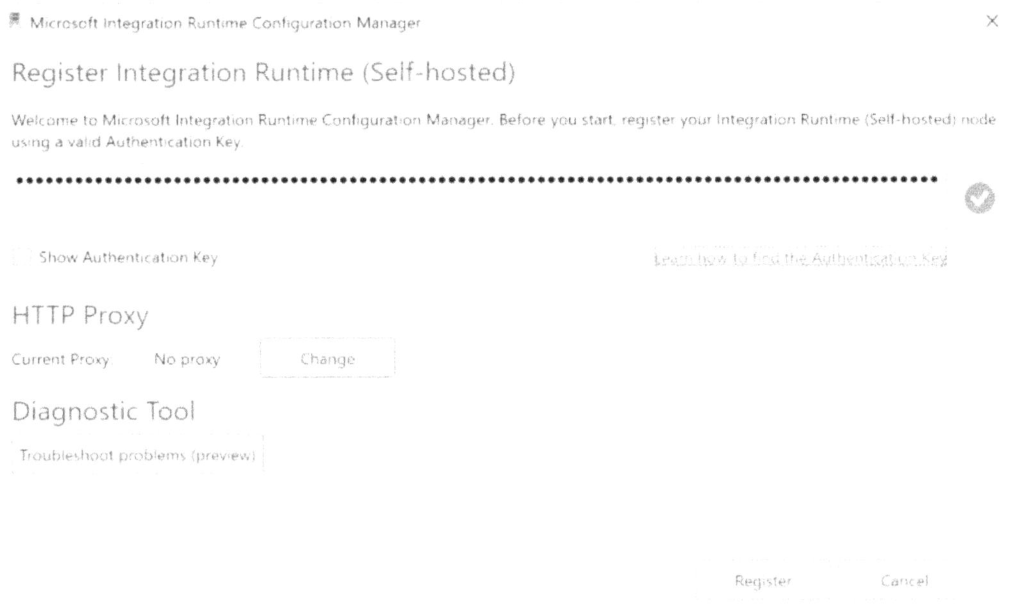

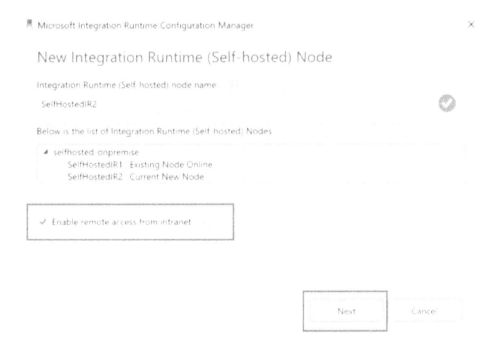

Figure 4.31 – Registering the key

10. Click on **Enable remote access from intranet** and click the **Next** button:

Figure 4.32 – Adding the secondary node

11. Hit the **Finish** button. Ensure to provide the same port number as the active node (the default is 8060). Ensure that port 8060 is open for communication between active and passive nodes and is not blocked by a firewall.

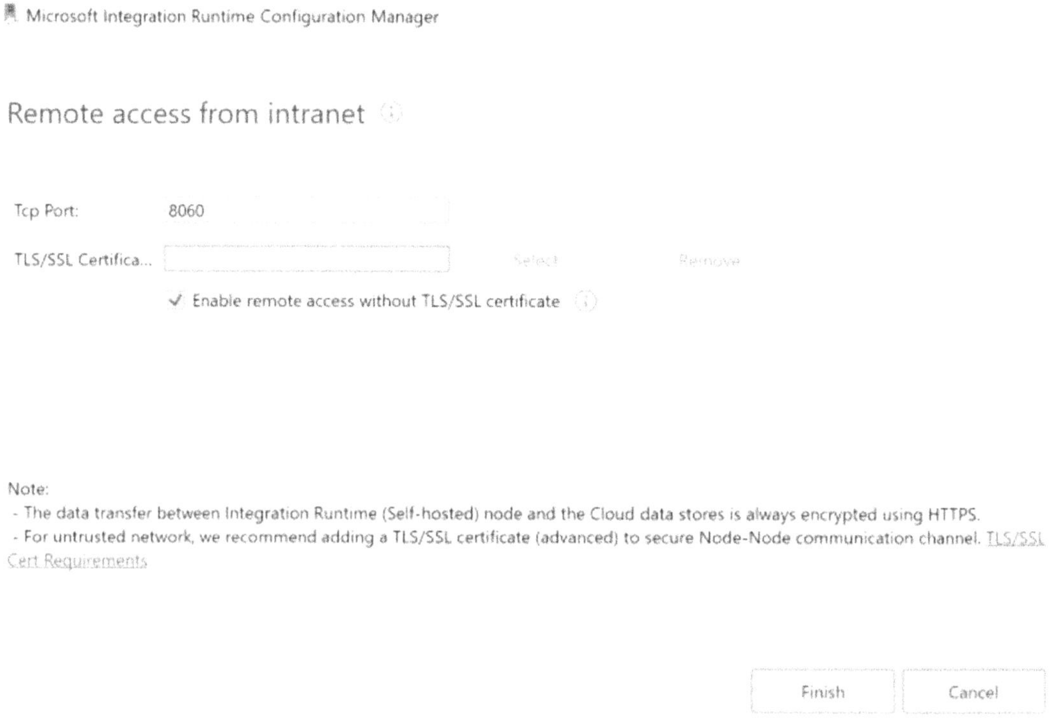

Microsoft Integration Runtime Configuration Manager

Remote access from intranet

Tcp Port: 8060

TLS/SSL Certifica... [   ]   Select   Remove

✓ Enable remote access without TLS/SSL certificate

Note:
- The data transfer between Integration Runtime (Self-hosted) node and the Cloud data stores is always encrypted using HTTPS.
- For untrusted network, we recommend adding a TLS/SSL certificate (advanced) to secure Node-Node communication channel. TLS/SSL Cert Requirements

Finish   Cancel

Figure 4.33 – Completing adding the secondary node

12. You should get the following screen after a successful registration. Hit the **Close** button to complete the process:

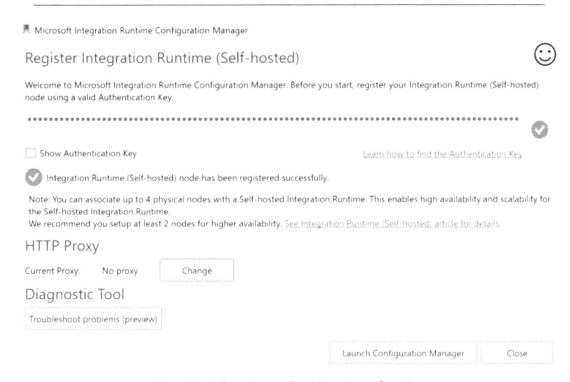

Figure 4.34 – Secondary node registration confirmation

13. You may verify the successful addition of a secondary node in `portal.azure.com`. Go to the Data Factory account created in this chapter. Click **Open Azure Data Factory Studio**. Click **Manage** -> **Integration runtimes** -> **selfhosted-onpremise**. Click the **Nodes** tab. Notice that the second node has been successfully added:

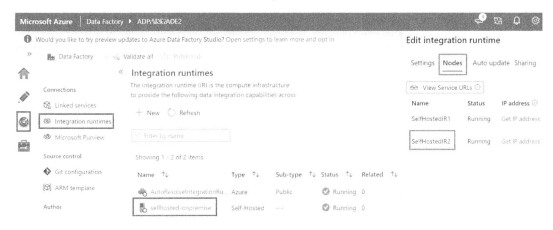

Figure 4.35 – Verifying the secondary node on the Azure portal

How it works...

Downloading the IR and installing it on a machine provides the libraries that are required for moving the data between an on-premises data source and Azure. Registering the key from the self-hosted IR that has already been provisioned in your Data Factory account helps the secondary node to identify the Data Factory account and the self-hosted IR, which it needs to register with and get provisioned as a secondary node to provide high availability and load-sharing capabilities. As of December 2021, you may add up to four nodes within a self-hosted IR. Having four nodes allows more data flows to be processed in parallel and offers greater high availability.

Patching a self-hosted IR

A self-hosted IR is a key component in Azure Data Factory, facilitating data movement between on-premises data sources and Azure. The self-hosted IR is configured by installing the Microsoft Integration Runtime installer on a machine that can connect to an on-premises data source and Azure. Keeping the self-hosted IR updated with the latest version is of the utmost importance to keep it bug-free and secure.

By default, a self-hosted IR has automatic updates enabled, which will apply the patches and keep the self-hosted IR updated. However, many users of Azure Data Factory would prefer to apply the patches themselves, as they would like to test their data transfer tasks and pipelines in a **User Acceptance Test** (**UAT**) environment with the latest patches, before running them in a production environment. So, in this recipe we will perform the following:

- Disabling the automatic patch update
- Installing the latest patches for a self-hosted IR

Getting ready

Perform the following steps before working on this recipe:

1. Log in to `https://portal.azure.com` using your Azure credentials.
2. You will need an existing Data Factory account. If you don't have one, create one by executing a PowerShell script. Open a new PowerShell prompt. Execute the following commands to log in to your Azure account from PowerShell and create a Data Factory account:

```
Connect-AzAccount
$resourceGroupName = "PacktADE";
$location = 'east us'
$dataFactoryName = "ADFPacktADE2";
```

```
$DataFactory = Set-AzDataFactoryV2 -ResourceGroupName
$resourceGroupName -Location $location -Name
$dataFactoryName
```

3. Configure the self-hosted IR, as explained in the *Configuring a self-hosted IR* section of this chapter.

How to do it...

We have to perform two major tasks – disabling the auto update, followed by applying the latest patch. Let's perform the following steps to complete the tasks:

1. Log in to `portal.azure.com`. Go to Data Factory. Click **Open Azure Data Factory Studio**.

2. Click **Manage -> Integration runtimes**. Make a note of the version number of the self-hosted IR. Click on **selfhosted-onpremise**:

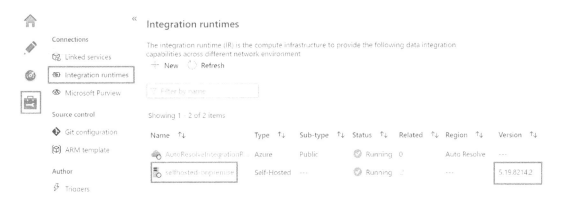

Figure 4.36 – Patching a self-hosted IR

3. Click **Auto update**. Set **Auto update** to **Disable** and hit the **Apply** button:

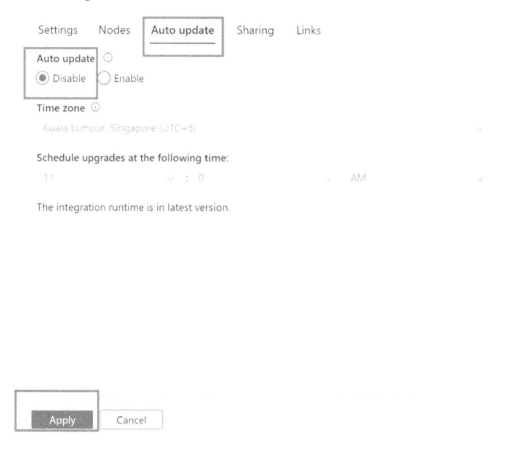

Figure 4.37– Disabling an auto update

4. To update the IR, click **selfhosted-onpremise** in **Integration runtimes** again.

5. Click **Download and install integration runtime** under **Option 2: Manual setup**:

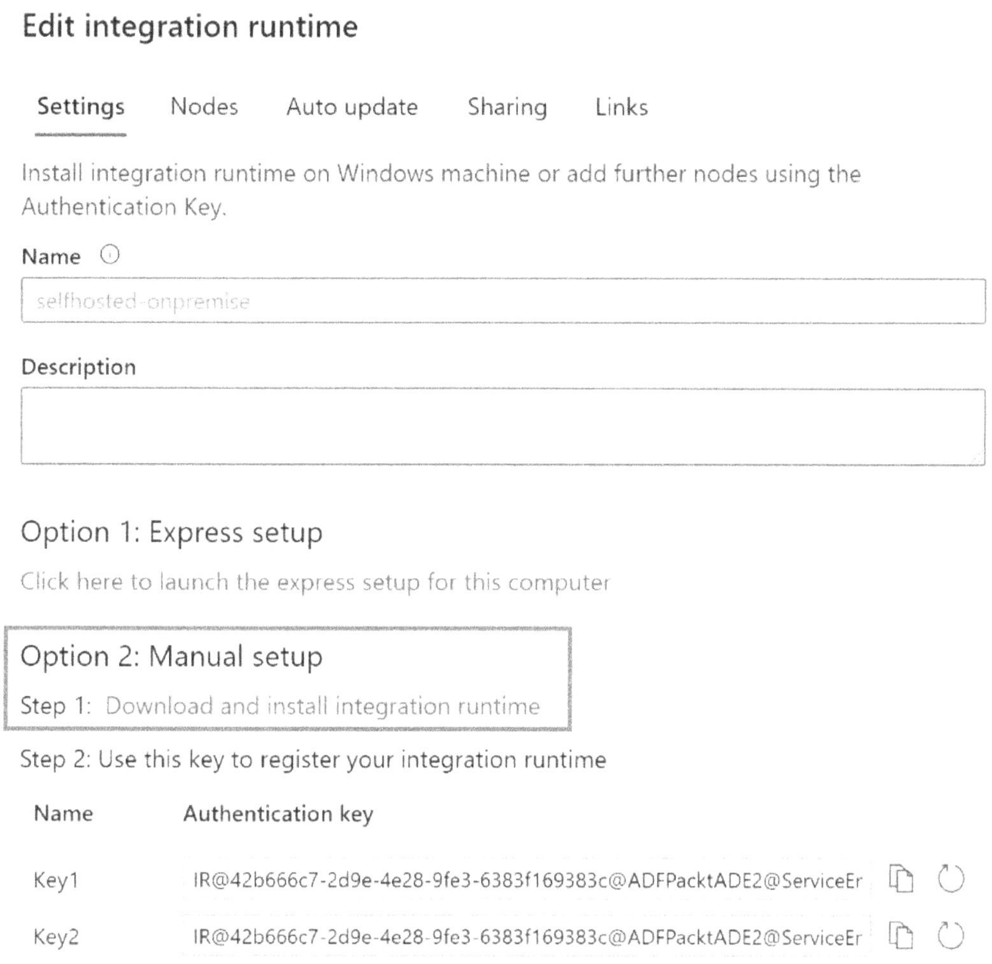

Edit integration runtime

Settings   Nodes   Auto update   Sharing   Links

Install integration runtime on Windows machine or add further nodes using the
Authentication Key.

Name ⓘ

selfhosted-onpremise

Description

Option 1: Express setup

Click here to launch the express setup for this computer

Option 2: Manual setup

Step 1: Download and install integration runtime

Step 2: Use this key to register your integration runtime

Name	Authentication key		
Key1	IR@42b666c7-2d9e-4e28-9fe3-6383f169383c@ADFPacktADE2@ServiceEr	📋	↻
Key2	IR@42b666c7-2d9e-4e28-9fe3-6383f169383c@ADFPacktADE2@ServiceEr	📋	↻

Figure 4.38 – Downloading the patches

6. Select the latest version to download, as long as it is higher than the version of self-hosted IR
you recorded in *step 2*:

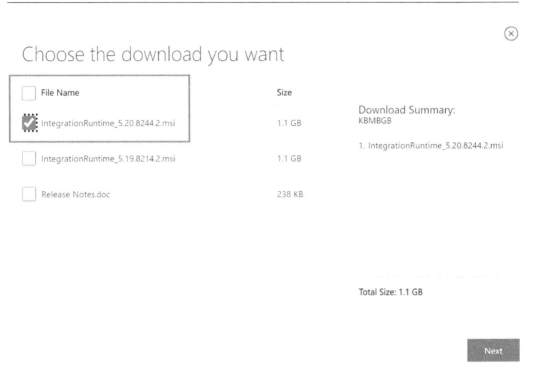

Figure 4.39 – Downloading the latest patches

7.   Copy the installer to the self-hosted runtime machine and follow the installation instructions on the screen to complete the installation of the `.msi` file:

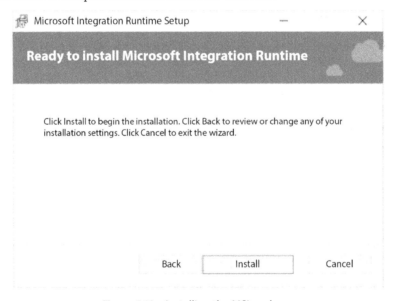

Figure 4.40 – Installing the MSI package

8. Upon completion of the installation, the **Microsoft Integration Runtime** application will automatically open and show that it is successfully connected to Azure Data Factory, as shown in the following screenshot:

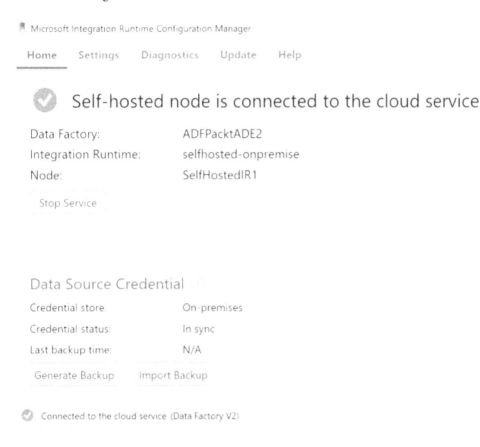

Figure 4.41 – Patch installation completion

9. To confirm the patch has been applied successfully, let's verify it from Azure portal. Go to portal.azure.com. Go to the Data Factory account that you created earlier in this chapter. Click **Open Azure Data Factory Studio**. Click the **Monitor** button on the left-hand side of the screen. Click **Integration runtimes** and then click **selfhosted-onpremise**:

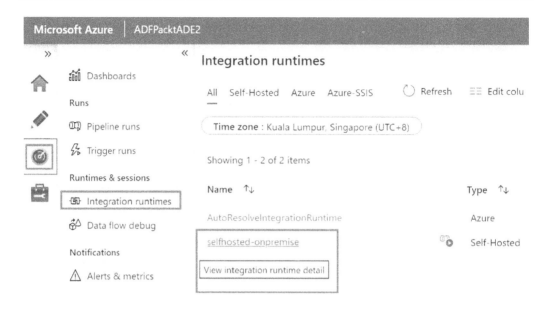

Figure 4.42 – Verifying the patch installation

In the **Node Details** section, verify whether the version number of the node has been updated to the version that was installed:

Figure 4.43 – Verifying patch installation with the version number

10. Repeat the preceding nine steps for each node in your self-hosted IR if your self-hosted integration setup contains multiple nodes.

How it works...

Disabling the auto update of the self-hosted IR ensures that patches are not rolled out automatically. While auto update is convenient, applying patches manually gives better control, especially when we are running mission-critical pipelines, where testing in a UAT environment is mandatory before the production rollout.

Downloading the latest patches and applying them on **SelfHostedIR1** works smoothly and doesn't demand providing the IR keys again, as the self-hosted IR is already connected to Azure Data Factory. Post-update, the version number of the node installed is also reflected in Data Factory Studio in the Azure portal in the **Monitoring** section for the self-hosted IR.

During the upgrade, active data flows may experience connectivity failures, and you are recommended to configure high availability, as explained in the *Configuring high availability for a self-hosted IR* section of this chapter.

Migrating an SSIS package to Azure Data Factory

SQL SSIS is a widely used on-premises **Extract, Transform, and Load** (ETL) tool. In this recipe, we'll learn how to migrate an existing SSIS package to Azure Data Factory.

We'll do this by configuring an Azure SSIS IR, uploading the SSIS package to the Azure SQL Database SSISDB, and then executing the package using the **Execute SSIS Package** activity.

Getting ready

To get started, do the following:

1. Open a new PowerShell prompt. Execute the following command to log in to your Azure account from PowerShell:

    ```
    Connect-AzAccount
    ```

2. You will need an existing Data Factory account. If you don't have one, create one by executing the following PowerShell script:

    ```
    $resourceGroupName = "PacktADE";
    $location = 'east us'
    $dataFactoryName = "ADFPacktADE2";
    $DataFactory = Set-AzDataFactoryV2 -ResourceGroupName
    $resourceGroupName -Location $location -Name
    $dataFactoryName
    ```

3. Download the `Azure-SSIS` folder from `https://github.com/PacktPublishing/`
    `Azure-Data-Engineering-Cookbook-2nd-edition/tree/main/Chapter04/`.
    You may fork the entire repository to your GitHub account and download it as a ZIP file to
    your local machine.

4. In PowerShell, go to the `Azure-SSIS` folder and run the following command:

    ```
    .\UploadDatatoAzureStorage "PacktADE" "packtadechapter4"
    "east us" "Data" 1 1
    ```

    This command creates a storage account named `packtadechapter4` and pushes all the
    files in the `Data` folder into the blob storage account. You should have an output similar to
    the one shown in the following screenshot:

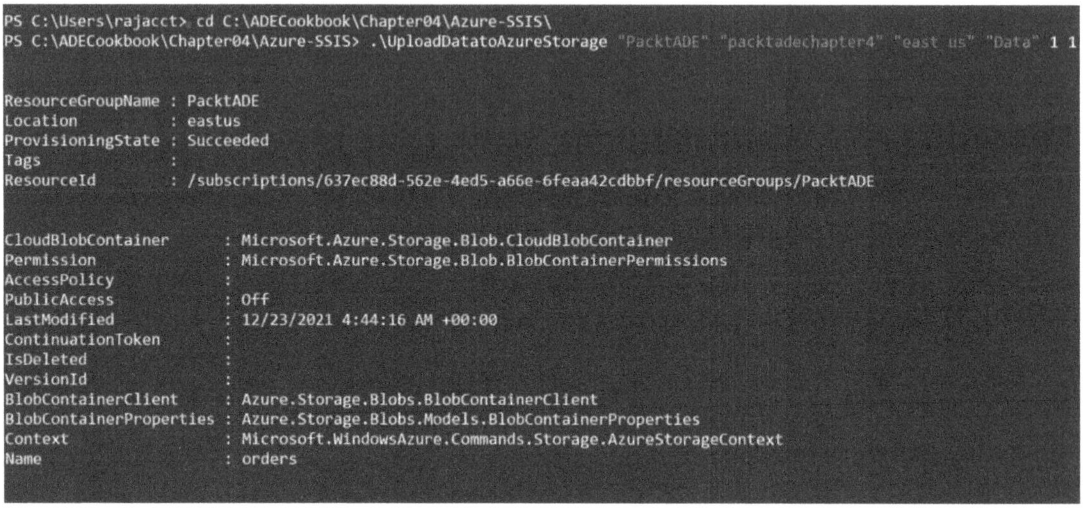

```
PS C:\Users\rajacct> cd C:\ADECookbook\Chapter04\Azure-SSIS\
PS C:\ADECookbook\Chapter04\Azure-SSIS> .\UploadDatatoAzureStorage "PacktADE" "packtadechapter4" "east us" "Data" 1 1


ResourceGroupName : PacktADE
Location          : eastus
ProvisioningState : Succeeded
Tags              :
ResourceId        : /subscriptions/637ec88d-562e-4ed5-a66e-6feaa42cdbbf/resourceGroups/PacktADE


CloudBlobContainer      : Microsoft.Azure.Storage.Blob.CloudBlobContainer
Permission              : Microsoft.Azure.Storage.Blob.BlobContainerPermissions
AccessPolicy            :
PublicAccess            : Off
LastModified            : 12/23/2021 4:44:16 AM +00:00
ContinuationToken       :
IsDeleted               :
VersionId               :
BlobContainerClient     : Azure.Storage.Blobs.BlobContainerClient
BlobContainerProperties : Azure.Storage.Blobs.Models.BlobContainerProperties
Context                 : Microsoft.WindowsAzure.Commands.Storage.AzureStorageContext
Name                    : orders
```

Figure 4.44 – A file upload to the blob storage

How to do it...

We'll start by creating a new Azure SSIS IR:

1. In the Azure portal, open the Data Factory **Manage** tab. Select **Integration runtimes**, and then
    select + **New**. Provide the name, location, node size, and node number. We have kept the node
    size and node number to the smallest available ones to save costs:

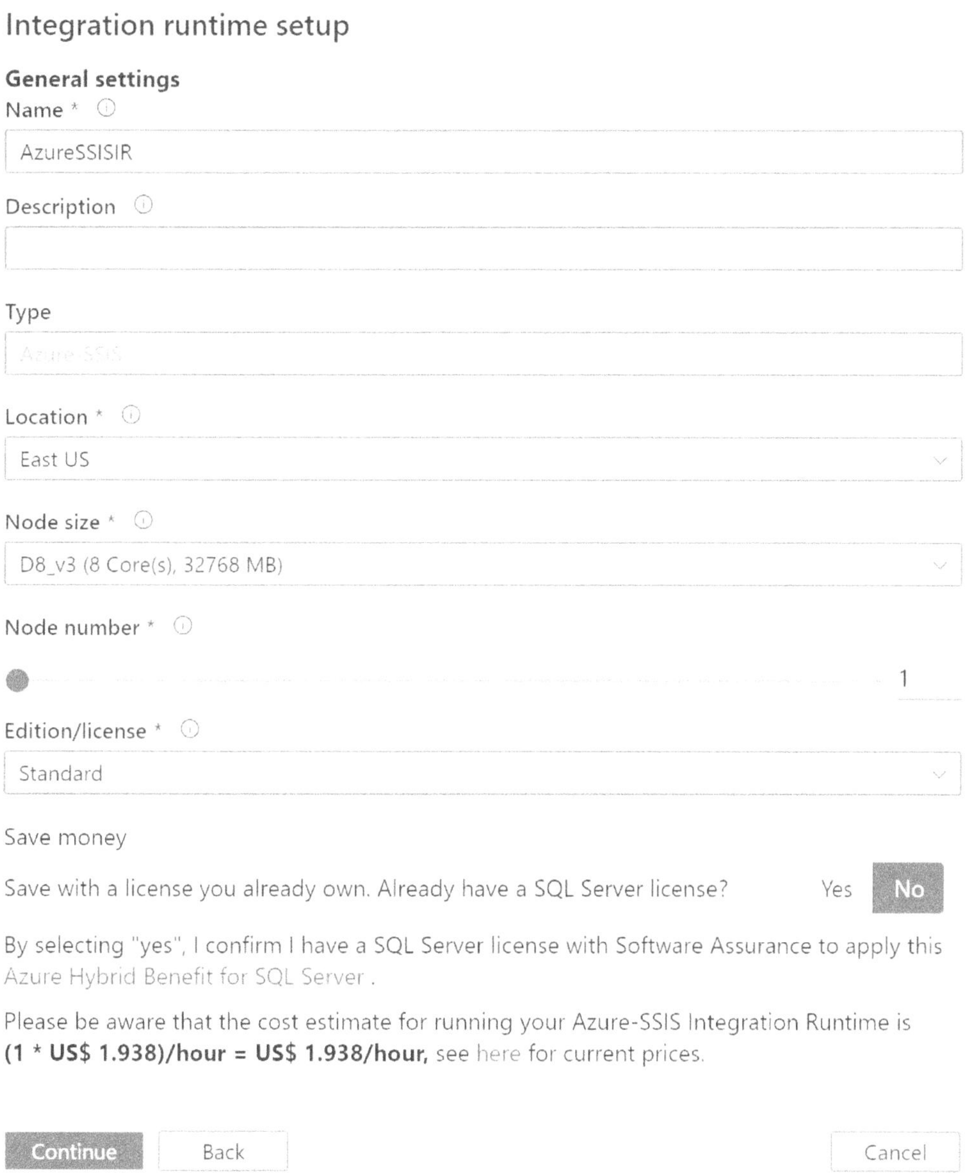

Integration runtime setup

General settings

Name * ⓘ

AzureSSISIR

Description ⓘ

Type

Azure-SSIS

Location * ⓘ

East US

Node size * ⓘ

D8_v3 (8 Core(s), 32768 MB)

Node number * ⓘ

●─────────────────────────────────────── 1

Edition/license * ⓘ

Standard

Save money

Save with a license you already own. Already have a SQL Server license?   Yes   No

By selecting "yes", I confirm I have a SQL Server license with Software Assurance to apply this Azure Hybrid Benefit for SQL Server .

Please be aware that the cost estimate for running your Azure-SSIS Integration Runtime is **(1 * US$ 1.938)/hour = US$ 1.938/hour,** see here for current prices.

Continue   Back   Cancel

Figure 4.45 – Configuring the Azure SSIS IR

Click **Continue** to go to the next step.

We can either host the SSIS package in an SSISDB database, hosted either on Azure SQL Database or SQL Managed Instance, or we can host the package on Azure Storage. In this recipe, we'll host the package on SSISDB.

2. In the **Deployment settings** section, check the **Create SSIS catalog...** checkbox and provide the Azure SQL Database details, as shown in the following screenshot. You may have to provision Azure SQL Database if you don't have these details. The PowerShell script to provision Azure SQL Database is provided at `https://github.com/PacktPublishing/Azure-Data-Engineering-Cookbook-2nd-edition/upload/main/Chapter04/Azure-SSIS/AzureSQLDB.ps1`. We have set **Catalog database service tier** to **Basic** to save on costs.

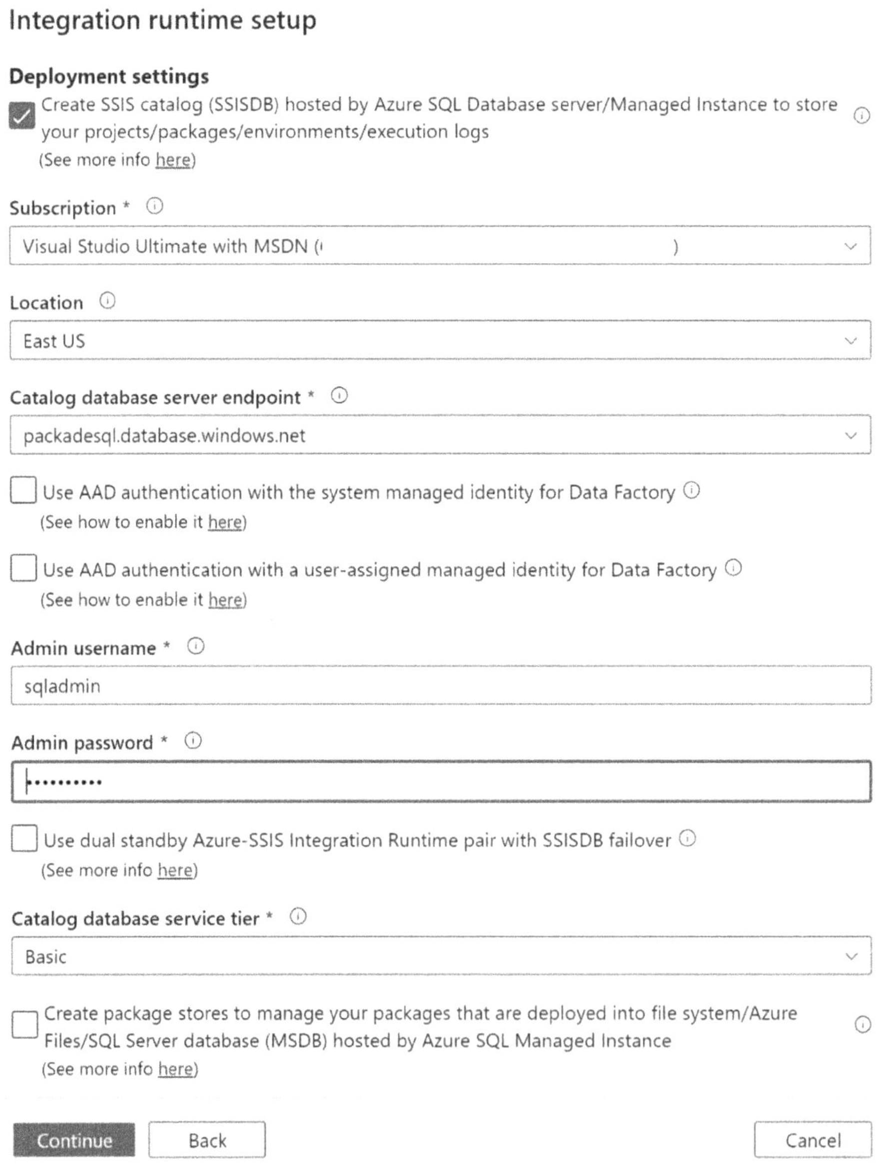

Integration runtime setup

Deployment settings

☑ Create SSIS catalog (SSISDB) hosted by Azure SQL Database server/Managed Instance to store ⓘ
your projects/packages/environments/execution logs
(See more info here)

Subscription * ⓘ

Visual Studio Ultimate with MSDN (   )   ∨

Location ⓘ

East US   ∨

Catalog database server endpoint * ⓘ

packadesql.database.windows.net   ∨

☐ Use AAD authentication with the system managed identity for Data Factory ⓘ
(See how to enable it here)

☐ Use AAD authentication with a user-assigned managed identity for Data Factory ⓘ
(See how to enable it here)

Admin username * ⓘ

sqladmin

Admin password * ⓘ

|··········

☐ Use dual standby Azure-SSIS Integration Runtime pair with SSISDB failover ⓘ
(See more info here)

Catalog database service tier * ⓘ

Basic   ∨

☐ Create package stores to manage your packages that are deployed into file system/Azure ⓘ
Files/SQL Server database (MSDB) hosted by Azure SQL Managed Instance
(See more info here)

Continue	Back		Cancel

Figure 4.46 – Configuring the SSISDB catalog

Click **Continue** to go to the next step.

3. On the **Advanced settings** page, set parallel executions to **1** and uncheck the **Select a VNet for your Azure-SSIS Integration Runtime to join...** option, as shown in the following screenshot, and click **Continue** to create the Azure SSIS IR:

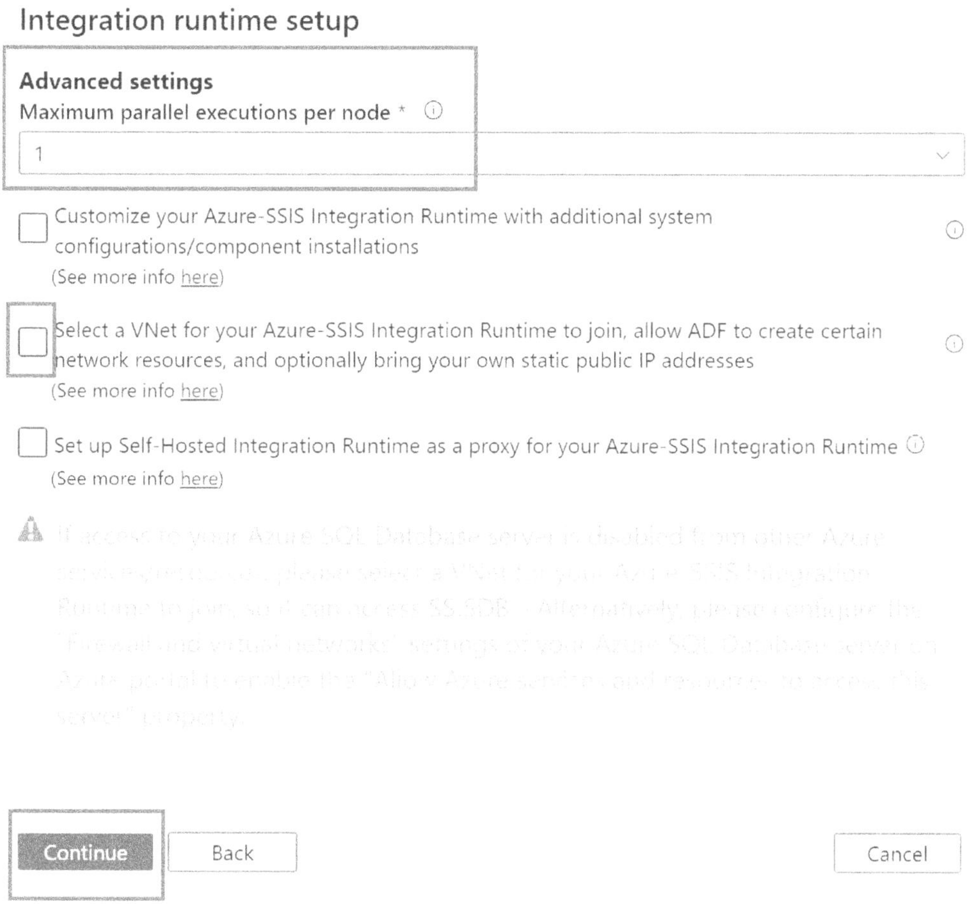

Integration runtime setup

Advanced settings

Maximum parallel executions per node *   ⓘ

| 1 | ⌄ |

☐ Customize your Azure-SSIS Integration Runtime with additional system
configurations/component installations   ⓘ
(See more info here)

☐ Select a VNet for your Azure-SSIS Integration Runtime to join, allow ADF to create certain
network resources, and optionally bring your own static public IP addresses   ⓘ
(See more info here)

☐ Set up Self-Hosted Integration Runtime as a proxy for your Azure-SSIS Integration Runtime ⓘ
(See more info here)

| Continue | Back | | Cancel |

Figure 4.47 – Configuring advanced settings

When created, the **AzureSSISIR** IR is now listed, as in the following screenshot:

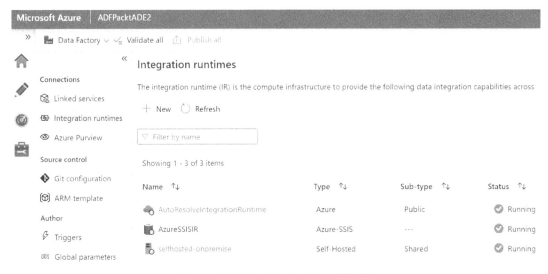

Figure 4.48 – Viewing the Azure SSIS IR

4. The next step is to deploy the SSIS package to the Azure SQL Database SSISDB catalog. To do that, open **SQL Server Management Studio** (**SSMS**), and connect to the **SSISDB** in Azure SQL Database (`packadesql.database.windows.net`) in **Object Explorer**. In the **Object Explorer** window, expand the **Integration Services Catalogs** folder, right-click **SSISDB**, and then select **Create Folder…**:

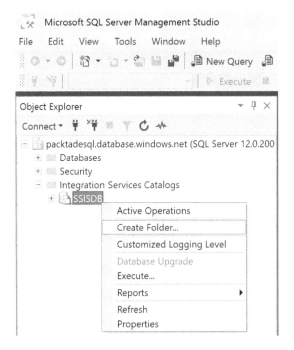

Figure 4.49 – Creating a folder in the SSISDB catalog

5. In the **Create Folder** dialog box, set **Folder name** to AzureSSIS:

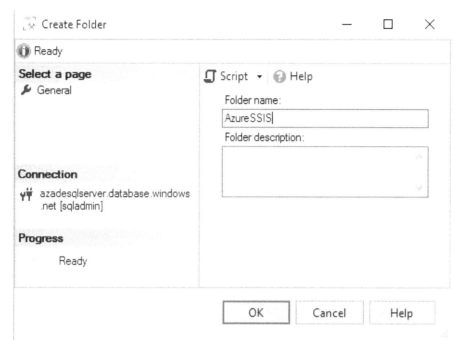

Figure 4.50 – Providing the folder settings

Click **OK** to create the folder.

6. In the **Object Explorer** window, expand the AzureSSIS folder, right-click **Project**, and select **Deploy Project...** from the context menu:

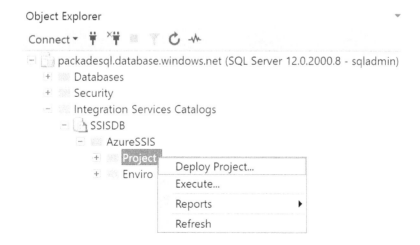

Figure 4.51 – Opening the SSIS deployment wizard

7. In the **Integration Services Deployment Wizard** window, select the **Select Source** tab, browse to the \chapter4\Azure-SSIS\ path, and select the CopyFiles.ispac file. If you haven't downloaded the CopyFiles.ispac file, download it from https://github.com/PacktPublishing/Azure-Data-Engineering-Cookbook-2nd-edition/tree/main/Chapter04/Azure-SSIS/:

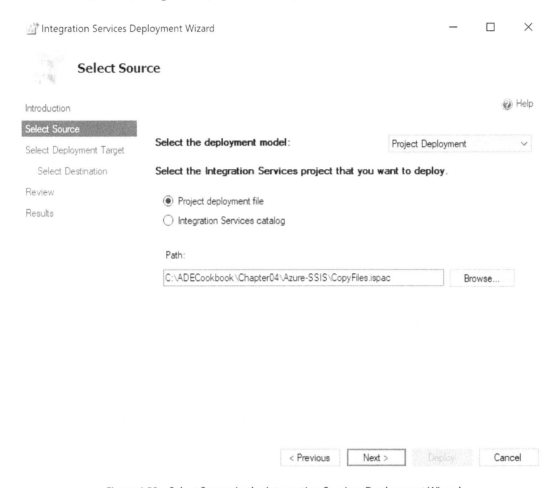

Figure 4.52 – Select Source in the Integration Services Deployment Wizard

> **Note**
> The SSIS package copies the files from an Azure storage account from the orders/datain folder to the orders/dataout folder. The Azure storage account name and key are passed as parameters.

Click **Next** to go to the next step.

8. In the **Select Deployment Target** tab, select **SSIS in Azure Data Factory**:

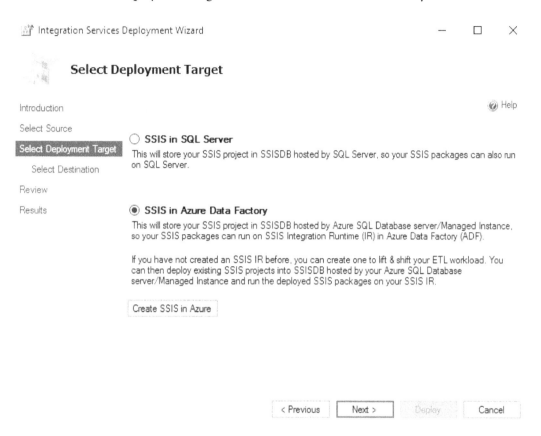

Figure 4.53 – Select Deployment Target in the Integration Services Deployment Wizard

Click **Next** to go to the next step.

9. In the **Select Destination** tab, provide the Azure SQL Server admin user and password, and click **Connect** to test the connection:

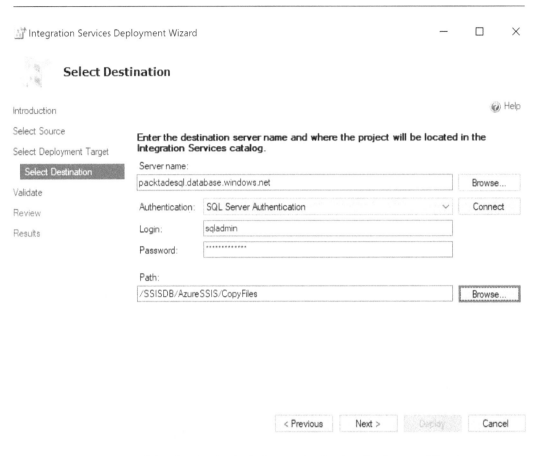

Figure 4.54 – Select Destination in the Integration Services Deployment Wizard

10. After a successful connection, click **Next**. The deployment wizard will validate that the package is without issues. After successful validation, click **Next** and then **Deploy** to complete the SSIS package deployment:

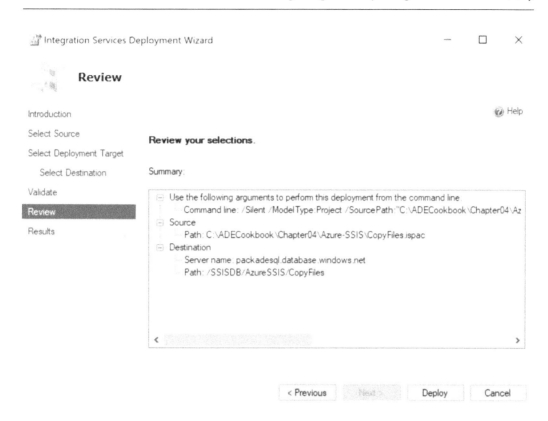

Figure 4.55 – Review and Deploy in the Integration Services Deployment Wizard

After a successful deployment, close the wizard.

11. In the **Object Explorer** window in SSMS, refresh the AzureSSIS folder. Observe that the CopyFiles package is now listed under the AzureSSIS | Projects folder.

We'll now create an Azure Data Factory pipeline to execute the package we deployed to the SSISDB catalog. To do this, switch to the Azure portal. Open the Data Factory **Author** tab. Create a new pipeline called Pipeline-AzureSSIS-IR. Drag and drop the **Execute SSIS package** activity from the **Activities** section to the **General** section. In the **General** tab, name the activity Execute CopyFiles SSIS Package. Switch to the **Settings** tab. Select **AzureSSISIR** in the **Azure-SSIS IR** drop-down list. Set the package location as SSISDB. Select AzureSSIS from the **Folder** drop-down list. If you don't see the folder name in the drop-down list, click **Refresh**. Select **CopyFiles** in the **Project** drop-down list. Select CopyFiles.dtsx in the **Package** drop-down list:

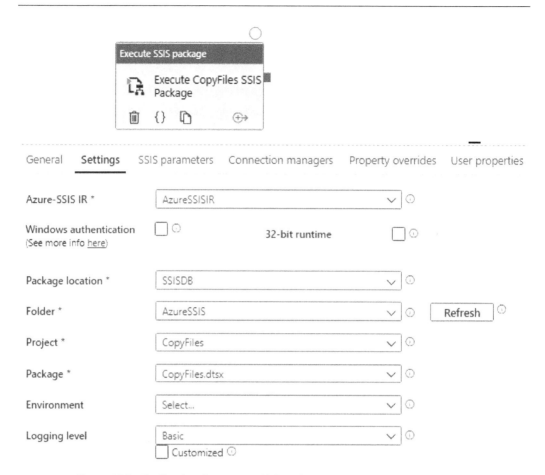

Figure 4.56 – Configuring the Execute SSIS package activity using the Settings tab

12. Switch to the **SSIS parameters** tab. Provide the **StorageAccountKey** value and the **StorageAccountName** value for the Azure Storage account:

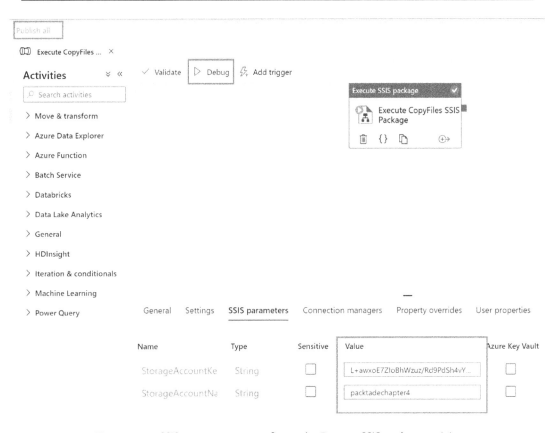

Figure 4.57 – SSIS parameters to configure the Execute SSIS package activity

> **Note**
> Make sure that the storage account you use has the `orders` container. You'll have to upload the
> `~/Chapter04/Azure-SSIS/Data/orders1.txt` file to the `orders/datain` folder.

13. Click **Publish all** to save your work. Click **Debug** to run the package. Once the package is complete, you should get an output similar to the one shown in the following screenshot:

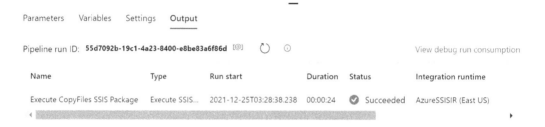

Figure 4.58 – Viewing the output

Observe that the IR used is `AzureSSISIR`.

> **Note**
>
> To stop the `Azure-SSIS` IR, navigate to **Data Factory** | **Manage** | **Integration runtimes**.
>
> Hover the mouse over the `Azure-SSIS` IR and click the **Pause** button.

How it works...

The Azure-SSIS IR lets us move on-premises SSIS packages to Azure Data Factory with few or no changes. We performed the following steps to move the `CopyFiles` SSIS package to Azure with no code change:

1. We created an Azure SSIS IR in our **ADFPacktADE2** Data Factory account.

2. In this recipe, we decided to store the package (`CopyFiles`) in Azure SQL Database's SSISDB. So, while configuring the Azure SSIS IR, we selected the option to create the SSISDB in Azure SQL Database (`packadesql.database.windows.net`). The SSISDB was provisioned while configuring the Azure SSIS IR.

3. We had to upload the `CopyFiles` package to the SSISDB. So, we connected to `packadesql.database.windows.net` using SSMS, created a project and folder inside the SSISDB, and uploaded the `CopyFiles` package into the folder using the `CopyFiles.ispac` package file.

4. We created the Data Factory pipeline, added the **Execute SSIS package** task, and configured it to use the `CopyFiles` package stored in the SSISDB. We also passed the storage account name (**packtadechapter4**) and its key as a parameter to the SSIS package so that it could move the files to the storage account.

5. We verified the successful execution by clicking the **Debug** button to run the package once. You can also verify it to see whether the `Order1.txt` file was successfully uploaded to the `orders/dataout` folder in **packtadechapter4**.

5

Configuring and Securing Azure SQL Database

Azure SQL Database, a fundamental relational database as a service offered in Azure, acts as a source, destination, or even as an intermediate storage layer in data engineering pipelines. Azure SQL Database can be used to consolidate data coming from several relational data sources and build mini data warehouses or data marts. With the introduction of Hyperscale tier in Azure SQL Database, the capacity of Azure SQL Database has increased leaps and bounds too. Securing Azure SQL Database is also pivotal in protecting access to the database. Having a strong understanding of Azure SQL Database's capabilities and security options is essential for any data engineer.

In this chapter, we will learn how to provision a serverless Azure SQL database, secure its connectivity to private links, integrate with Azure Key Vault to secure its credentials, configure a wake-up script to start a serverless Azure SQL database, and also configure the Hyperscale tier of Azure SQL Database.

In this chapter, we'll cover the following recipes:

- Provisioning and connecting to an Azure SQL database using PowerShell
- Implementing an Azure SQL Database elastic pool using PowerShell
- Configuring a virtual network and private endpoints for Azure SQL Database
- Configuring Azure Key Vault for Azure SQL Database
- Provisioning and configuring a wake-up script for a serverless SQL database
- Configuring the Hyperscale tier of Azure SQL Database

Technical requirements

For this chapter, the following are required:

- A Microsoft Azure subscription.

- PowerShell 7 or above.

- Microsoft Azure PowerShell – the installation instructions can be found at `https://docs.microsoft.com/en-us/powershell/azure/install-az-ps?view=azps-8.0.0`.

Provisioning and connecting to an Azure SQL database using PowerShell

In this recipe, we'll learn how to create and connect to an Azure SQL database instance. Azure SQL Database comes in three flavors: **standalone Azure SQL Database**, **Azure SQL Database elastic pools**, and **managed instances**. In this recipe, we'll create a standalone Azure SQL database.

Getting ready

In a new PowerShell window, execute the `Connect-AzAccount` command to log in to your Microsoft Azure account.

How to do it...

Let's begin by provisioning an Azure SQL database.

Provisioning an Azure SQL database

Execute the following steps to provision an Azure SQL database:

1. Execute the following PowerShell command to create a new resource group:

   ```
   New-AzResourceGroup -Name packtadesql -Location "eastus"
   ```

2. Execute the following command to create a new Azure SQL server:

   ```
   #create credential object for the Azure SQL Server admin
   credential
   $sqladminpassword = ConvertTo-SecureString 'Sql@
   Server@1234' -AsPlainText -Force
   $sqladmincredential = New-Object System.Management.
   Automation.PSCredential('sqladmin', $sqladminpassword)
   ```

```
# create the Azure SQL Server
New-AzSqlServer -ServerName azadesqlserver
-SqlAdministratorCredentials $sqladmincredential
-ResourceGroupName packtadesql -Location "eastus"
```

You should get an output similar to the one in the following screenshot:

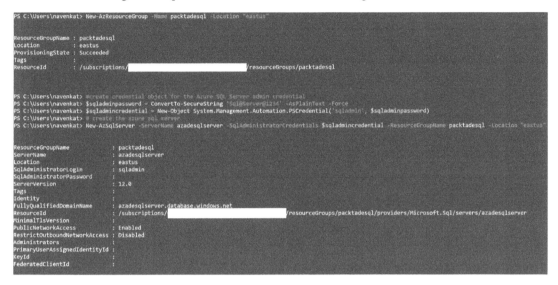

Figure 5.1 – Creating a new Azure SQL server

3. Execute the following command to create a new Azure SQL database:

```
New-AzSqlDatabase -DatabaseName azadesqldb -Edition basic
-ServerName azadesqlserver -ResourceGroupName packtadesql
```

You should get an output similar to the one shown in the following screenshot:

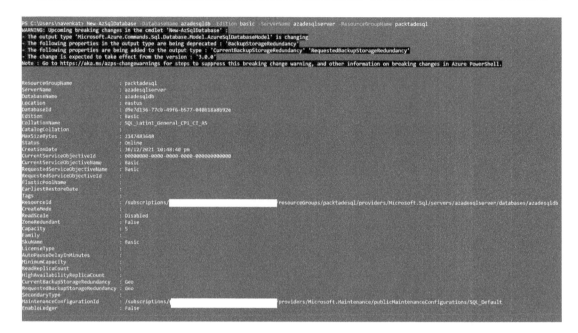

Figure 5.2 – Creating a new Azure SQL database

Connecting to an Azure SQL database

To connect to an Azure SQL database, let's first whitelist the IP address in the Azure SQL Server firewall:

1. Execute the following command to whitelist the public IP address of the machine to connect to an Azure SQL database (this recipe assumes that you are connecting from your local system. To connect from a system other than your local system, change the IP address in the following command). Execute the following command in the PowerShell window to whitelist the machine's public IP address in the Azure SQL Server firewall:

    ```
    $clientip = (Invoke-RestMethod -Uri https://ipinfo.io/
    json).ip
    New-AzSqlServerFirewallRule -FirewallRuleName "home"
    -StartIpAddress $clientip -EndIpAddress $clientip
    -ServerName azadesqlserver -ResourceGroupName packtadesql
    ```

    You will get an output similar to the one shown in the following screenshot:

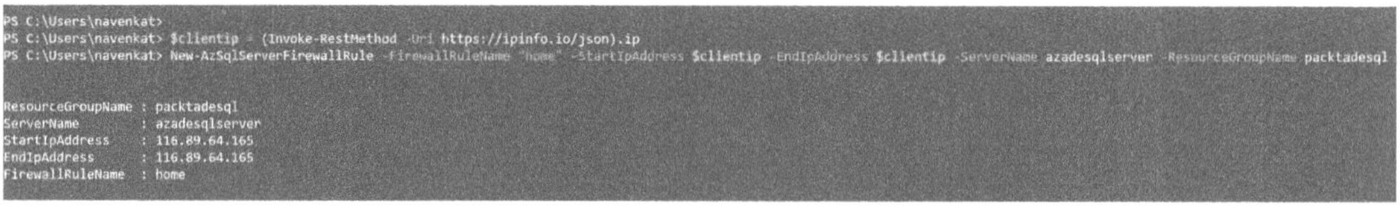

```
PS C:\Users\navenkat>
PS C:\Users\navenkat> $clientip = (Invoke-RestMethod -Uri https://ipinfo.io/json).ip
PS C:\Users\navenkat> New-AzSqlServerFirewallRule -FirewallRuleName "home" -StartIpAddress $clientip -EndIpAddress $clientip -ServerName azadesqlserver -ResourceGroupName packtadesql

ResourceGroupName : packtadesql
ServerName        : azadesqlserver
StartIpAddress    : 116.89.64.165
EndIpAddress      : 116.89.64.165
FirewallRuleName  : home
```

Figure 5.3 – Creating a new Azure SQL Server firewall rule

2. Execute the following command to connect to an Azure SQL database from SQLCMD (SQLCMD comes with the SQL Server installation, or you can download the SQLCMD utility from `https://docs.microsoft.com/en-us/sql/tools/sqlcmd-utility?view=sql-server-ver15`):

```
sqlcmd -S "azadesqlserver.database.windows.net" -U
sqladmin -P "Sql@Server@1234" -d azadesqldb -Q "Select
name from sys.databases"
```

Here's the output:

Figure 5.4 – Connecting to an Azure SQL database

How it works...

We first execute the `New-AzSQLServer` command to provision a new Azure SQL Server. The command accepts the server name, location, resource group, and login credentials.

An Azure SQL Server, unlike an on-premises SQL Server, is not a physical machine or a **virtual machine (VM)** that is accessible to customers.

We then execute the `New-AzSQLDatabase` command to create an Azure SQL database. This command accepts the database name, the Azure SQL Server name, the resource group, and the edition. There are multiple SQL database editions to choose from based on the application workload. However, for the sake of this demo, we will create a basic edition.

To connect to an Azure SQL database, we first need to whitelist the machine's IP address in the Azure SQL Server firewall. Only whitelisted IPs are allowed to connect to the database.

To whitelist the client's public IP, we use the `New-AzSQLServerFirewallRule` command. This command accepts the server name, resource group, and start and end IPs. We can whitelist either a single IP address or a range of IP addresses.

We can connect to an Azure SQL database from **SQL Server Management Studio (SSMS)**, SQLCMD, or Azure Data Studio, or with a programming language using the appropriate SQL Server drivers. When connecting to an Azure SQL database, we need to specify the server name as `azuresqlservername.database.windows.net`, and then specify the Azure SQL database to connect to.

Implementing an Azure SQL Database elastic pool using PowerShell

An elastic pool is a cost-effective mechanism to group single Azure SQL databases with varying peak usage times. For example, consider 20 different SQL databases with varying usage patterns, each S3 Standard storage class requiring 100 **database throughput units** (**DTUs**) to run. We need to pay for 100 DTUs separately. However, we can group all of them in an elastic pool of S3 Standard storage classes. In this case, we only need to pay for elastic pool pricing and not for each individual SQL database.

In this recipe, we'll create an elastic pool of multiple single Azure databases.

Getting ready

In a new PowerShell window, execute the `Connect-AzAccount` command and follow the steps to log in to your Azure account.

How to do it...

The steps for this recipe are as follows:

1. Execute the following query on an Azure SQL Server:

    ```
    #create credential object for the Azure SQL Server admin
    credential
    $sqladminpassword = ConvertTo-SecureString 'Sql@
    Server@1234' -AsPlainText -Force
    $sqladmincredential = New-Object System.Management.
    Automation.PSCredential('sqladmin', $sqladminpassword)
    # create the Azure SQL Server
    New-AzSqlServer -ServerName azadesqlserver
    -SqlAdministratorCredentials $sqladmincredential
    -Location "eastus" -ResourceGroupName packtadesql
    #Execute the following query to create an elastic pool.
    #Create an elastic pool
    New-AzSqlElasticPool -ElasticPoolName adepool -ServerName
    azadesqlserver -Edition standard -Dtu 100 -DatabaseDtuMin
    20 -DatabaseDtuMax 100 -ResourceGroupName packtadesql
    ```

You should get an output similar to the one shown in the following screenshot:

Figure 5.5 – Creating a new Azure elastic pool

2. Execute the following query to create and add an Azure SQL database to an elastic pool:

```
New-AzSqlDatabase -DatabaseName azadedb1 -ElasticPoolName
adepool -ServerName azadesqlserver -ResourceGroupName
packtadesql
```

You should get an output similar to the one shown in the following screenshot:

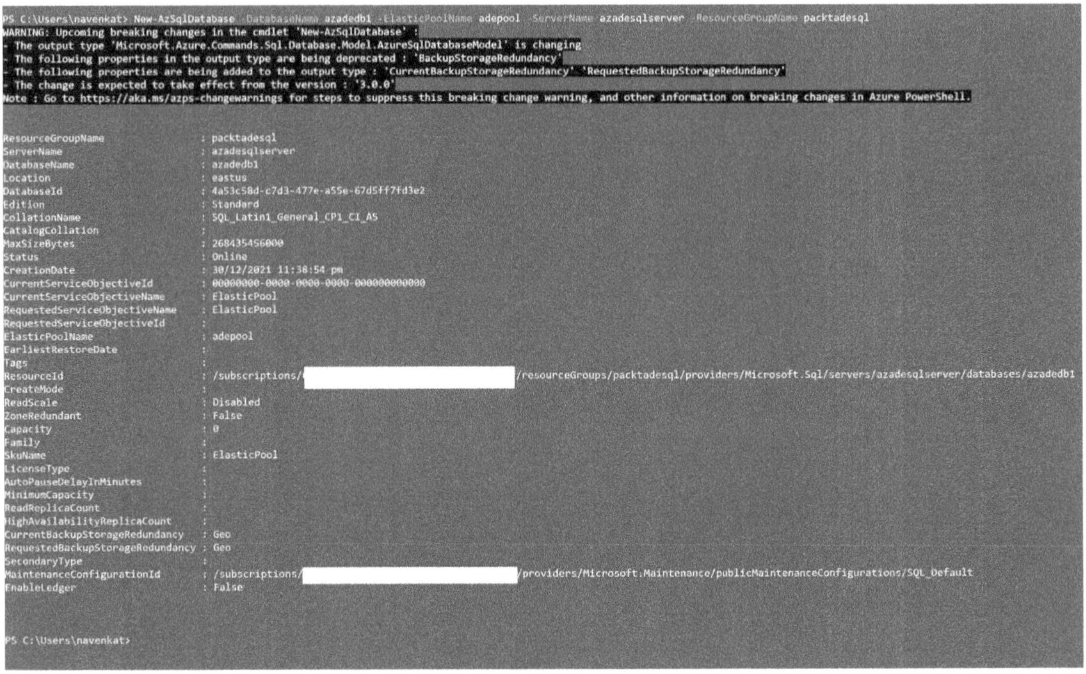

```
PS C:\Users\navenkat> New-AzSqlDatabase -DatabaseName azadedb1 -ElasticPoolName adepool -ServerName azadesqlserver -ResourceGroupName packtadesql
WARNING: Upcoming breaking changes in the cmdlet 'New-AzSqlDatabase' :
 - The output type 'Microsoft.Azure.Commands.Sql.Database.Model.AzureSqlDatabaseModel' is changing
 - The following properties in the output type are being deprecated : 'BackupStorageRedundancy'
 - The following properties are being added to the output type : 'CurrentBackupStorageRedundancy','RequestedBackupStorageRedundancy'
 - The change is expected to take effect from the version : '3.0.0'
Note : Go to https://aka.ms/azps-changewarnings for steps to suppress this breaking change warning, and other information on breaking changes in Azure PowerShell.


ResourceGroupName                 : packtadesql
ServerName                        : azadesqlserver
DatabaseName                      : azadedb1
Location                          : eastus
DatabaseId                        : 4a53c58d-c7d3-477e-a55e-67d5ff7fd3e2
Edition                           : Standard
CollationName                     : SQL_Latin1_General_CP1_CI_AS
CatalogCollation                  :
MaxSizeBytes                      : 268435456000
Status                            : Online
CreationDate                      : 30/12/2021 11:38:54 pm
CurrentServiceObjectiveId         : 00000000-0000-0000-0000-000000000000
CurrentServiceObjectiveName       : ElasticPool
RequestedServiceObjectiveName     : ElasticPool
RequestedServiceObjectiveId       :
ElasticPoolName                   : adepool
EarliestRestoreDate               :
Tags                              :
ResourceId                        : /subscriptions/                              /resourceGroups/packtadesql/providers/Microsoft.Sql/servers/azadesqlserver/databases/azadedb1
CreateMode                        :
ReadScale                         : Disabled
ZoneRedundant                     : False
Capacity                          : 0
Family                            :
SkuName                           : ElasticPool
LicenseType                       :
AutoPauseDelayInMinutes           :
MinimumCapacity                   :
ReadReplicaCount                  :
HighAvailabilityReplicaCount      :
CurrentBackupStorageRedundancy    : Geo
RequestedBackupStorageRedundancy  : Geo
SecondaryType                     :
MaintenanceConfigurationId        : /subscriptions/                              /providers/Microsoft.Maintenance/publicMaintenanceConfigurations/SQL_Default
EnableLedger                      : False


PS C:\Users\navenkat>
```

Figure 5.6 – Creating a new SQL database in an elastic pool

3.　Execute the following query to create a new Azure SQL database outside of the elastic pool:

```
New-AzSqlDatabase -DatabaseName azadedb2 -Edition
Standard -RequestedServiceObjectiveName S3 -ServerName
azadesqlserver -ResourceGroupName packtadesql
```

You should get an output similar to the one shown in the following screenshot:

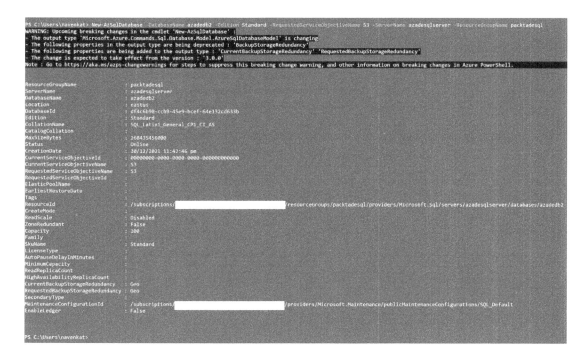

Figure 5.7 – Creating a new SQL database

4. Execute the following query to add the `azadedb2` database created in the preceding step to the elastic pool:

```
#Add an existing database to the elastic pool
$db = Get-AzSqlDatabase -DatabaseName azadedb2
-ServerName azadesqlserver -ResourceGroupName packtadesql
$db | Set-AzSqlDatabase -ElasticPoolName adepool
```

You should get an output similar to the one shown in the following screenshot:

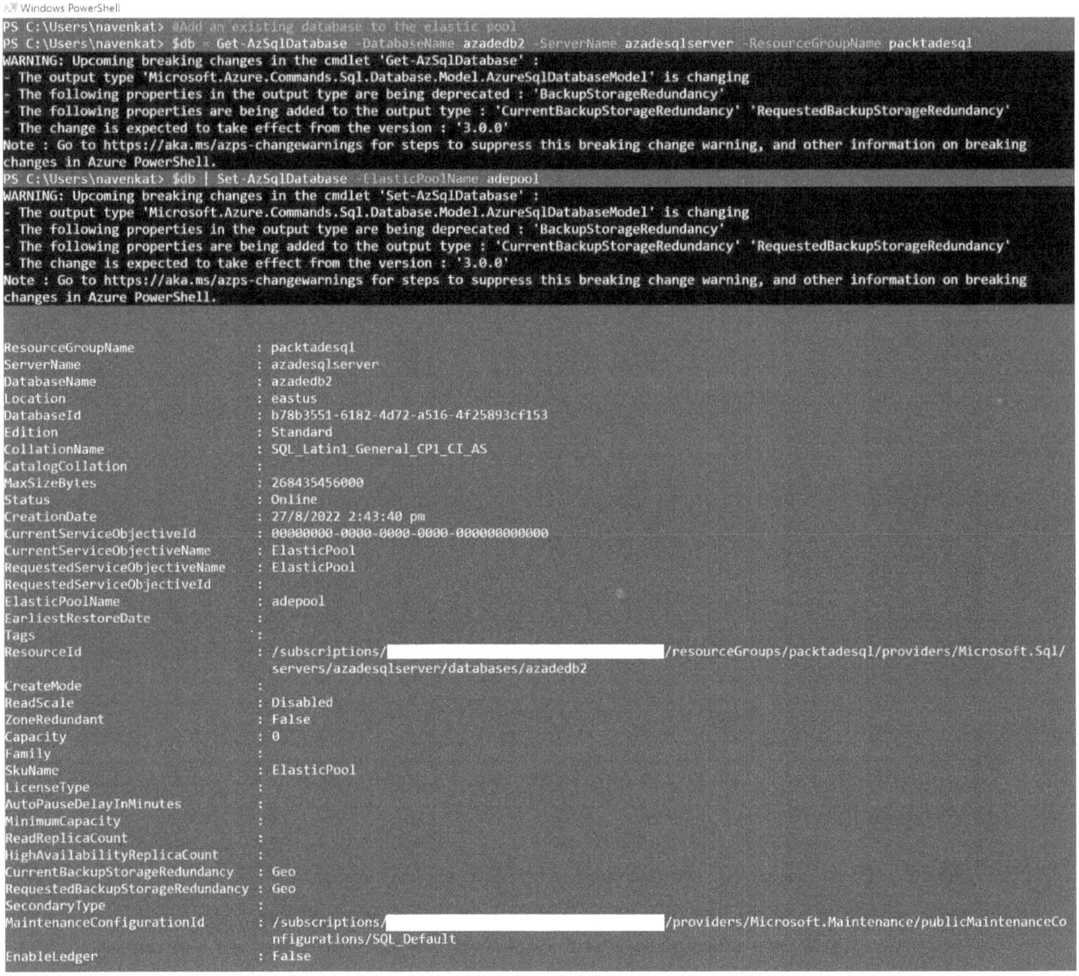

Figure 5.8 – Adding an existing SQL database to an elastic pool

5. To verify this in the Azure portal, log in with your Azure account. Navigate to **All resources | azadesqlserver | SQL elastic pools | adepool | Configure** and click on the **Databases** tab:

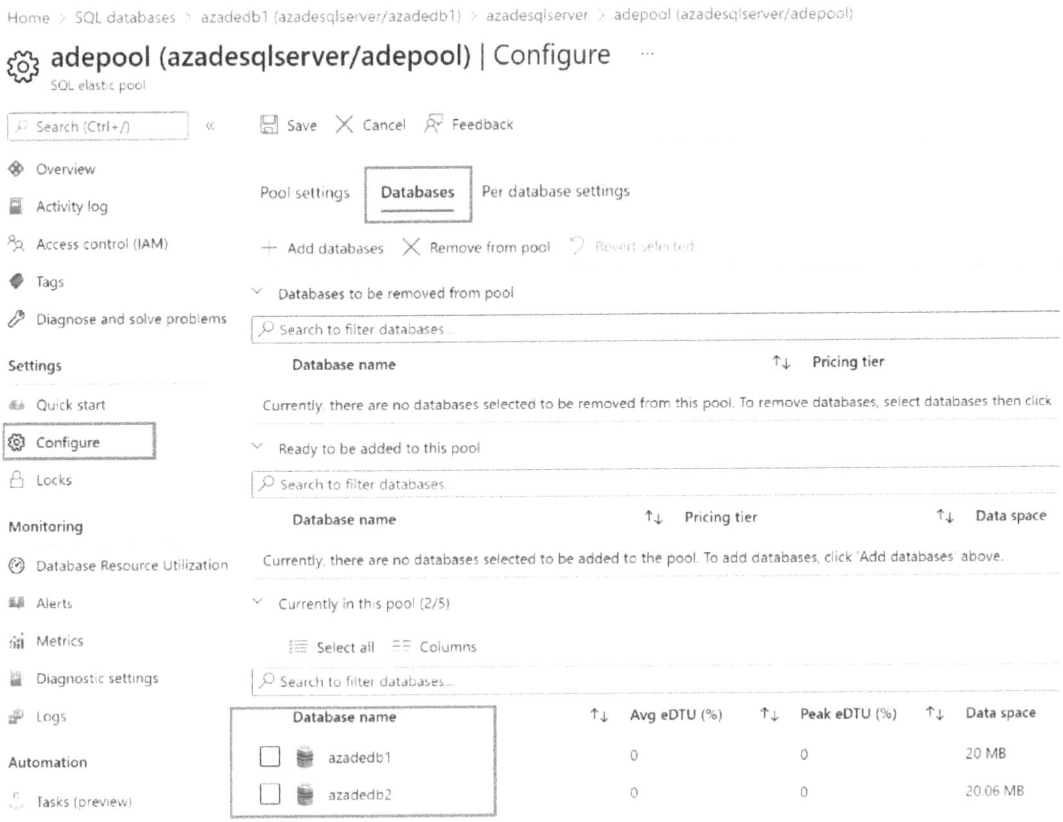

Home > SQL databases > azadedb1 (azadesqlserver/azadedb1) > azadesqlserver > adepool (azadesqlserver/adepool)

⚙ **adepool (azadesqlserver/adepool) | Configure** ⋯
   SQL elastic pool

🔍 Search (Ctrl+/) «	🖫 Save ✕ Cancel ⤺ Feedback

❖ Overview

🗐 Activity log

🔏 Access control (IAM)

🏷 Tags

🖉 Diagnose and solve problems

Settings

⚙ Quick start

⚙ Configure

🔒 Locks

Monitoring

⊘ Database Resource Utilization

📊 Alerts

📊 Metrics

🗎 Diagnostic settings

📜 Logs

Automation

⚙ Tasks (preview)

Pool settings | **Databases** | Per database settings

➕ Add databases ✕ Remove from pool ⤺ Revert selected

∨ Databases to be removed from pool

🔍 Search to filter databases...

Database name	↑↓ Pricing tier

Currently, there are no databases selected to be removed from this pool. To remove databases, select databases then click

∨ Ready to be added to this pool

🔍 Search to filter databases...

Database name	↑↓ Pricing tier	↑↓ Data space

Currently, there are no databases selected to be added to the pool. To add databases, click 'Add databases' above.

∨ Currently in this pool (2/5)

☰ Select all ☰ Columns

🔍 Search to filter databases...

Database name	↑↓ Avg eDTU (%)	↑↓ Peak eDTU (%)	↑↓ Data space
☐ 🗄 azadedb1	0	0	20 MB
☐ 🗄 azadedb2	0	0	20.06 MB

Figure 5.9 – Viewing the elastic pool in the Azure portal

6. Execute the following command to remove an Azure SQL database from an elastic pool. To move a database out of an elastic pool, we need to set the edition and the service objective explicitly:

```
#remove a database from an elastic pool
$db = Get-AzSqlDatabase -DatabaseName azadedb2
-ServerName azadesqlserver -ResourceGroupName packtadesql
$db | Set-AzSqlDatabase -Edition Standard
-RequestedServiceObjectiveName S3
```

You should get an output similar to the one shown in the following screenshot:

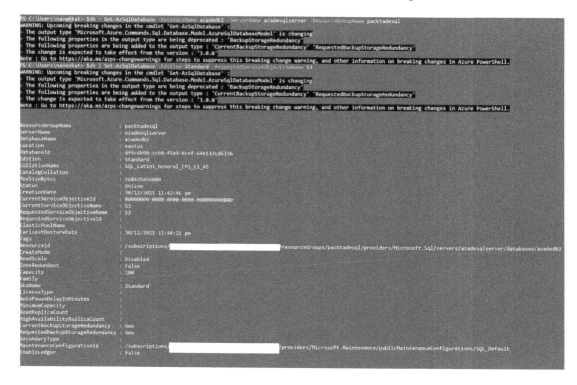

Figure 5.10 – Removing a SQL database from an elastic pool

7. Execute the following command to remove an elastic pool. An elastic pool has to be empty before it can be removed. Execute the following query to remove all the databases in an elastic pool:

```
# get elastic pool object
$epool = Get-AzSqlElasticPool -ElasticPoolName adepool
-ServerName azadesqlserver -ResourceGroupName packtadesql
# get all databases in an elastic pool
$epdbs = $epool | Get-AzSqlElasticPoolDatabase
# change the edition of all databases in an elastic pool
to standard S3
foreach($db in $epdbs) {
$db | Set-AzSqlDatabase -Edition Standard
-RequestedServiceObjectiveName S3
}
# Remove an elastic pool
$epool | Remove-AzSqlElasticPool
```

> **Note**
>
> The command sets the edition of the SQL databases to **Standard**. This is for demo purposes only. If this is to be done in production, modify the edition and the service objective accordingly.

How it works...

We create an elastic pool using the `New-AzSqlElasticPool` command. In addition to the parameters, such as the server name, resource group name, compute model, compute generation, and edition, which are the same as when we created a new Azure SQL database, we can also specify `DatabaseMinDtu` and `DatabaseMaxDtu`. `DatabaseMinDtu` specifies the minimum amount of DTUs that all the databases in an elastic pool can have. `DatabaseMaxDtu` is the maximum amount of DTUs that a database can consume in an elastic pool.

Similarly, for the vCore-based purchasing model, we can specify `DatabaseVCoreMin` and `DatabaseVCoreMax`.

To add a new database to an elastic pool, specify the elastic pool name at the time of database creation using the `New-AzSqlDatabase` command.

To add an existing database to an elastic pool, modify the database using `Set-AzSqlDatabase` to specify the elastic pool name.

To remove a database from an elastic pool, modify the database using the `Set-AzSqlDatabase` command to specify a database edition explicitly.

To remove an elastic pool, first, empty it by moving all of the databases out of the elastic pool, and then remove it using the `Remove-AzSqlElasticPool` command.

Configuring a virtual network and private endpoints for Azure SQL Database

Securing the connectivity to an Azure SQL database is important to limit the exposure of the database to external attacks such as **distributed denial-of-service (DDOS)** attacks and **SQL injection**. Using private endpoints for connecting to Azure SQL Database ensures that the database connectivity flows through Azure's backbone network and does not use the public internet. Placing the SQL endpoint behind a virtual network prevents Azure SQL Database from being exposed to a connection request from the public internet. In this recipe, we will explore how we can configure private endpoints using virtual networks for Azure SQL Database.

Getting ready

1. Log in to `portal.azure.com`.
2. Create an Azure SQL Database, as explained in the *Provisioning and connecting to an Azure SQL database using PowerShell* recipe in this chapter.

How to do it...

Perform the following steps to create an Azure SQL database with a private link and virtual network:

1. **Creating a virtual network**: Go to `portal.azure.com` and click **Create a Resource**. Search for `Virtual Network` and click **Create**, as shown in the following screenshot:

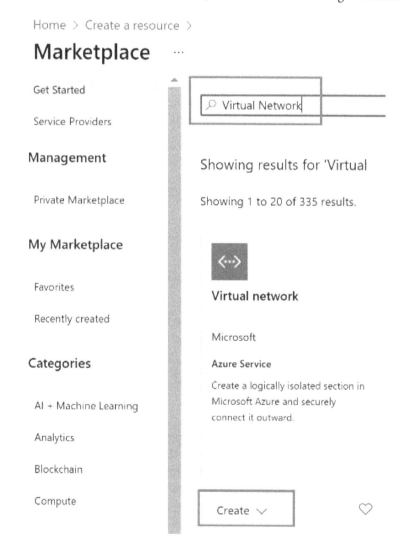

Figure 5.11 – Creating a virtual network

2. Provide the resource group name as `packtadesql` (create one if it doesn't exist using the **Create new** button). Provide the virtual network instance name as `packtadesqlvnet`, as shown in the following screenshot. Click **Review + create**:

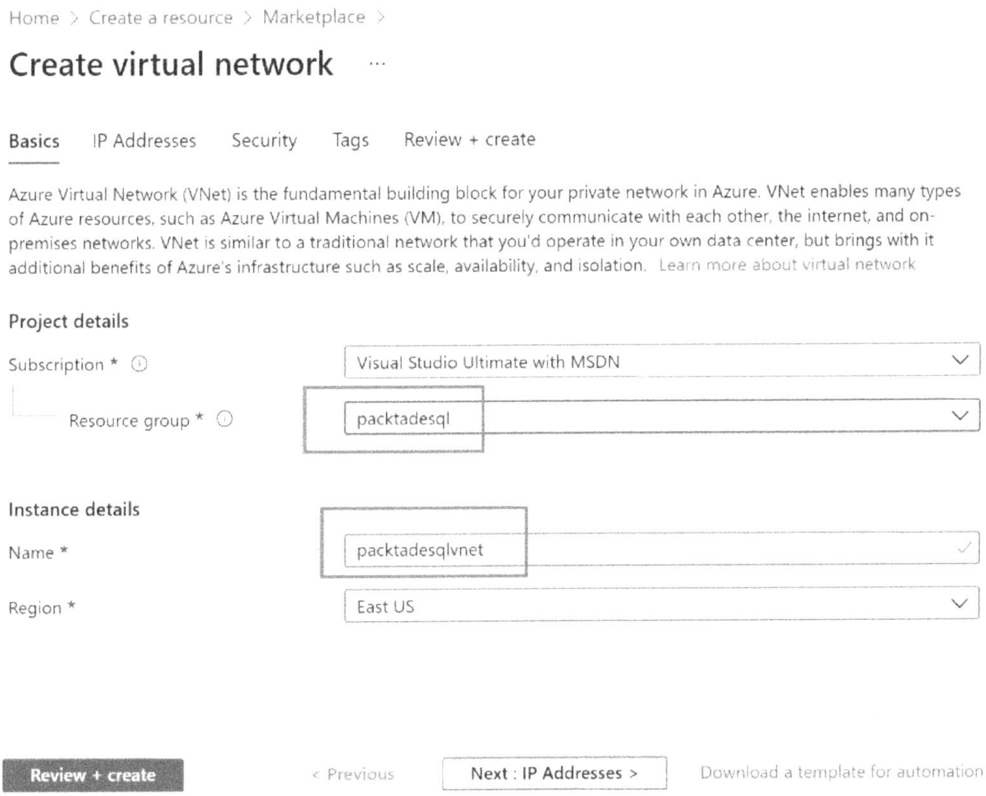

Home > Create a resource > Marketplace >

Create virtual network   ...

Basics   IP Addresses   Security   Tags   Review + create

Azure Virtual Network (VNet) is the fundamental building block for your private network in Azure. VNet enables many types of Azure resources, such as Azure Virtual Machines (VM), to securely communicate with each other, the internet, and on-premises networks. VNet is similar to a traditional network that you'd operate in your own data center, but brings with it additional benefits of Azure's infrastructure such as scale, availability, and isolation.   Learn more about virtual network

Project details

Subscription * ⓘ   Visual Studio Ultimate with MSDN   ⌄

          Resource group * ⓘ   packtadesql   ⌄

Instance details

Name *   packtadesqlvnet   ✓

Region *   East US   ⌄

[Review + create]   < Previous   [Next : IP Addresses >]   Download a template for automation

Figure 5.12 – Create virtual network

3. Once the virtual network has been created, go to **All resources**. Find **azadesqlserver**. Click **azadesqlserver** and click **Firewalls and virtual networks**, as shown here:

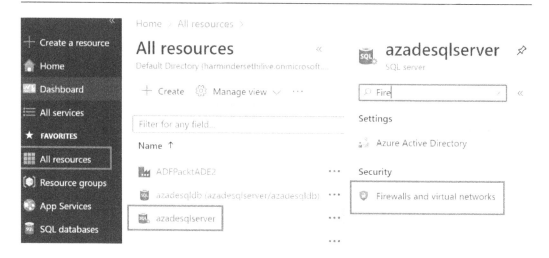

Figure 5.13 – Firewalls and virtual networks

4.   Check the **Deny public network access** checkbox and click the **Save** button:

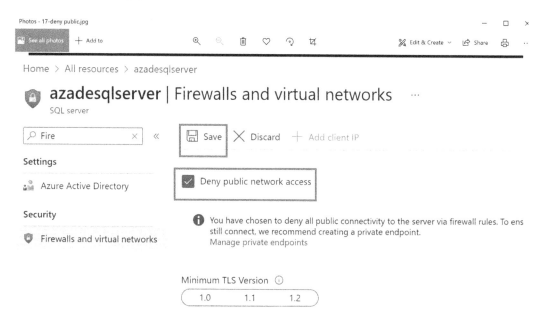

Figure 5.14 – Firewalls and virtual networks

5. Now, on **azadesqlserver**, search for `Private endpoint` and click the + **Private endpoint** button:

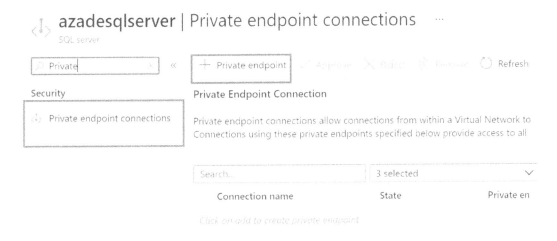

Figure 5.15 – Private endpoint

6. Create a private endpoint named `pepsqlserver`. Ensure that the **Region** detail is the same as that of Azure SQL Database (**East US**):

Figure 5.16 – Create a private endpoint

7. On the **Resource** tab, set **Resource type** to **Microsoft.Sql/servers**, **Resource** to **azadesqlserver**, and **Target sub-resource** to **sqlServer**, as shown here:

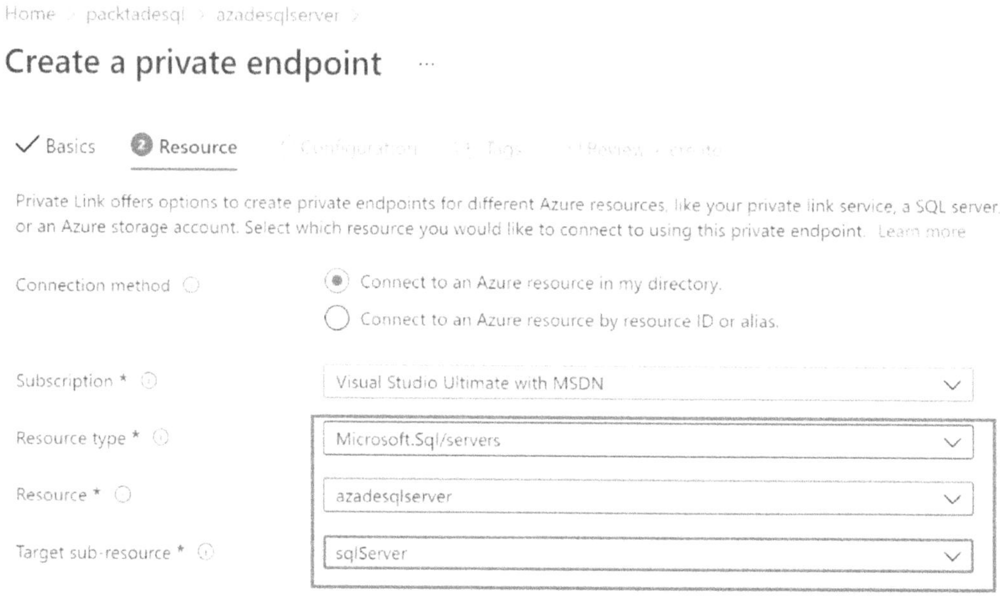

Figure 5.17 – Resource type

8. Move to the **Configuration** tab. Set **Virtual network** to **packtadesqlvnet**, which we created in *step 1*:

Home > packtadesql > azadesqlserver >

Create a private endpoint ···

✓ Basics   ✓ Resource   ❸ Configuration   (4) Tags   (5) Review + create

Networking

To deploy the private endpoint, select a virtual network subnet. Learn more

Virtual network * ⓘ	packtadesqlvnet ∨
Subnet * ⓘ	packtadesqlvnet/default (10.0.0.0/24) ∨

ⓘ If you have a network security group (NSG) enabled for the subnet above, it will be disabled for private endpoints on this subnet only. Other resources on the subnet will still have NSG enforcement.

Private DNS integration

To connect privately with your private endpoint, you need a DNS record. We recommend that you integrate your private endpoint with a private DNS zone. You can also utilize your own DNS servers or create DNS records using the host files on your virtual machines. Learn more

Integrate with private DNS zone   ⦿ Yes   ◯ No

Configuration name	Subscription	Resource group	Private DNS zone
privatelink-database-windows-net	Visual Studio Ultimate with MSDN ∨	packtadesql ∨	(new) privatelink.database.windows.n...

Figure 5.18 – Create a private endpoint

9. Hit **Review + create** and then the **Create** button to create the Azure SQL database with a private endpoint. With this, we have created a private endpoint to connect to Azure SQL Database.

10. To test it, let's try to connect to the database from our local machine. It is expected to fail, as it's not connected to the virtual network. Install SSMS and test the connectivity to azadesqlserver. database.windows.net. It fails with an error message, as shown in the following screenshot:

Connect to Server   ✕

❌   Cannot connect to azadesqlserver.database.windows.net.

Additional information:

⤷ Reason: An instance-specific error occurred while establishing a connection to SQL Server. Connection was denied since Deny Public Network Access is set to Yes (https://docs.microsoft.com/azure/azure-sql/database/connectivity-settings#deny-public-network-acce ss). To connect to this server, use the Private Endpoint from inside your virtual network (https://docs.microsoft.com/azure/sql-database/sql-database-private-endpoint-overview#how-to-set-u p-private-link-for-azure-sql-database). (Microsoft SQL Server, Error: 47073)

ⓗ Help ▾   ⬲ Copy message   📄 Show details | OK |

Figure 5.19 – Connection error

11. Create a new VM in the Azure portal. Ensure to allow remote desktop connection to the VM, as shown here:

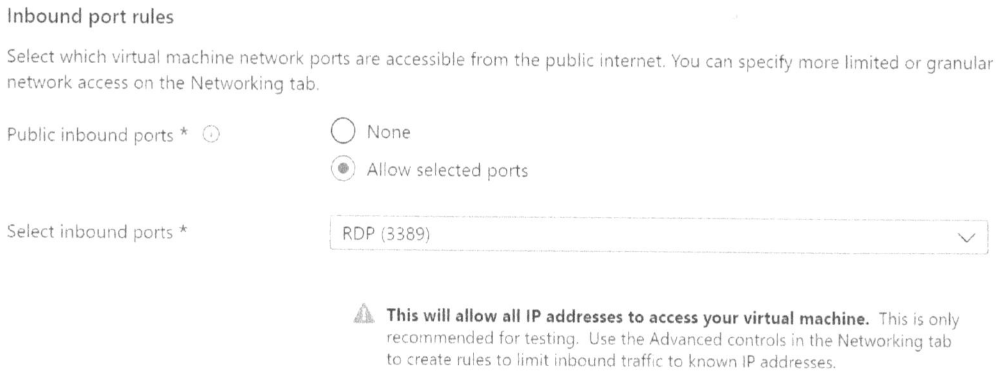

Inbound port rules

Select which virtual machine network ports are accessible from the public internet. You can specify more limited or granular network access on the Networking tab.

Public inbound ports * ⓘ   ◯ None
                               ⦿ Allow selected ports

Select inbound ports *   RDP (3389)   ⌄

⚠ **This will allow all IP addresses to access your virtual machine.** This is only
recommended for testing. Use the Advanced controls in the Networking tab
to create rules to limit inbound traffic to known IP addresses.

Figure 5.20 – Creating a new VM and allowing remote desktop connection

12. While creating the VM, under **Networking**, set **Virtual network** to **packtadesqlvnet**. A VM will be created inside the virtual network:

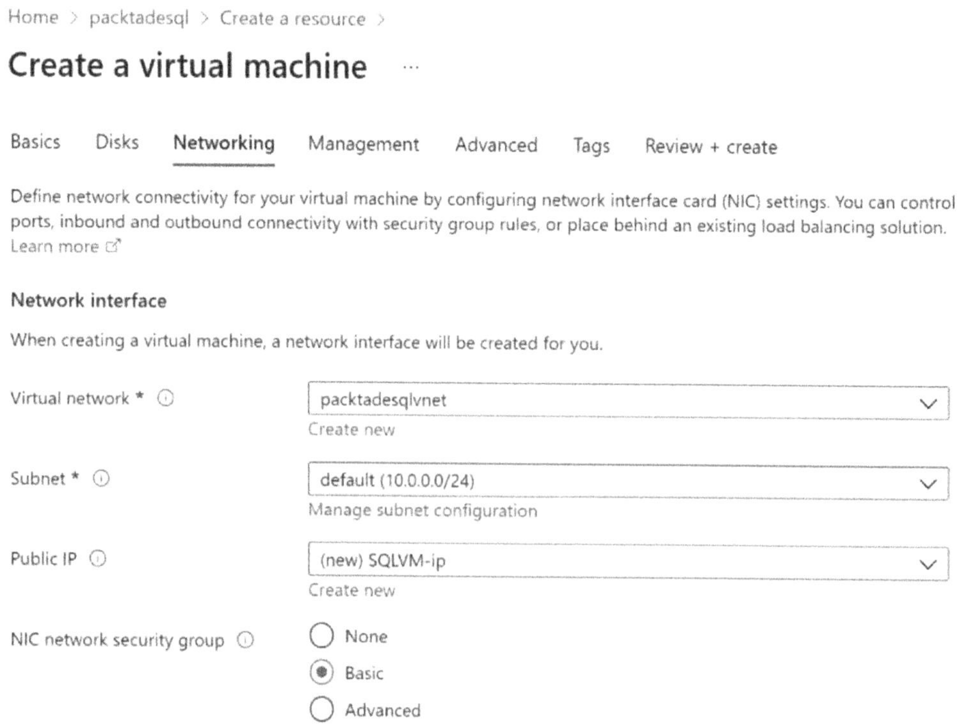

Home > packtadesql > Create a resource >

Create a virtual machine   ···

Basics   Disks   **Networking**   Management   Advanced   Tags   Review + create

Define network connectivity for your virtual machine by configuring network interface card (NIC) settings. You can control ports, inbound and outbound connectivity with security group rules, or place behind an existing load balancing solution. Learn more ☐

Network interface

When creating a virtual machine, a network interface will be created for you.

Virtual network * ⓘ   packtadesqlvnet   ⌄
                             Create new

Subnet * ⓘ   default (10.0.0.0/24)   ⌄
                             Manage subnet configuration

Public IP ⓘ   (new) SQLVM-ip   ⌄
                             Create new

NIC network security group ⓘ   ◯ None
                                ⦿ Basic
                                ◯ Advanced

Figure 5.21 – Configuring the VM to use the virtual network

13. Once the VM is created, open SSMS and test the connectivity to `azadesqlserver.database.windows.net`. It will successfully connect to the database, as it is now part of the virtual network.

How it works...

The objective of the recipe was to ensure that the database can only be connected using Azure's backbone network and no connection from the public internet is allowed. To achieve this, we performed the following:

1. We created a virtual network called **packtadesqlvnet**.

2. The private endpoint acts as a private tunnel between the client connection and the resource (Azure SQL Database). So, we created a private endpoint connection called **pepsqlserver** inside the virtual network, **packtadesqlvnet**. We configured the private endpoint, **pepsqlserver**, to connect to our Azure SQL database, **azadesqlserver**. This ensures that the endpoint can be used to connect to the database.

3. We shut down any public connections to the Azure SQL database, **azadesqlserver**, by setting **Deny public network access**. This ensures that no connection through a public network is possible to our Azure SQL database, **azadesqlserver**.

4. A private endpoint connection only works from devices joined to the Azure virtual network. We tested this by trying to connect from a public network (our machine), which failed. We then created a VM inside the virtual network (**packtadesqlvnet**) and tested the connectivity to the database, which was successful.

Configuring Azure Key Vault for Azure SQL Database

Azure SQL Database is encrypted at rest by default using Microsoft-managed keys. However, many customers prefer to encrypt Azure SQL Database using keys that are managed by them, as it offers more control over the encryption keys. This recipe will show how you can use customer-managed keys to encrypt Azure SQL Database by integrating it with Azure Key Vault.

Getting ready

Create an Azure SQL database, as explained in the *Provisioning and connecting to an Azure SQL database using PowerShell* recipe in this chapter.

How to do it...

Perform the following steps to configure Azure Key Vault:

1. Go to `portal.azure.com`, click **All resources**, and find the SQL server, **azadesqlserver**.

2. Find **Transparent data encryption** under **Security**. Click **Customer-managed key** and click **Change key**:

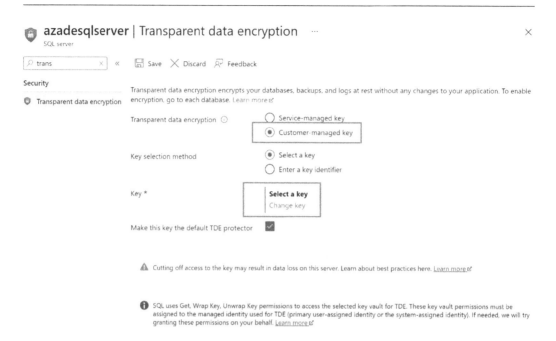

Figure 5.22 – Configuring Azure Key Vault

3. Set **Key store type** to **Key vault**. Click **Create new key vault**:

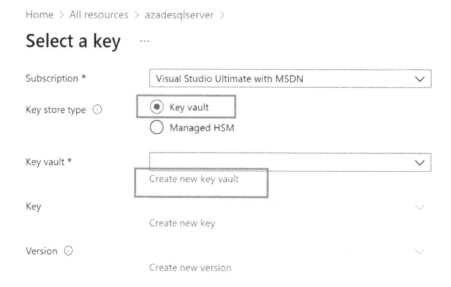

Figure 5.23 – Configuring Azure Key Vault

4. Provide `azadekeyvault` (or any new name) as the **Key vault name** and hit the **Review + create** button:

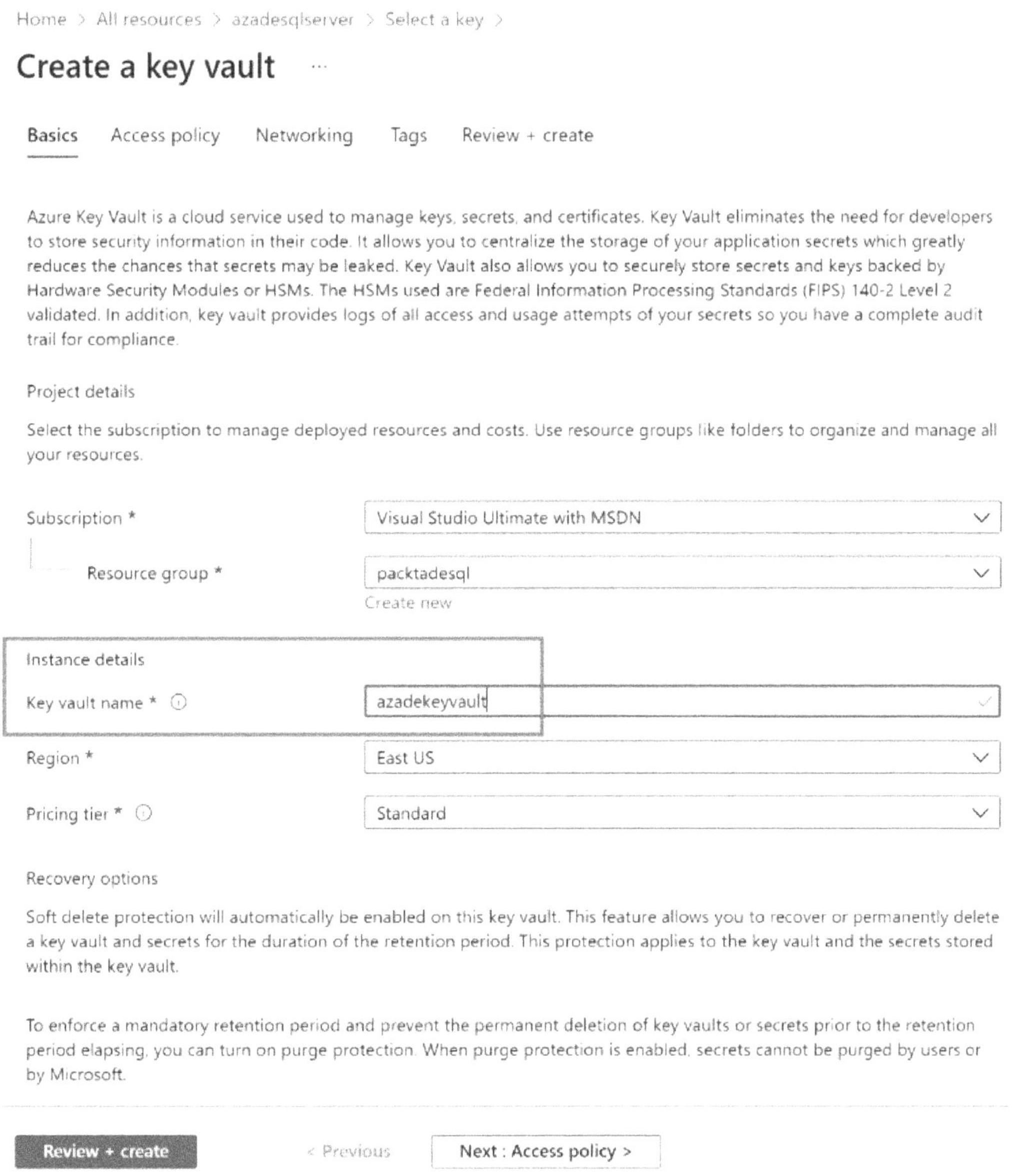

Home > All resources > azadesqlserver > Select a key >

Create a key vault   ...

Basics   Access policy   Networking   Tags   Review + create

Azure Key Vault is a cloud service used to manage keys, secrets, and certificates. Key Vault eliminates the need for developers to store security information in their code. It allows you to centralize the storage of your application secrets which greatly reduces the chances that secrets may be leaked. Key Vault also allows you to securely store secrets and keys backed by Hardware Security Modules or HSMs. The HSMs used are Federal Information Processing Standards (FIPS) 140-2 Level 2 validated. In addition, key vault provides logs of all access and usage attempts of your secrets so you have a complete audit trail for compliance.

Project details

Select the subscription to manage deployed resources and costs. Use resource groups like folders to organize and manage all your resources.

Subscription *	Visual Studio Ultimate with MSDN ⌄
Resource group *	packtadesql ⌄
	Create new

Instance details

Key vault name * ⓘ	azadekeyvault ⌄
Region *	East US ⌄
Pricing tier * ⓘ	Standard ⌄

Recovery options

Soft delete protection will automatically be enabled on this key vault. This feature allows you to recover or permanently delete a key vault and secrets for the duration of the retention period. This protection applies to the key vault and the secrets stored within the key vault.

To enforce a mandatory retention period and prevent the permanent deletion of key vaults or secrets prior to the retention period elapsing, you can turn on purge protection. When purge protection is enabled, secrets cannot be purged by users or by Microsoft.

[Review + create]   < Previous   [Next : Access policy >]

Figure 5.24 – New key vault creation

5. After the key vault is created in *step 4*, you will be taken back to the **Select a key** screen again. Click **Create new key**:

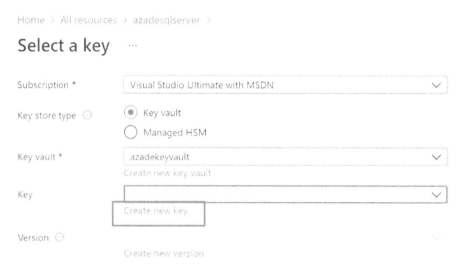

Figure 5.25 – Create a new key

6. Provide a key name. It is recommended to use a name that makes it easy to identify its purpose.
    Then, click **Create**:

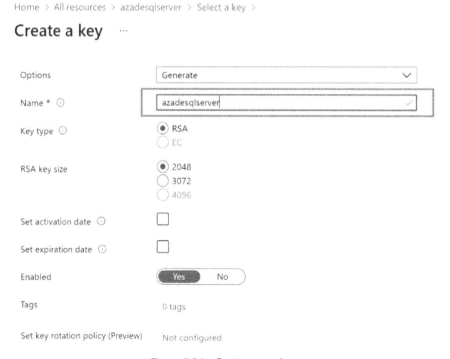

Figure 5.26 – Create a new key

7. You will be taken back to the **Select a key** screen, but this time, all the values for **Key vault**, **Key**, and **Version** will automatically be filled in. Click the **Select** button:

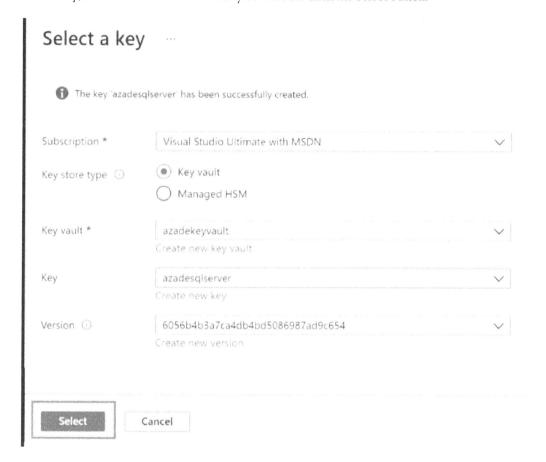

Figure 5.27 – Key selection completion

8. Check the **Auto-rotate key** checkbox. Hit the **Save** button to complete the configuration:

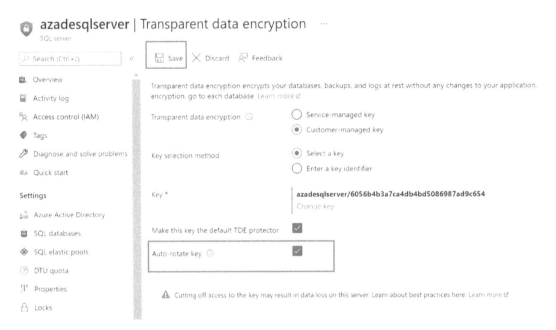

Figure 5.28 – Save the transparent data encryption configuration

How it works...

Turning on **Transparent data encryption** for Azure SQL Database using Azure Key Vault ensures that the database is always encrypted when at rest. By turning on the **Auto-rotate key** option while configuring the Key vault integration, we have ensured keys used for encryption are rotated automatically by Azure. Customers can still access the keys used in Key Vault using the Azure portal. This allows us to have the best of both worlds – enjoying the power of automation of the cloud while having control over your resources. During creation of Key Vault, Azure automatically grants permission for SQL Server to access Key Vault. You can verify this by going to **All resources** in the Azure portal, searching for **azadekeyvault**, and clicking **Access policies** under **Settings**:

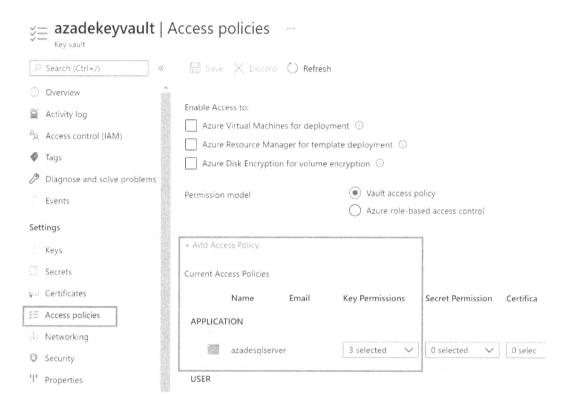

Figure 5.29 – Verify permissions

Azure SQL Database will use its service account, **azadesqlserver**, access Key Vault using the permission granted, and perform operations such as encryption, decryption, and key rotation automatically with full transparency to users.

Provisioning and configuring a wake-up script for a serverless SQL database

Azure SQL Database is offered in two compute tiers: provisioned, and serverless. While a provisioned SQL database will be always running, a serverless database can pause when not in use and start again when connections flow in. Automatically pausing an Azure SQL database helps to save costs when the database is not being used.

While pausing the database is automatic, the database only starts when the first connection request comes in. Starting a database can take a few minutes of waiting time and cause inconvenience to customers. Configuring a wake-up script to start a paused database is a proactive method to reduce this waiting time. The following recipe will configure a serverless database and deploy a wake-up script to start the database at a specific schedule. At a high level, we will be doing the following:

1. Configuring the serverless compute tier for an existing database

2. Creating an **Azure Automation** account

3. Provisioning a PowerShell script to start the database

4. Scheduling the PowerShell script using a runbook in the Azure Automation account

Getting ready

Create an Azure SQL database, as explained in the *Provisioning and connecting to an Azure SQL database using PowerShell* section of this chapter.

How to do it...

Perform the following steps to configure a serverless database and schedule a wake-up script for the database:

1. Go to portal.azure.com, click **All resources**, and find the SQL database, **azadesqldb**. Click **Compute + storage** under **Settings**. Pick **General Purpose (Scalable compute and storage options)** for **Service tier** and **Serverless** under **Computer tier,** as shown in the following screenshot:

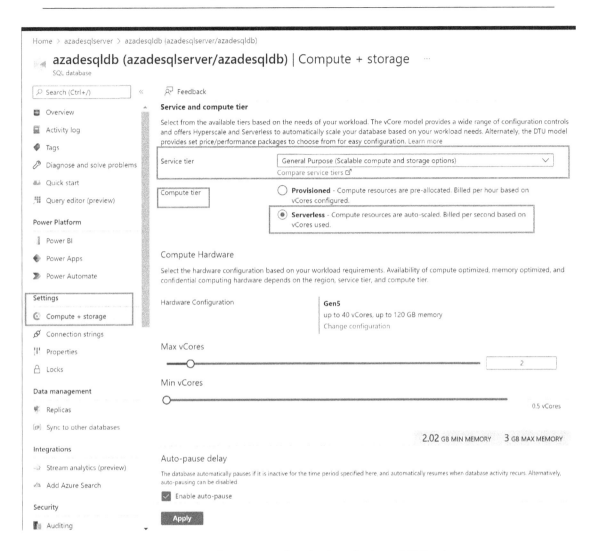

Figure 5.30 – Configure a serverless database for Azure SQL Database

2. Scroll down and make note of the **Auto-pause delay** duration of **1** hour and click the **Apply** button:

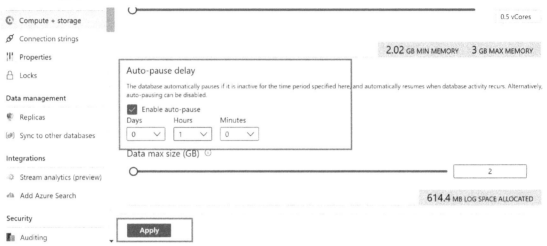

Figure 5.31 – Set auto-pause duration

3. To create an Automation account, go to `portal.azure.com`, click **Create a resource**, and search for `Automation Account`. Click **Create** for the **Automation** service listed by Microsoft:

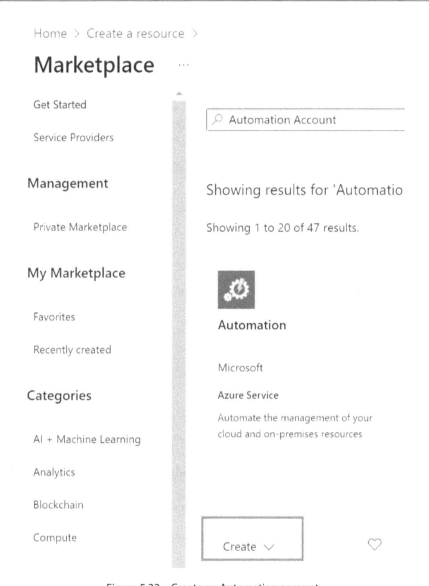

Figure 5.32 – Create an Automation account

4. Provide the **Resource group** name as packtadesql (the same as the one given to Azure SQL Database) and set **Automation account name** as azadeautomation (or any other new name). Click **Review + Create**:

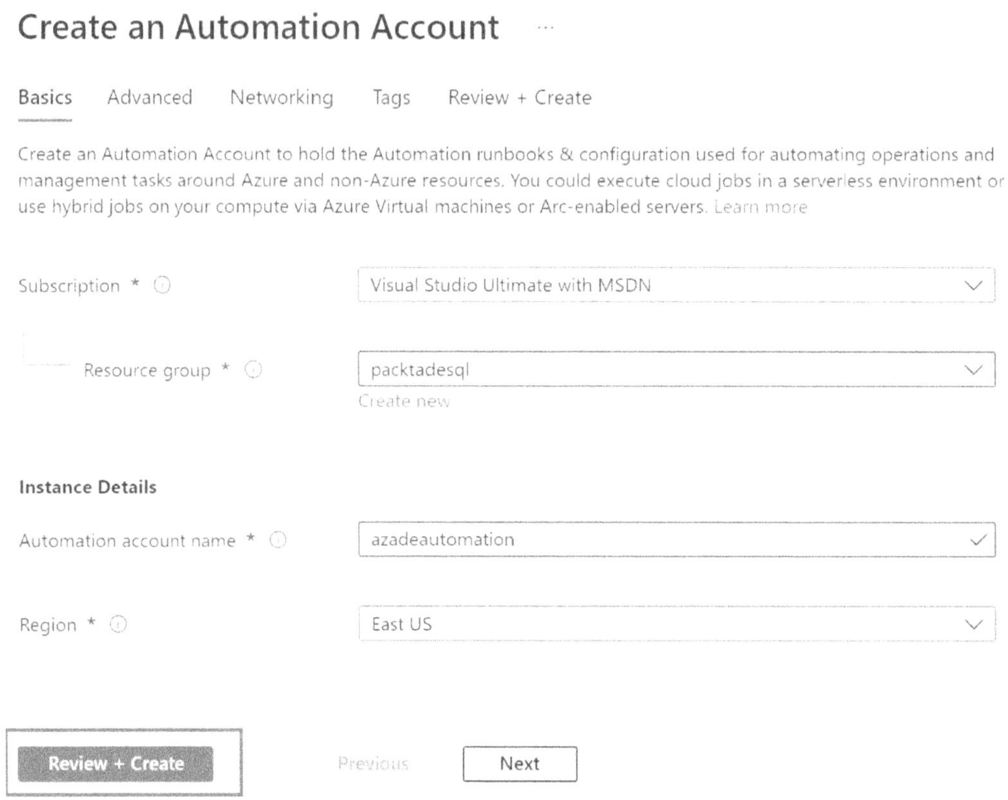

Create an Automation Account   ...

Basics   Advanced   Networking   Tags   Review + Create

Create an Automation Account to hold the Automation runbooks & configuration used for automating operations and management tasks around Azure and non-Azure resources. You could execute cloud jobs in a serverless environment or use hybrid jobs on your compute via Azure Virtual machines or Arc-enabled servers. Learn more

Subscription * ⓘ

> Visual Studio Ultimate with MSDN   ⌄

Resource group * ⓘ

> packtadesql   ⌄

Create new

Instance Details

Automation account name * ⓘ

> azadeautomation   ✓

Region * ⓘ

> East US   ⌄

[Review + Create]   Previous   [Next]

Figure 5.33 – Automation account creation

5.   Open the Azure Automation account and click **Modules**. Click the **Browse gallery** button:

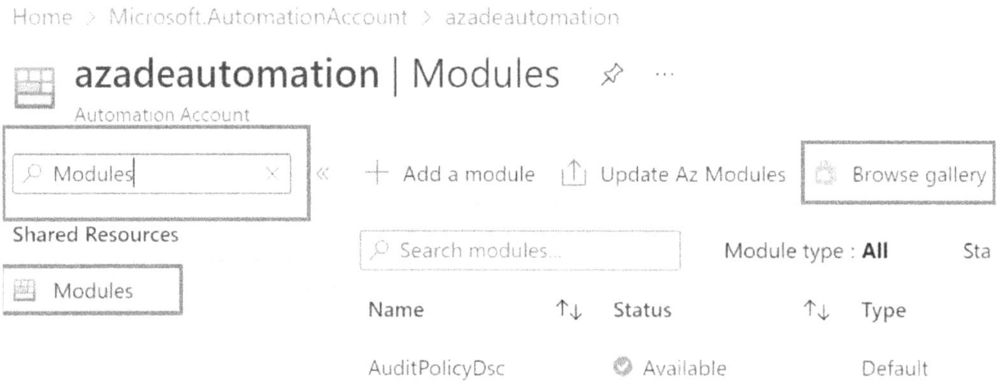

Home > Microsoft.AutomationAccount > azadeautomation

🗔 azadeautomation | Modules   📌 ...
Automation Account

🔍 Modules   ✕   «   + Add a module   ⬆ Update Az Modules   🗔 Browse gallery

Shared Resources

🔍 Search modules...   Module type : **All**   Sta

🗔 Modules

Name	↑↓	Status	↑↓	Type
AuditPolicyDsc		🕑 Available		Default

Figure 5.34 – Module installation in an Automation account

6. Search for `sqlServer`. Select the **SqlServer** module. Hit the **Select** button on the next screen:

Figure 5.35 – sqlserver module installation in an Azure Automation account

7. Set **Runtime version** to **7.1** on the **Add a module** screen and hit the **Import** button:

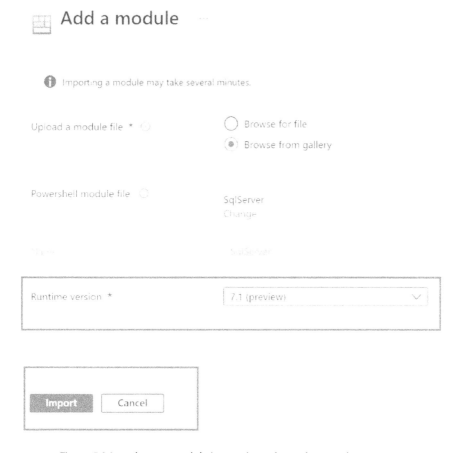

Figure 5.36 – sqlserver module import in an Azure Automation account

8. After adding the module, go to the Azure Automation account **Credentials** in the **Shared Resources** section. We will be providing the SQL credentials required to run the wake-up script. Click + **Add a credential**:

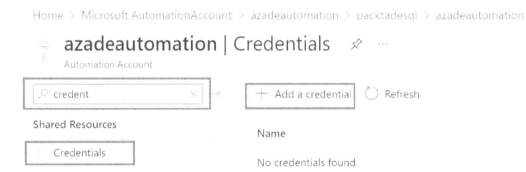

Figure 5.37 – Add a credential to an Azure Automation account

9. Provide the credential name as `sqlcredentials`. Provide the user ID and password (`sqladmin/Sql@Server@1234`) for **azadesqldb**:

Figure 5.38 – Add a credential to an Azure Automation account

10. After adding the credential, go to the Azure Automation account and then **Runbooks** in the **Process Automation** section. We will be creating the PowerShell script to wake up the database. Click **Create a runbook**:

Figure 5.39 – Add a runbook to an Azure Automation account

11. Provide the script name as `sqlwakeupscript`. Set **Runbook type** to **PowerShell** and **Runtime version** to **7.1 (preview)**, and click the **Create** button:

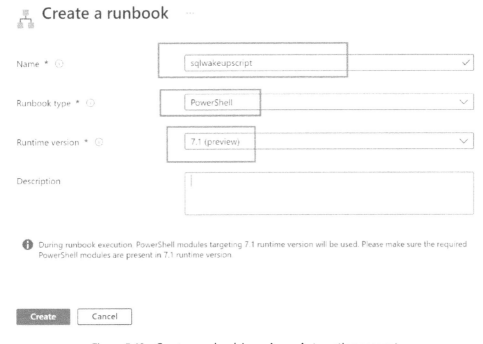

Figure 5.40 – Create a runbook in an Azure Automation account

12. Copy the following script in the **Edit PowerShell Runbook** window:

```
$SqlCredential = Get-AutomationPSCredential -Name
"sqlcredentials"
# Query to execute
$Query = "select getdate()"


# Execute query
 "----- Running SQL Command "
 invoke-sqlcmd -ServerInstance "azadesqlserver.
database.windows.
net" -Database "azadesqldb" -Credential $SqlCredential
-Query "$Query" -Encrypt
 "`n ----- END SQL Command"
```

Ensure to provide the credential name (`sqlcredentials`) created in *step 9* for the `-Name` parameter of the `Get-AutomationPSCredential` command. Provide the SQL Server name and database name to be woken up as the values for the `ServerInstance` parameter and the `Database` parameter in the `invoke-sqlcmd` command, as shown in the following screenshot. Hit **Save** and then hit the **Publish** button:

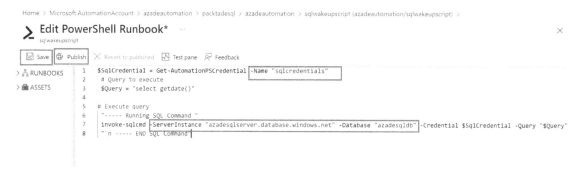

Figure 5.41 – Azure Automation runbook script creation

13. After publishing the script, go to the Azure Automation account, and under the **Resources** section, click **Schedules**. Click on + **Add a schedule**:

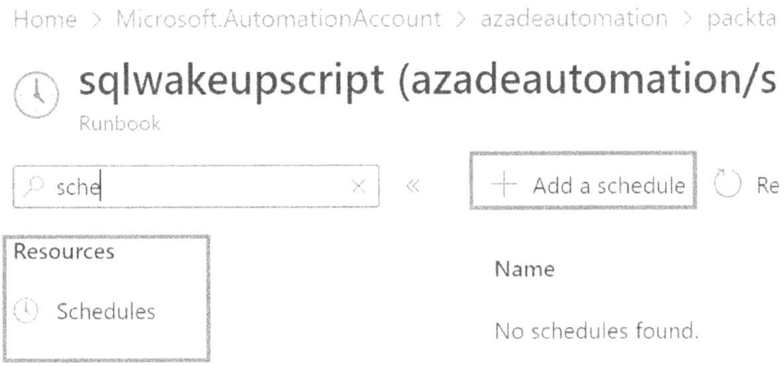

Home > Microsoft.AutomationAccount > azadeautomation > packta

sqlwakeupscript (azadeautomation/s
Runbook

sche   ×   «   + Add a schedule   ⟳ Re

Resources

Schedules

Name

No schedules found.

Figure 5.42 – Azure Automation schedule

14. Provide the schedule name and set the **Recurrence** option to **Recurring**. Set the **Recur every** option as **1 Day** and hit the **Create** button:

Figure 5.43 – The Azure Automation schedule

15. After adding the schedule, go to **All resources** in the Azure portal. Search for the `sqlwakeupscript` runbook. Click **Schedules** under **Resources**. Click **+ Add a schedule**:

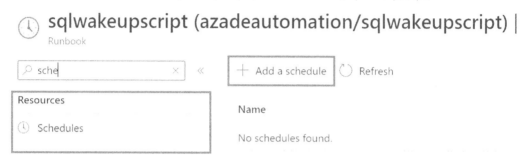

Figure 5.44 – Add a schedule to the runbook

16. Select **Link a schedule to your runbook**:

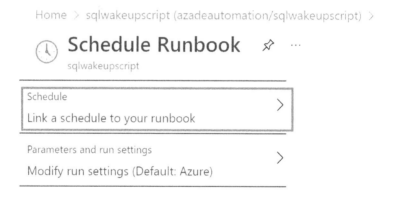

Figure 5.45 – Link a schedule to your runbook

17. Click on the **wakescript-schedule** schedule created earlier and click **OK**:

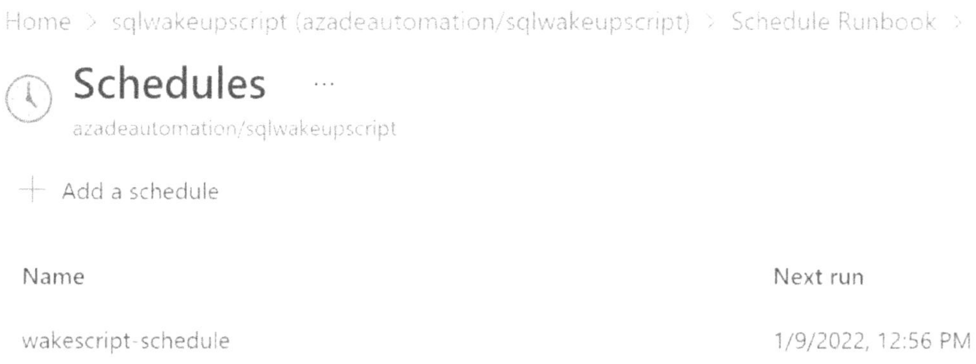

Home > sqlwakeupscript (azadeautomation/sqlwakeupscript) > Schedule Runbook >

Schedules ...

azadeautomation/sqlwakeupscript

+ Add a schedule

Name	Next run
wakescript-schedule	1/9/2022, 12:56 PM

Figure 5.46 – Linking a schedule to a runbook

18. Upon creation, the runbook will now have the schedule linked to it, as shown in the following screenshot:

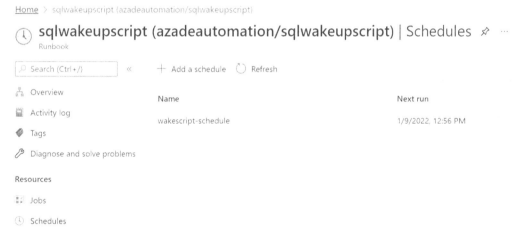

Figure 5.47 – Link the schedule to the runbook

19. Go to `portal.azure.com`, click on **All resources**, and find the SQL database, **azadesqldb**. On the **Overview** page, check the status of the database. If it is not paused, close all connections to the database to ensure that it pauses:

Figure 5.48 – A paused database

20. Go to **All resources**, search for `sqlwakeupscript`, and go to the runbook. Hit the **Start** button:

Home >

➤ sqlwakeupscript (azadeautomation/sqlwakeupscript)
Runbook

🔍 Search (Ctrl+/)	«	▷ Start </> View ✎ Edit ⏲ Link to schedule

∧ Essentials

🖧 Overview

🖼 Activity log

🏷 Tags

🔧 Diagnose and solve problems

Resources

📊 Jobs

🕐 Schedules

Resource gro...   : packtadesql

Account   : azadeautomation

Location   : East US

Subscription   : Visual Studio Ultimate with MSDN

Tags (Edit)   : Click here to add tags

Figure 5.49 – Start the runbook

21. Wait until the job completes. It will take a few minutes. Upon completion, the job status changes to **Completed**, as shown in the following screenshot:

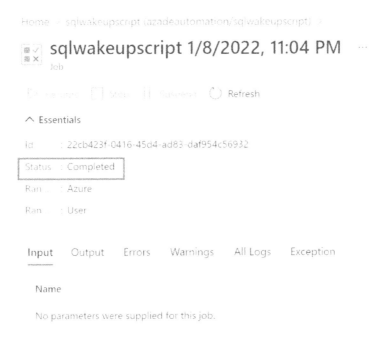

Figure 5.50 – The completed runbook

22. Go to portal.azure.com, click on **All resources**, and find the SQL database, **azadesqldb**. On the **Overview** page, check the status of the database. The database status should be either **Resuming** or **Online**, as shown in the following screenshot:

Figure 5.51 – The database status

How it works...

We performed the following steps to configure the serverless tier and provision a wake-up script for the Azure SQL database, **azadesqldb**:

1. As the first step, we configured **azadesqldb** on the serverless compute tier, so that it pauses when not in use for 1 hour or longer.

2. As a next step, to configure a wake-up script for **azadesqldb**, we provisioned an Azure Automation account named **azadeautomation**. An Automation account is required for running any scheduled task in Azure. We can leverage the same automation account for running any other maintenance task too.

3. Next, we created a runbook called `sqlwakeupscript` in our **azadeautomation** account. A runbook is used to run any PowerShell script.

4. On the runbook, we added a PowerShell script that connects to the database and runs a simple query against the database. The connection request from the runbook will start the database if it is paused.

5. We added a shared schedule called **wakescript-schedule** to the Automation account and linked it to the `sqlwakeupscript` runbook. This schedules the runbook to be executed daily. Having the schedule created in the Automation account as a shared schedule allows us to reuse the schedule if any other job needs to be run at the same frequency.

6. Finally, we started the runbook manually and noticed that the paused database came online.

This recipe is typically useful for UAT and development environments in organizations where users or developers work at certain times during the day. The database can be automatically paused and then resume just before working hours. Pausing the database when not in use and resuming it only when it is needed reduces the cost incurred from Azure SQL Database.

Configuring the Hyperscale tier of Azure SQL Database

Azure SQL Database offers three service tiers – General Purpose, Business Critical, and Hyperscale – when purchased in the vCore model. General Purpose is the most commonly used service tier for medium-sized applications, while the Business Critical tier offers enterprise-class performance and enhanced high-availability options. However, both the General Purpose and Business Critical tiers have a maximum database size limit of 4 TB. A Hyperscale database allows the database to scale up to 100 TB, as well as offering enterprise-class performance and high-availability capabilities. One of the aspects of Hyperscale databases is that we don't need to specify an upper limit for the database size; the database automatically grows as you use it and you pay for the storage you have used.

This recipe will show how to provision the Hyperscale tier of Azure SQL Database.

Getting ready

Log in to `portal.azure.com`.

How to do it...

Perform the following steps to provision the Hyperscale tier:

1. Go to **Create a resource** in the Azure portal, find **SQL Database**, and click on the **Create** button. On the **Create SQL Database** screen, provide a new resource group if you don't have one already. Provide any database name (a sample one is used in this example). Click **Create new** under **Server** if you don't have a SQL Server created. If you have followed the *Provisioning and connecting to an Azure SQL database using PowerShell* recipe, you would already have a SQL Server named **azadesqlserver**. You can reuse this one:

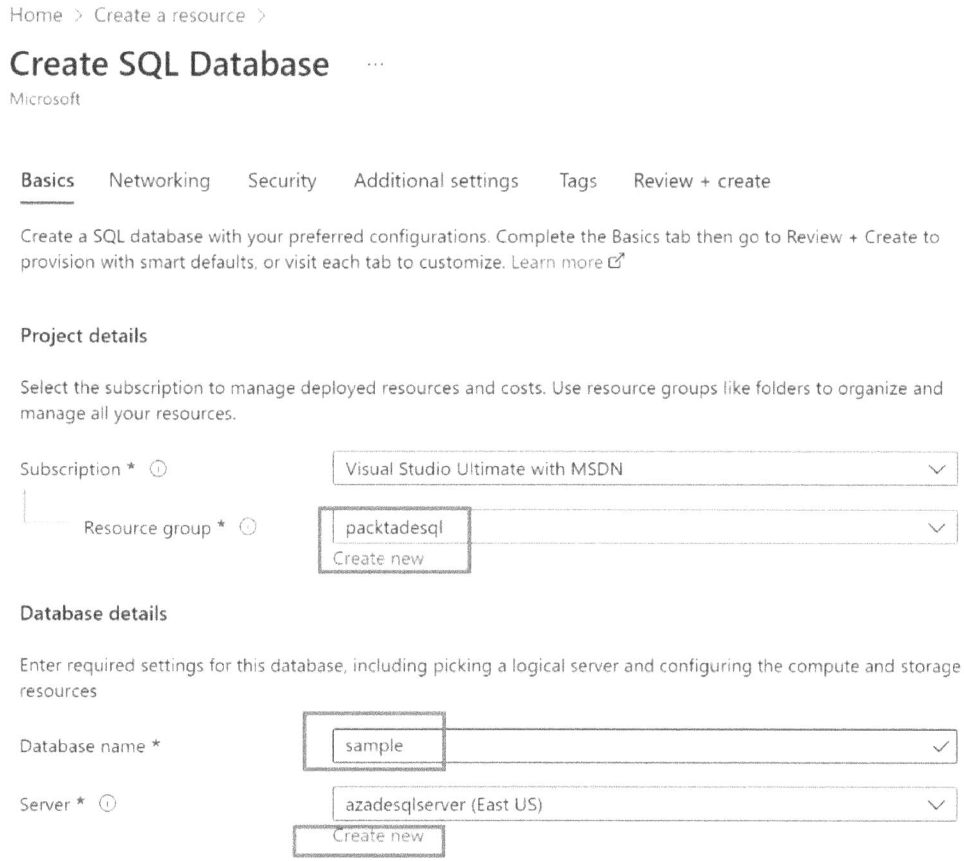

Home > Create a resource >

Create SQL Database   ...
Microsoft

Basics   Networking   Security   Additional settings   Tags   Review + create

Create a SQL database with your preferred configurations. Complete the Basics tab then go to Review + Create to provision with smart defaults, or visit each tab to customize. Learn more ☑

Project details

Select the subscription to manage deployed resources and costs. Use resource groups like folders to organize and manage all your resources.

Subscription * ⓘ | Visual Studio Ultimate with MSDN   ∨ |

    Resource group * ⓘ | packtadesql   ∨ |
                          | Create new |

Database details

Enter required settings for this database, including picking a logical server and configuring the compute and storage resources

Database name * | sample   ✓ |

Server * ⓘ | azadesqlserver (East US)   ∨ |
                        | Create new |

Figure 5.52 – Create a new database

2.   Provide new server details, as shown in the following screenshot. Provide the user ID and password as `sqladmin/Sql@Server@1234`:

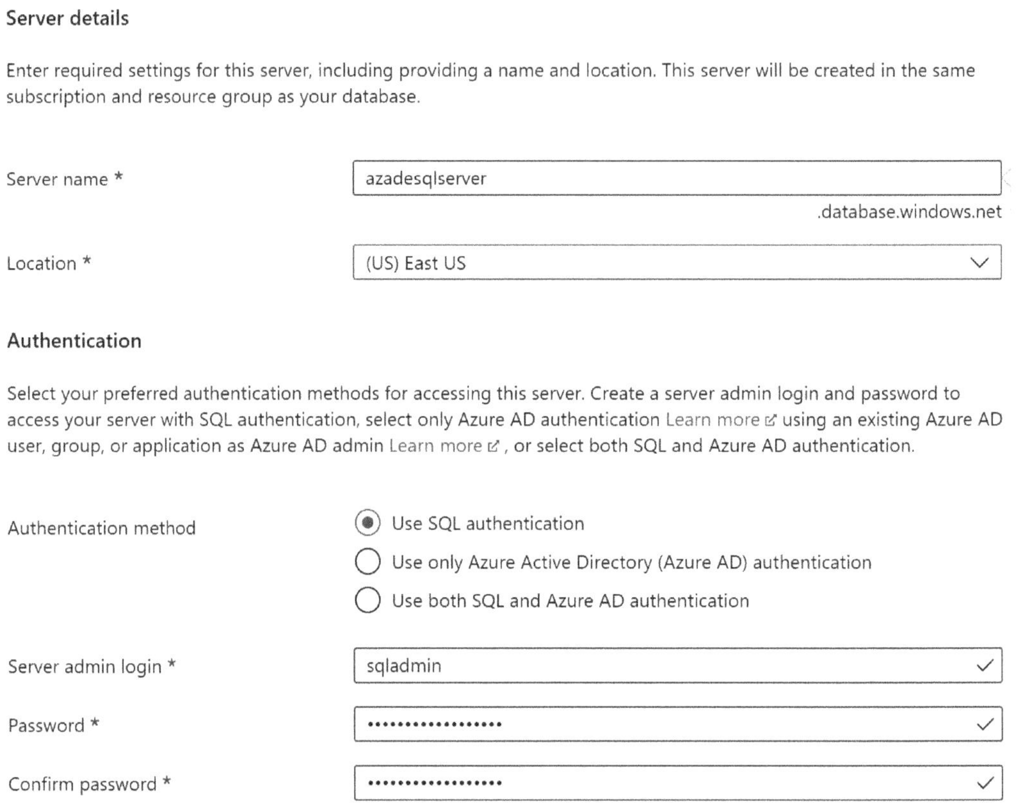

Server details

Enter required settings for this server, including providing a name and location. This server will be created in the same subscription and resource group as your database.

Server name *	azadesqlserver
	.database.windows.net
Location *	(US) East US ⌄

Authentication

Select your preferred authentication methods for accessing this server. Create a server admin login and password to access your server with SQL authentication, select only Azure AD authentication Learn more ⧉ using an existing Azure AD user, group, or application as Azure AD admin Learn more ⧉ , or select both SQL and Azure AD authentication.

Authentication method	⦿ Use SQL authentication
	◯ Use only Azure Active Directory (Azure AD) authentication
	◯ Use both SQL and Azure AD authentication
Server admin login *	sqladmin ✓
Password *	•••••••••••••••• ✓
Confirm password *	•••••••••••••••• ✓

Figure 5.53 – Create a new server

3.   Click **Configure database**:

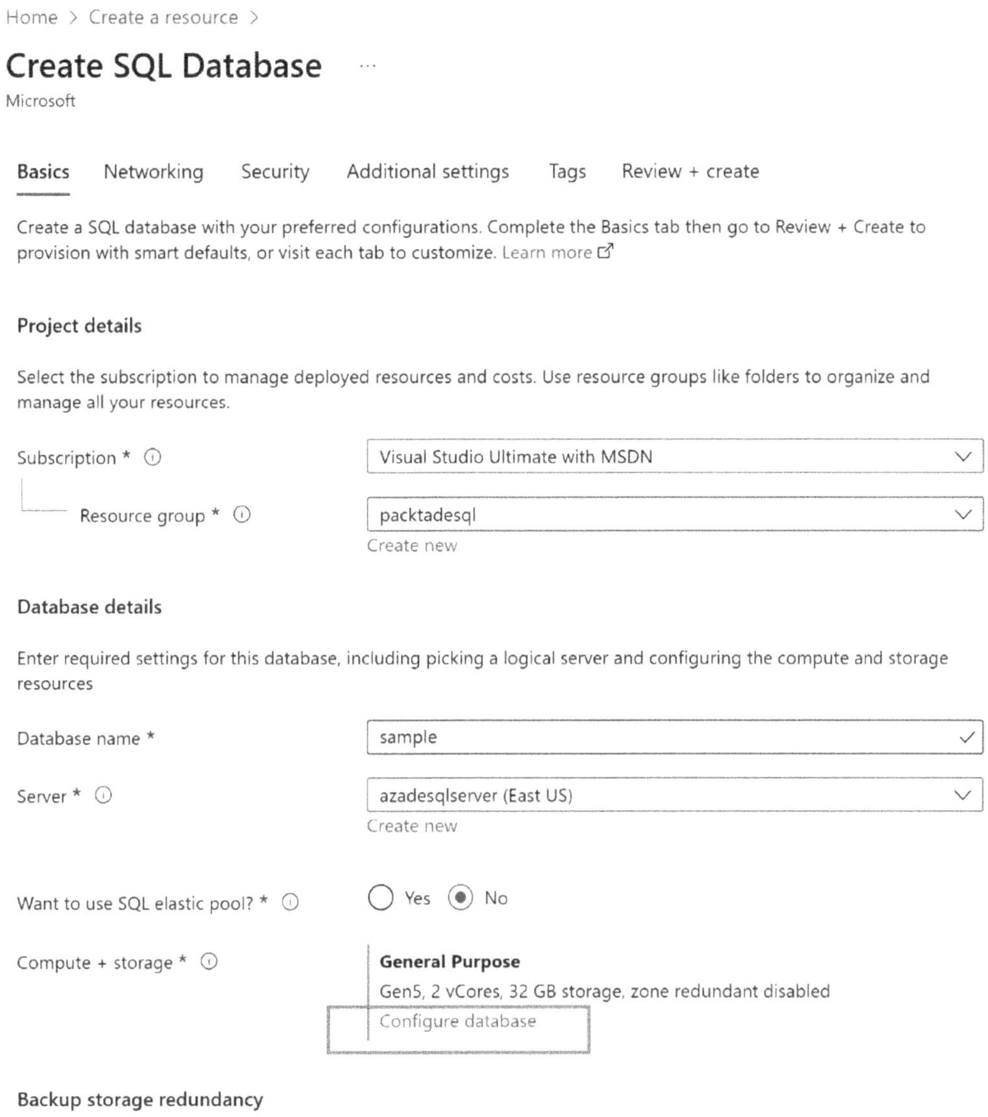

Home > Create a resource >

Create SQL Database   ...
Microsoft

Basics   Networking   Security   Additional settings   Tags   Review + create

Create a SQL database with your preferred configurations. Complete the Basics tab then go to Review + Create to provision with smart defaults, or visit each tab to customize. Learn more ☐

Project details

Select the subscription to manage deployed resources and costs. Use resource groups like folders to organize and manage all your resources.

Subscription * ⓘ

| Visual Studio Ultimate with MSDN | ⌄ |

  └── Resource group * ⓘ

| packtadesql | ⌄ |

Create new

Database details

Enter required settings for this database, including picking a logical server and configuring the compute and storage resources

Database name *

| sample | ✓ |

Server * ⓘ

| azadesqlserver (East US) | ⌄ |

Create new

Want to use SQL elastic pool? * ⓘ   ◯ Yes ⦿ No

Compute + storage * ⓘ

General Purpose
Gen5, 2 vCores, 32 GB storage, zone redundant disabled

Configure database

Backup storage redundancy

Figure 5.54 – Configure the Hyperscale tier

4. Under **Service and compute tier**, select **Hyperscale (On-demand scalable storage)**. Check the **I understand that scaling from Hyperscale to another service tier is not possible.** checkbox and click the **Apply** button:

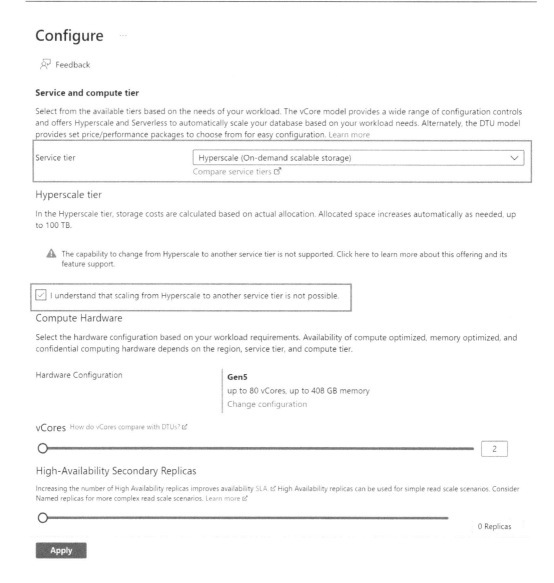

Figure 5.55 – Set the Hyperscale service tier

5. Click **Review + create** on the **Create SQL Database** screen. Hit the **Create** button to create the database:

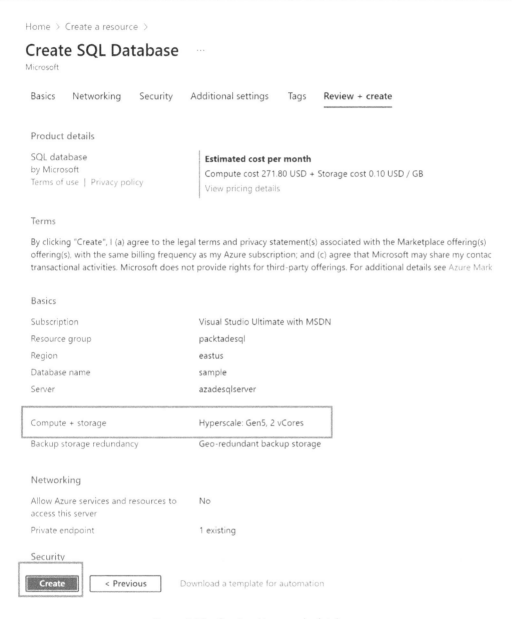

Figure 5.56 – Create a Hyperscale database

The preceding steps will create a Hyperscale SQL database that can scale up to 100 TB. As you may have noticed, we didn't specify an upper limit for the database size when creating the database, unlike we would need to with the General Purpose and Business Critical service tiers. The database created would have a starting size of 40 GB, grow if required, and you would only pay for the storage that you consumed.

6

Implementing High Availability and Monitoring in Azure SQL Database

Azure SQL Database, a fundamental relational database as a service offering in Azure acts as a source, destination, or even an intermediate storage layer in data engineering pipelines. Azure SQL Database can be used to consolidate data coming from several relational data sources and build mini data warehouses or data marts. Keeping Azure SQL Database in good health and available at all times is of the utmost importance to any organization. In this chapter, we will explore recipes that will help us monitor the health of a database and ensure database availability. By the end of the chapter, you will have learned how to configure replication, an auto-failover group, auditing, monitoring for Azure SQL Database, and high availability for the Hyperscale tier of Azure SQL Database. This chapter has the following recipes:

- Implementing active geo-replication for an Azure SQL database using PowerShell

- Implementing an auto-failover group for an Azure SQL database using PowerShell

- Configuring high availability to the Hyperscale tier of Azure SQL Database

- Implementing vertical scaling for an Azure SQL database using PowerShell

- Monitoring an Azure SQL database using the Azure portal

- Configuring auditing for Azure SQL Database

Implementing active geo-replication for an Azure SQL database using PowerShell

The active geo-replication feature allows you to create up to four readable databases for a primary Azure SQL Database. Active geo-replication uses **SQL Server AlwaysOn** to asynchronously replicate transactions to the secondary databases. The secondary database can be in the same or a different region from the primary database.

Active geo-replication can be used for the following cases:

- To provide business continuity by failing over to the secondary database in case of a disaster. The failover is manual.

- To offload reads to the readable secondary database.

- To migrate a database to a different server in another region.

In this recipe, we'll configure active geo-replication for an Azure SQL database and perform a manual failover.

Getting ready

In a new PowerShell window, execute the `Connect-AzAccount` command and follow the steps to log in to your Azure account.

You need an existing Azure SQL database for this recipe. If you don't have one, create an Azure SQL database by following the steps mentioned in the *Provisioning and connecting to an Azure SQL database using PowerShell* recipe in *Chapter 5, Configuring and Securing Azure SQL Database*.

How to do it...

First, let's create a readable secondary database.

Creating a readable secondary

The steps are as follows:

1. Execute the following command to provision a new Azure SQL Server to host the secondary replica:

```
#create credential object for the Azure SQL Server admin credential
$sqladminpassword = ConvertTo-SecureString 'Sql@Server@1234' -AsPlainText -Force
$sqladmincredential = New-Object System.Management.
```

```
Automation.PSCredential ('sqladmin', $sqladminpassword)
New-AzSQLServer -ServerName azadesqlsecondary
-SqlAdministratorCredentials $sqladmincredential
-Location westus -ResourceGroupName packtadesql
```

2. Execute the following command to configure the geo-replication from the primary server to the secondary server:

```
$primarydb = Get-AzSqlDatabase -DatabaseName azadesqldb
-ServerName azadesqlserver -ResourceGroupName packtadesql
$primarydb | New-AzSqlDatabaseSecondary
-PartnerResourceGroupName packtadesql -PartnerServerName
azadesqlsecondary -AllowConnections "All"
```

You should get an output as shown in the following screenshot:

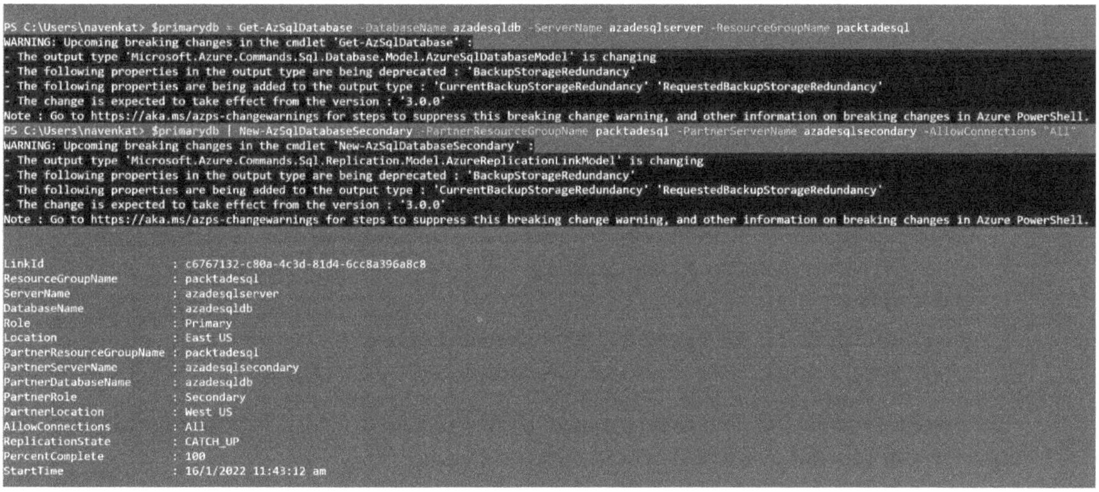

Figure 6.1 – Configuring geo-replication

Moreover, we can also check geo-replication configuration on the Azure portal, as shown in the following screenshot. Go to the **azadesqldb** database and, under **Data management**, click **Replicas**:

Figure 6.2 – Verifying geo-replication in the Azure portal

Performing manual failover to the secondary

The steps are as follows:

1. Execute the following command to manually fail over to the secondary database:

    ```
    $secondarydb = Get-AzSqlDatabase -DatabaseName azadesqldb
    -ServerName azadesqlsecondary -ResourceGroupName
    packtadesql
    $secondarydb | Set-AzSqlDatabaseSecondary
    -PartnerResourceGroupName packtadesql -Failover
    ```

    The preceding command performs a planned failover without data loss. To perform a manual failover with data loss, use the Allowdataloss switch.

    If we check the Azure portal, we'll see that **azadesqlsecondary/azadesqldb** in **West US** is the primary database:

Home > packtadesql > azadesqldb (azadesqlserver/azadesqldb) > azadesqldb (azadesqlserver/azadesqldb)

🌐 azadesqldb (azadesqlserver/azadesqldb) | Replicas ⋯
SQL database

| 🔍 Search (Ctrl+/) | ≪ | + Create replica 🔄 Refresh 🗣 Feedback |

🖉 Connection strings

Geo replicas for your database are listed below. Geo replicas reside on a different logical server from the primary and protect against regional
Learn more ↗

‖' Properties

🔒 Locks

Name ↑↓	Server ↑↓	Region ↑↓	Failover policy ↑↓	Pricing tier ↑↓
∨ Primary				
azadesqldb	azadesqlsecondary	West US	None	Basic

Data management

🌐 Replicas

∨ Geo replicas

azadesqldb	azadesqlserver	East US		Basic

[🖉] Sync to other databases

Integrations

Figure 6.3 – Failing over to the secondary server

2.   We can also get the active geo-replication information by executing the following command:

```
$Get-AzSqlDatabaseReplicationLink -DatabaseName
azadesqldb -PartnerResourceGroupName packtadesql
-PartnerServerName azadesqlsecondary -ServerName
azadesqlserver -ResourceGroupName packtadesql
```

You should get an output as shown in the following screenshot:

Figure 6.4 – Getting the geo-replication status

Removing active geo-replication

Execute the following command to remove the active geo-replication link between the primary and secondary databases:

```
$primarydb = Get-AzSqlDatabase -DatabaseName azadesqldb
-ServerName azadesqlserver -ResourceGroupName packtadesql

$primarydb | Remove-AzSqlDatabaseSecondary
-PartnerResourceGroupName packtadesql -PartnerServerName
azadesqlsecondary
```

You should get an output as shown in the following screenshot:

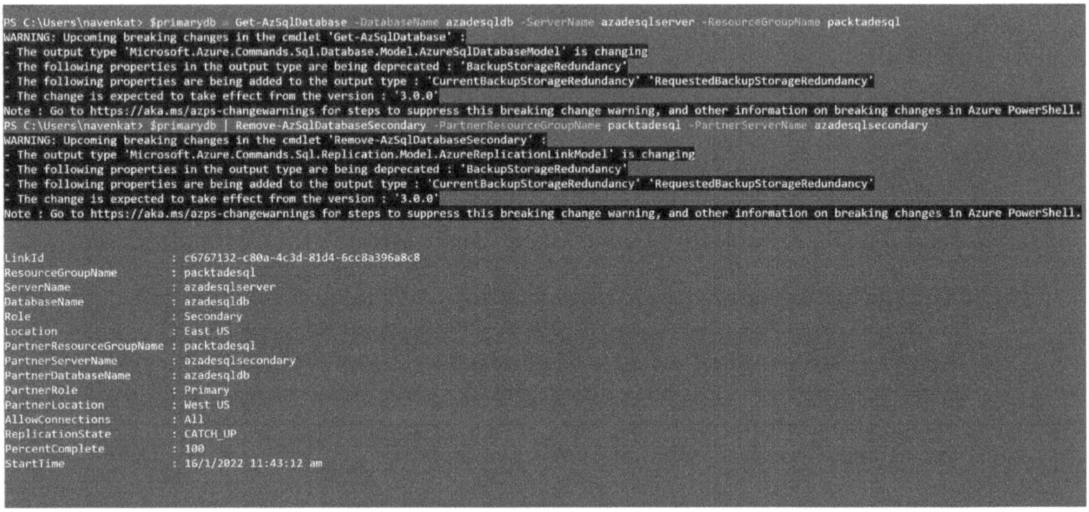

Figure 6.5 – Removing active geo-replication

How it works...

To configure active geo-replication, we use the New-AzSqlDatabaseSecondary command. This command expects the primary database name, server name, and resource group name; the secondary resource group name and server name; and the **Allow connections** parameter. If we want a readable secondary, then we set **Allow connections** to **All**; otherwise, we set it to **No**.

The active geo-replication provides manual failover with and without data loss. To perform a manual failover, we use the Set-AzSqlDatabaseSecondary command. This command expects the secondary server name, the database name, the resource group name, a failover switch, and the Allowdataloss switch in case of failover with data loss.

To remove active geo-replication, we use the `Remove-AzSqlDatabaseSecondary` command. This command expects the secondary server name, secondary database name, and resource name to remove the replication link between the primary and secondary databases.

Removing active geo-replication doesn't remove the secondary database.

Implementing an auto-failover group for an Azure SQL database using PowerShell

An auto-failover group allows a group of databases to fail to a secondary server in another region if the SQL database service in the primary region fails. Unlike active geo-replication, the secondary server should be in a different region from the primary. The secondary databases can be used to offload read workloads. The failover can be manual or automatic.

In this recipe, we'll create an auto-failover group, add databases to the auto-failover group, and perform a failover to the secondary server.

Getting ready

In a new PowerShell window, execute the `Connect-AzAccount` command and follow the steps to log in to your Azure account.

You will need an existing Azure SQL database for this recipe. If you don't have one, create an Azure SQL database by following the steps mentioned in the *Provisioning and connecting to an Azure SQL database using PowerShell* recipe of *Chapter 5, Configuring and Securing Azure SQL Database*.

How to do it...

First, let's create an auto-failover group.

Creating an auto-failover group

The steps are as follows:

1. Execute the following PowerShell command to create a secondary server. The server should be in a different region than the primary server:

    ```
    #create credential object for the Azure SQL Server admin
    credential
    $sqladminpassword = ConvertTo-SecureString 'Sql@
    Server@1234' -AsPlainText -Force
    $sqladmincredential = New-Object System.Management.
    Automation.PSCredential ('sqladmin', $sqladminpassword)
    ```

```
New-AzSQLServer -ServerName azadesqlsecondary
-SqlAdministratorCredentials $sqladmincredential
-Location westus -ResourceGroupName packtadesql
```

2. Execute the following command to create the auto-failover group:

```
New-AzSqlDatabaseFailoverGroup -ServerName azadesqlserver
-FailoverGroupName adefg -PartnerResourceGroupName
packtadesql -PartnerServerName azadesqlsecondary
-FailoverPolicy Automatic -ResourceGroupName packtadesql
```

You should get an output as shown in the following screenshot:

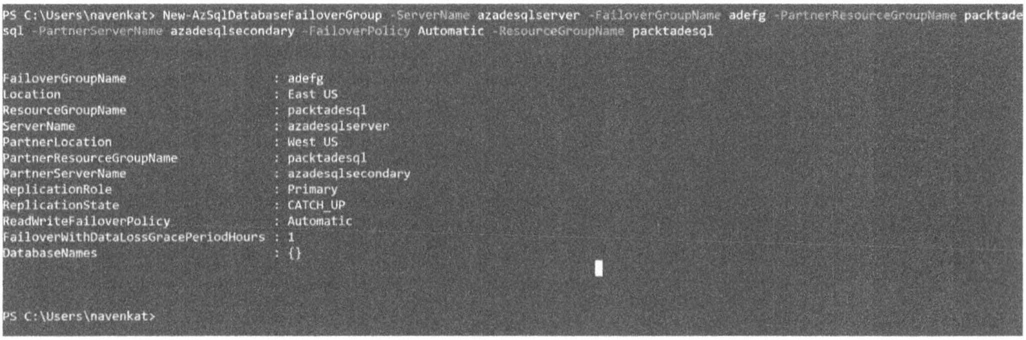

```
PS C:\Users\navenkat> New-AzSqlDatabaseFailoverGroup -ServerName azadesqlserver -FailoverGroupName adefg -PartnerResourceGroupName packtade
sql -PartnerServerName azadesqlsecondary -FailoverPolicy Automatic -ResourceGroupName packtadesql

FailoverGroupName                     : adefg
Location                              : East US
ResourceGroupName                     : packtadesql
ServerName                            : azadesqlserver
PartnerLocation                       : West US
PartnerResourceGroupName              : packtadesql
PartnerServerName                     : azadesqlsecondary
ReplicationRole                       : Primary
ReplicationState                      : CATCH_UP
ReadWriteFailoverPolicy               : Automatic
FailoverWithDataLossGracePeriodHours  : 1
DatabaseNames                         : {}

PS C:\Users\navenkat>
```

Figure 6.6 – Creating an auto-failover group

3. Execute the following command to add an existing database to the auto-failover group:

```
$db = Get-AzSqlDatabase -DatabaseName azadesqldb
-ServerName azadesqlserver -ResourceGroupName packtadesql
```
```
$db|Add-AzSqlDatabaseToFailoverGroup -ResourceGroupName
packtadesql -ServerName azadesqlserver -FailoverGroupName
adefg
```

4. Execute the following command to add a new Azure SQL database to the auto-failover group:

```
$db = New-AzSqlDatabase -DatabaseName azadesqldb2
-Edition basic -ServerName azadesqlserver
-ResourceGroupName packtadesql
```
```
$db | Add-AzSqlDatabaseToFailoverGroup -FailoverGroupName
adefg
```

5. Execute the following PowerShell command to get the details about the auto-failover group:

```
Get-AzSqlDatabaseFailoverGroup -ServerName azadesqlserver
-FailoverGroupName adefg -ResourceGroupName packtadesql
```

You should get an output as shown in the following screenshot:

```
PS C:\Users\navenkat> Get-AzSqlDatabaseFailoverGroup -ServerName azadesqlserver -FailoverGroupName adefg -ResourceGroupName packtadesql_

FailoverGroupName                          : adefg
Location                                   : East US
ResourceGroupName                          : packtadesql
ServerName                                 : azadesqlserver
PartnerLocation                            : West US
PartnerResourceGroupName                   : packtadesql
PartnerServerName                          : azadesqlsecondary
ReplicationRole                            : Primary
ReplicationState                           : CATCH_UP
ReadWriteFailoverPolicy                    : Automatic
FailoverWithDataLossGracePeriodHours       : 1
DatabaseNames                              : {azadesqldb, azadesqldb2}
```

Figure 6.7 – Getting the auto-failover group details

The endpoint used to connect to the primary server of an auto-failover group is in the `<auto-failover group name>.database.windows.net` form. In our case, this will be `adefg.database.windows.net`.

To connect to a readable secondary in an auto-failover group, the endpoint used is in the `<auto-failover group name>.secondary.database.windows.net` form. In our case, the endpoint will be `adefg.secondary.database.windows.net`. In addition to this, we need to specify **ApplicationIntent** as **readonly** in the connection string when connecting to the secondary.

6. In the Azure portal, the failover groups can be found on the Azure SQL Server page, as shown in the following screenshot. Go to **azadesqlserver** and, under **Data management**, click on **Failover groups**:

Figure 6.8 – Viewing an auto-failover group in the Azure portal

7. To open the failover group details, click on the failover group name, **adefg**:

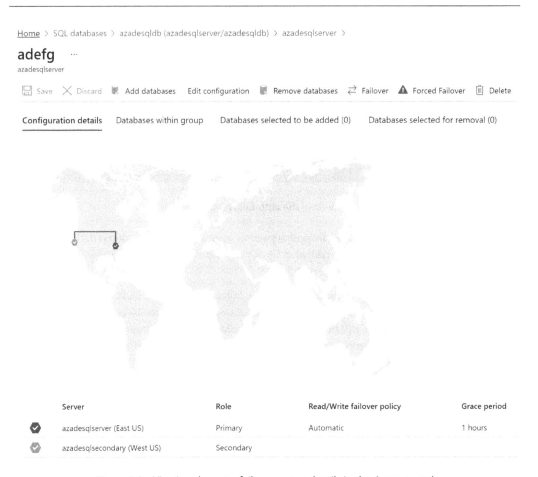

Home > SQL databases > azadesqldb (azadesqlserver/azadesqldb) > azadesqlserver >

adefg   ...
azadesqlserver

🖫 Save   ✕ Discard   🖳 Add databases   Edit configuration   🖳 Remove databases   ⇄ Failover   ⚠ Forced Failover   🗑 Delete

Configuration details   Databases within group   Databases selected to be added (0)   Databases selected for removal (0)

	Server	Role	Read/Write failover policy	Grace period
✅	azadesqlserver (East US)	Primary	Automatic	1 hours
✅	azadesqlsecondary (West US)	Secondary		

Figure 6.9 – Viewing the auto-failover group details in the Azure portal

Performing a failover to the secondary server

The steps are as follows:

1. Execute the following command to fail over to the secondary server:

    ```
    $secondarysqlserver = Get-AzSqlServer -ResourceGroupName
    packtadesql -ServerName azadesqlsecondary

    $secondarysqlserver | Switch-AzSqlDatabaseFailoverGroup
    -FailoverGroupName adefg
    ```

    If we check in the Azure portal, the primary server is now **azadesqlsecondary** and the secondary server is **azadesqlserver**, as shown in the following screenshot:

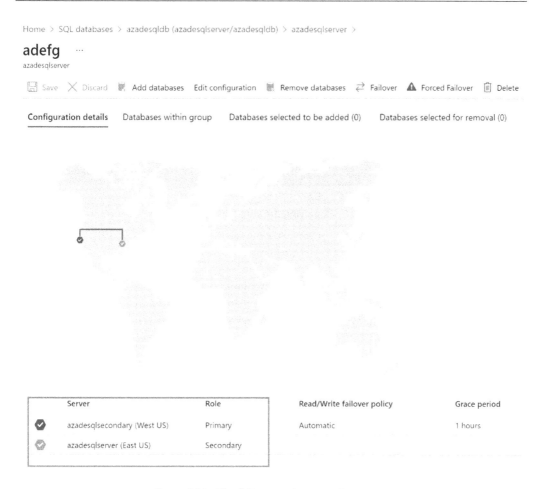

Figure 6.10 – The failover to the secondary server

2. Execute the following command to remove the auto-failover group. Removing the auto-failover group doesn't remove the secondary or primary SQL databases:

```
Remove-AzSqlDatabaseFailoverGroup -ServerName
azadesqlsecondary -FailoverGroupName adefg
-ResourceGroupName packtadesql
```

You should get an output as shown in the following screenshot:

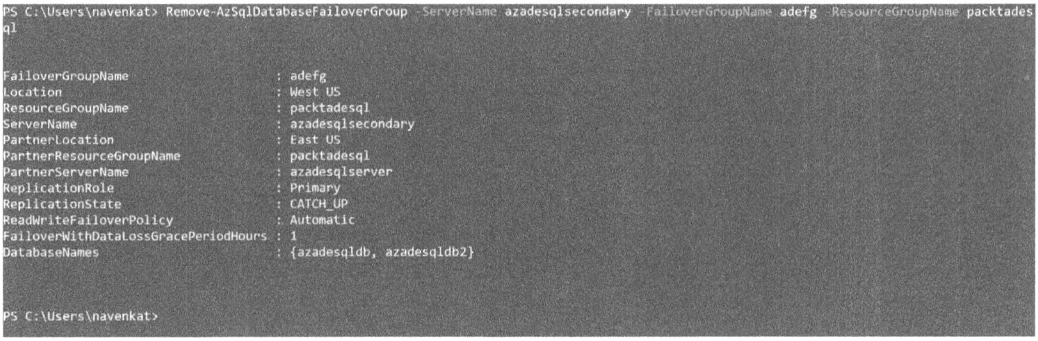

```
PS C:\Users\navenkat> Remove-AzSqlDatabaseFailoverGroup -ServerName azadesqlsecondary -FailoverGroupName adefg -ResourceGroupName packtades
ql

FailoverGroupName                       : adefg
Location                                : West US
ResourceGroupName                       : packtadesql
ServerName                              : azadesqlsecondary
PartnerLocation                         : East US
PartnerResourceGroupName                : packtadesql
PartnerServerName                       : azadesqlserver
ReplicationRole                         : Primary
ReplicationState                        : CATCH_UP
ReadWriteFailoverPolicy                 : Automatic
FailoverWithDataLossGracePeriodHours    : 1
DatabaseNames                           : {azadesqldb, azadesqldb2}


PS C:\Users\navenkat>
```

Figure 6.11 – Removing the auto-failover group

How it works...

The `New-AzSqlDatabaseFailoverGroup` command is used to create an auto-failover group. We need to specify the auto-failover group name, the primary and secondary server names, the resource group name, and the failover policy (either automatic or manual). In addition to this, we can also specify `GracePeriodWithDataLossHours`. As the replication between the primary and secondary is asynchronous, the failover may result in data loss.

The `GracePeriodWithDataLossHours` value specifies how many hours the system should wait before initiating the automatic failover. This can therefore limit the data loss that happens because of a failover.

After the auto-failover group creation, we can add the databases to the auto-failover group by using the `Add-AzSqlDatabaseToFailoverGroup` command. The database to be added should exist on the primary server and not on the secondary server.

We can perform a failover by executing the `Switch-AzSqlDatabaseFailoverGroup` command. We need to provide the primary server's name, the auto-failover group name, and the primary server resource group name.

To remove an auto-failover group, execute the `Remove-AzSqlDatabaseFailoverGroup` command by specifying the primary server's name and resource group, and the auto-failover group name.

Configuring high availability to the Hyperscale tier of Azure SQL Database

Azure SQL Database in the Hyperscale tier supports databases with up to 100 TB of storage size, compared to the 4 TB offered in the General Purpose and Business Critical tiers. In addition to supporting large databases, the Hyperscale tier offers high-availability capabilities too, with the three following types of replica:

- **Geo-replica**: Allows the creation of asynchronous replicas in the same or a different region, allowing users to have a read-only copy of the database

- **High-availability replica**: Acts as a hot standby server, offering automatic failover

- **Named replica**: Allows the creation of replicas with a different database name from the primary database, mainly used for supporting read scale-out scenarios

Both high-availability and named replicas are configured on synchronous mode, and can be created in the same location as a primary database.

In this recipe, we will configure a named replica for a Hyperscale database and execute read-only queries against the secondary database. We will also showcase how a secondary replica will continue to function even when a primary replica is under maintenance due to a scale-up activity.

Getting ready

Create a Hyperscale Azure SQL database, as mentioned in the *Configuring the Hyperscale tier of Azure SQL Database* recipe of *Chapter 5, Configuring and Securing Azure SQL Database*.

Download and install **SQL Server Management Studio (SSMS)** on your local machine. SSMS can be downloaded from `https://docs.microsoft.com/en-us/sql/ssms/download-sql-server-management-studio-ssms?view=sql-server-ver15`.

How to do it...

Follow these steps to create a secondary replica for a Hyperscale database:

1. Log in to `portal.azure.com`, click **All resources**, and find **azadesqlserver**. Go to the **HyperScale Azure SQL database** created. Under **Data management**, click **Replicas** and then click **Create replica**:

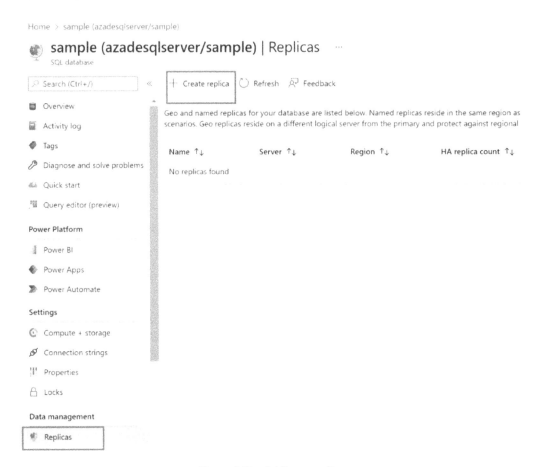

Figure 6.12 – Adding a replica

2. Select **Named replica** for **Replica type**. **Named replica** allows you to create another database that can be used as a read-only replica in the same region as the primary. Our primary database is named **sample** and our secondary is named **sample_NamedReplica**. You may choose to create a new logical SQL server if you wish but, in this example, we will reuse the same logical server. Hit the **Review + create** button:

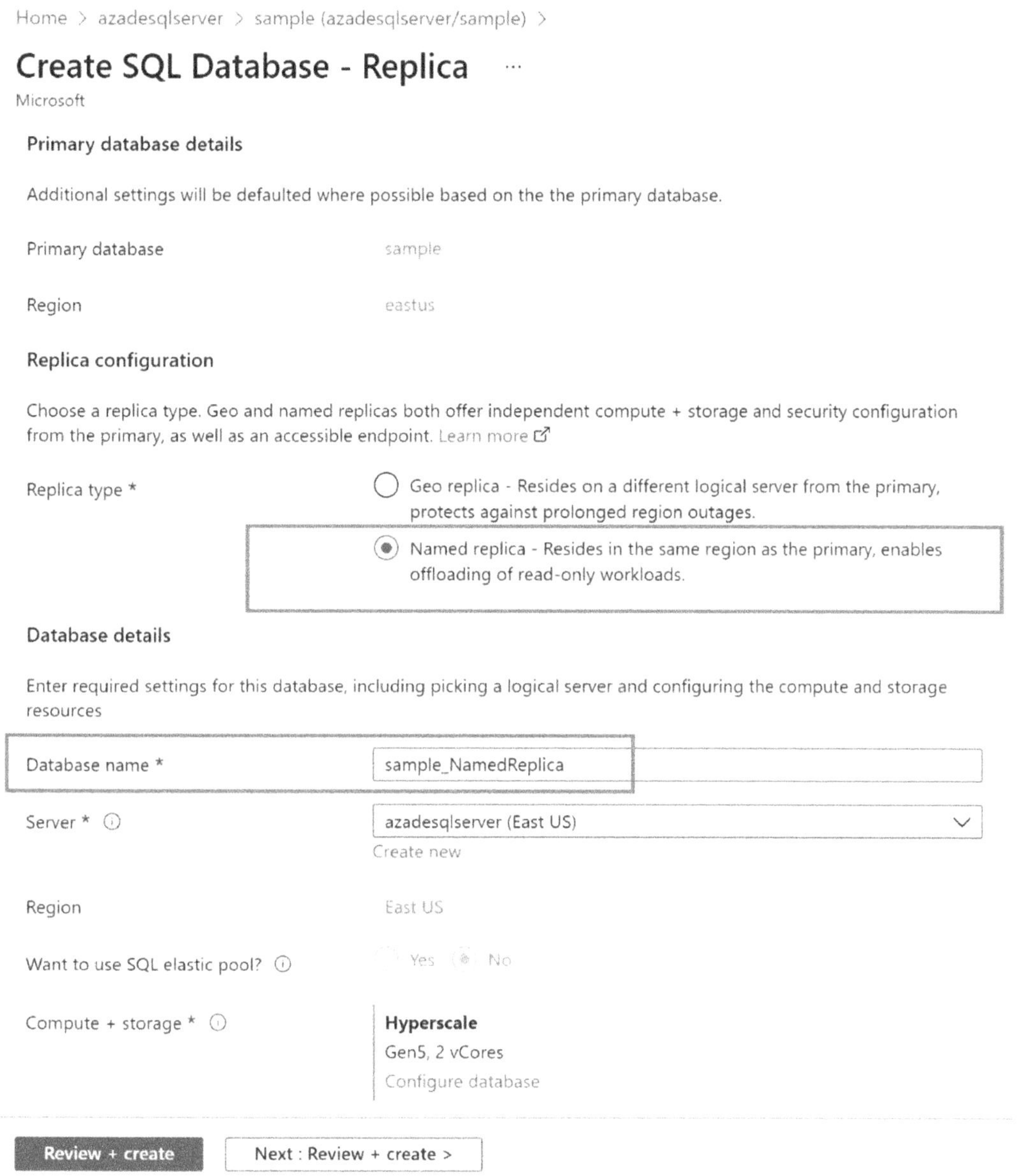

Home > azadesqlserver > sample (azadesqlserver/sample) >

Create SQL Database - Replica   ...
Microsoft

Primary database details

Additional settings will be defaulted where possible based on the the primary database.

Primary database   sample

Region   eastus

Replica configuration

Choose a replica type. Geo and named replicas both offer independent compute + storage and security configuration from the primary, as well as an accessible endpoint. Learn more ☐

Replica type *   ◯ Geo replica - Resides on a different logical server from the primary,
                                   protects against prolonged region outages.

                                ◉ Named replica - Resides in the same region as the primary, enables
                                   offloading of read-only workloads.

Database details

Enter required settings for this database, including picking a logical server and configuring the compute and storage resources

Database name *   sample_NamedReplica

Server * ⓘ   azadesqlserver (East US)   ⌄
                                Create new

Region   East US

Want to use SQL elastic pool? ⓘ   Yes ◉ No

Compute + storage * ⓘ   **Hyperscale**
                                Gen5, 2 vCores
                                Configure database

[Review + create]   [Next : Review + create >]

Figure 6.13 – Configuring the replica

3. Once the replica has been created, go to **azadesqlserver** in `portal.azure.com` and click **Firewalls and virtual networks**. Click **Add client IP**. After the IP is listed as a rule, click on the **Save** button. We connect to the primary and secondary database from a local client machine, and hence the local machine IP needs to be whitelisted:

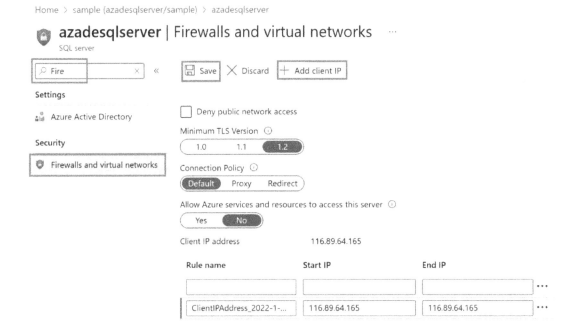

Figure 6.14 – Firewalls and virtual networks

4. Connect to **azadesqlserver.database.windows.net** using SSMS with the user ID and the password set to `sqladmin/Sql@Server@1234`. Right-click on the **sample_NamedReplica** database and click **New Query** to open a connection to it:

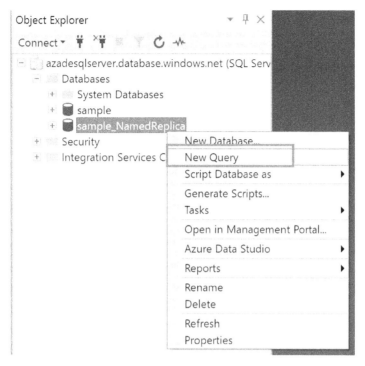

Figure 6.15 – New Query

5. Execute the following command:

```
Select db_name() as database_name
GO
```

Notice that the query executes successfully on **sample_NamedReplica**, as it just performs a read-only operation:

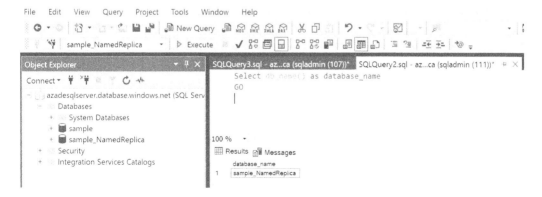

Figure 6.16 – A read-only query

6. Notice that when you perform a read-write operation such as table creation on **sample_ NamedReplica**, it fails:

Figure 6.17 – A read-write failure

7. Let's perform the following experiment. We will perform a scale-up operation on the primary database (**sample**), and while the scale-up is in progress, we will perform read-only queries against the secondary database. Due to Azure SQL Database's high-availability capabilities, there will be no connection drop in the secondary database when the primary database is scaled up.

Execute the following script against the secondary database (**sample_NamedReplica**). The script will continuously run read-only queries against the secondary database at an interval of 1 second:

```
While 1=1
Begin
Select * from sys.objects;
WAITFOR DELAY '0:00:01'
END
```

The following is a screenshot of the script execution. As expected, the script reads the secondary database continuously:

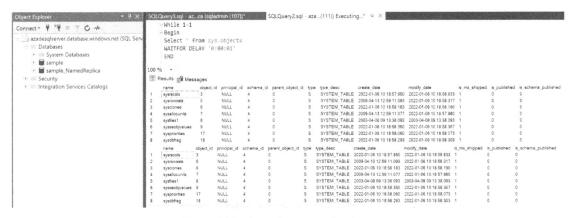

Figure 6.18 – A continuous read-only query

8. Go to portal.azure.com, find **azadesqlserver**, and go to the sample database. Click **Compute + storage**. Under **vCores**, scale up from 2 vCores to 8 and hit the **Apply** button:

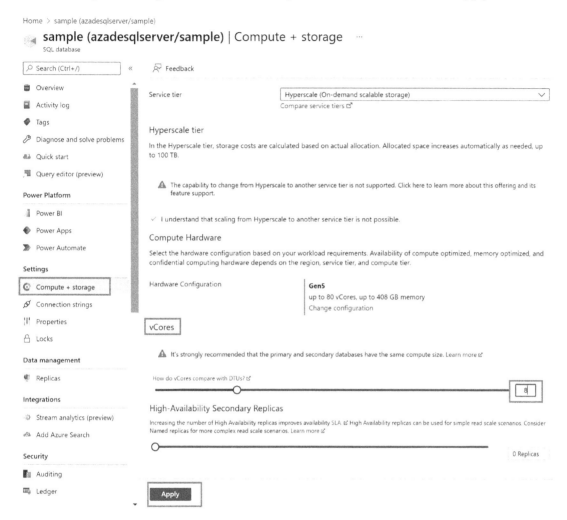

Figure 6.19 – Scaling up the primary database

9. While scale-up is in progress, go back to SSMS and check whether the query is running without errors against the secondary database (**sample_NamedReplica**). You will notice that the query was running just fine when the primary was being scaled up:

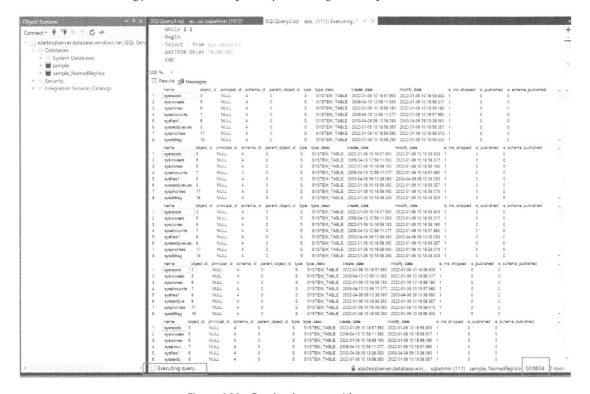

Figure 6.20 – Read-only query with no errors

How it works...

For any operation to be performed in a database, we need two key components. They are the compute, which is responsible for processing the query, and the storage, which actually stores the data. While creating replicas for a Hyperscale database, Azure doesn't copy any data or provision any additional storage. Data stored is common for both primary and secondary replicas. For a secondary replica, a logical database is created, and additional vCores are provisioned to handle compute operations required for read-only queries. Even though the database name of a read-only replica is different, it will query the same data files and facilitate read-only workloads. As the compute and storage are decoupled components in Azure SQL Database, there is no disruption to the connections of the secondary replica, even when the compute is scaled up (or scaled down) in the primary database.

Having a secondary replica offers read scale-out, excellent resilience, and high availability to mission-critical applications.

Implementing vertical scaling for an Azure SQL database using PowerShell

An Azure SQL Database has multiple purchase models and service tiers for different workloads. There are two purchasing models: **DTU-based** and **vCore-based**. There are multiple service tiers within these purchasing models.

Having multiple service tiers provides us with the flexibility to scale up or scale down based on the workload or activity in an Azure SQL database.

In this recipe, we'll learn how to automatically scale up an Azure SQL Database whenever the CPU percentage is above 40%.

Getting ready

In a new PowerShell window, execute the `Connect-AzAccount` command and follow the steps to log in to your Azure account.

You will need an existing Azure SQL database for this recipe. If you don't have one, create an Azure SQL database by following the steps mentioned in the *Provisioning and connecting to an Azure SQL database using PowerShell* recipe of *Chapter 5, Configuring and Securing Azure SQL Database*.

How to do it...

The steps for this recipe are as follows:

1. Execute the following PowerShell command to create an Azure Automation account:

```
#Create an Azure automation account
$automation = New-AzAutomationAccount -ResourceGroupName
packtadesql -Name azadeautomation -Location eastus -Plan
Basic
```

2. Execute the following command to create an Automation runbook of the PowerShell workflow type:

```
#Create a new automation runbook of type PowerShell
workflow
$runbook = New-AzAutomationRunbook -Name rnscalesql
-Description "Scale up sql azure when CPU is 40%" -Type
PowerShellWorkflow -ResourceGroupName packtadesql
-AutomationAccountName $automation.AutomationAccountName
```

3. Execute the following command to create Automation credentials. The credentials are passed as a parameter to the runbook and are used to connect to the Azure SQL database from the runbook:

```
#Create automation credentials.
$sqladminpassword = ConvertTo-SecureString 'Sql@
Server@1234' -AsPlainText -Force
$sqladmincredential = New-Object System.Management.
Automation.PSCredential('sqladmin', $sqladminpassword)
$creds = New-AzAutomationCredential -Name sqlcred
-Description "sql azure creds" -ResourceGroupName
packtadesql -AutomationAccountName $automation.
AutomationAccountName -Value $sqladmincredential
```

4. Go to portal.azure.com. Under **All resources**, search for and open the **azadeautomation** automation:

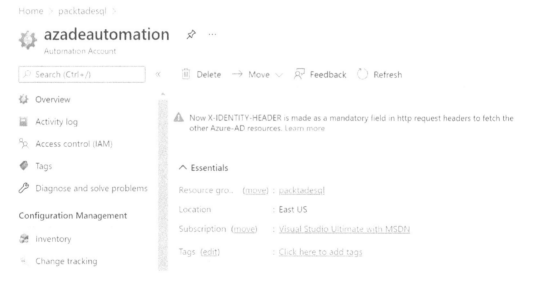

Figure 6.21 – Open Automation account

5. Search for and click on **Modules**. Click on the **Browse gallery** button:

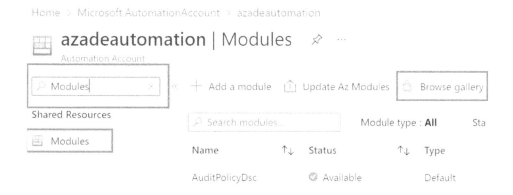

Figure 6.22 – Module installation in an Automation account

6. Search for sqlServer. Select the **SqlServer** module and hit the **Select** button on the next screen:

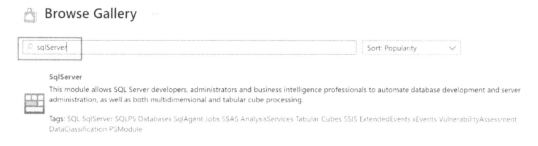

Figure 6.23 – An Azure Automation account SqlServer module installation

7. Set **Runtime version** to **7.1** on the **Add a module** screen and hit the **Import** button:

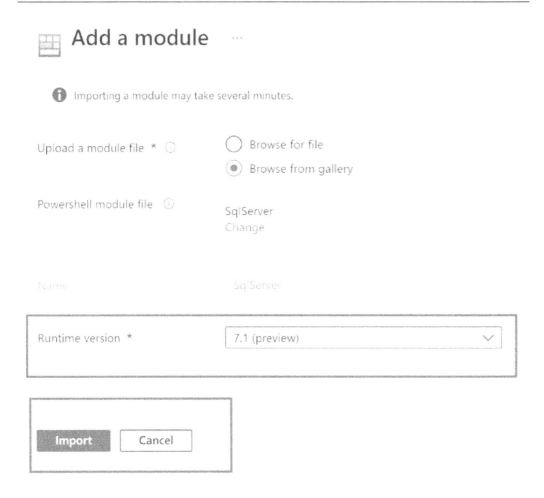

Figure 6.24 – Importing the SqlServer module

8. The next step is to edit the runbook and provide the PowerShell script to modify the service tier of an Azure SQL database. To do that, open https://portal.azure.com and log in to your Azure account. Under **All resources**, search for and open the **azadeautomation** Automation account:

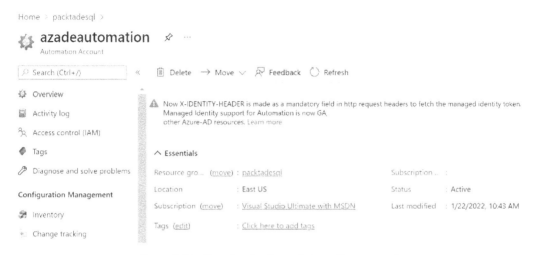

Figure 6.25 – Opening the Azure Automation account

9.   On the Azure Automation page, locate and select **Runbooks**:

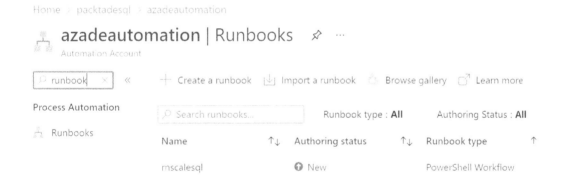

Figure 6.26 – Opening Runbooks in Azure Automation

10. Select the **rnscalesql** runbook to open the **Runbook** page. On the **Runbook** page, select **Edit**:

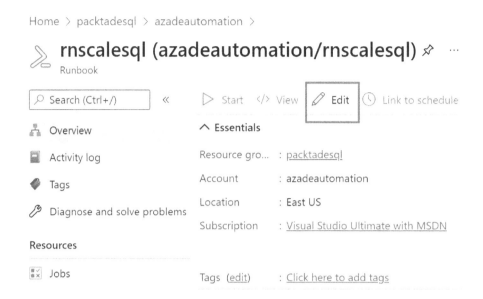

Figure 6.27 – Editing the runbook in your Azure Automation account

11. On the **Edit PowerShell Workflow Runbook** page, copy and paste the following PowerShell code into the canvas. The following section of the code receives parameters passed by the webhook:

```
workflow rnscalesql
{
param
(
# Name of the Azure SQL Database server (Ex:bzb98er9bp)
[parameter(Mandatory=$true)]
[string] $SqlServerName,
# Target Azure SQL Database name
[parameter(Mandatory=$true)]
[string] $DatabaseName,
# When using in the Azure Automation UI, please enter
the name of the credential asset for the "Credential"
parameter
[parameter(Mandatory=$true)]
[PSCredential] $Credential
)
```

The following section of the code uses the parameters passed and establishes a connection with the database:

```
inlinescript
{
$ServerName = $Using:SqlServerName + ".database.windows.
net"
$db = $Using:DatabaseName
$UserId = $Using:Credential.UserName
$Password = ($Using:Credential).GetNetworkCredential().
Password
$MasterDatabaseConnection = New-Object System.Data.
SqlClient.SqlConnection
$MasterDatabaseConnection.ConnectionString = "Server
= $ServerName; Database = Master; User ID = $UserId;
Password = $Password;"
$MasterDatabaseConnection.Open();
```

The following section of the code uses the ALTER DATABASE command and modifies the service tier of the Azure SQL Database to Standard S0:

```
$MasterDatabaseCommand = New-Object System.Data.
SqlClient.SqlCommand
$MasterDatabaseCommand.Connection
=$MasterDatabaseConnection
$MasterDatabaseCommand.CommandText = "ALTER DATABASE $db
MODIFY (EDITION = 'Standard', SERVICE_OBJECTIVE = 'S0');"
$MasterDbResult = $MasterDatabaseCommand.
ExecuteNonQuery();
}
}
```

12. Click **Save**, and then click **Publish** to publish the runbook:

Figure 6.28 – Saving and publishing the runbook

13. The next step is to create a webhook to trigger the runbook. Execute the following PowerShell command to create the webhook:

```
# define the runbook parameters
$Params = @
{"SQLSERVERNAME"="azadesqlserver";"DATABASENAME"
="azadesqldb";"CREDENTIAL"="sqlcred"}
# Create a webhook
$expiry = (Get-Date).AddDays(1)
New-AzAutomationWebhook -Name rnscaleazure -RunbookName
$runbook.Name -Parameters $Params -ResourceGroupName
packtadesql -AutomationAccountName $automation.
AutomationAccountName -IsEnabled $true -ExpiryTime
$expiry
```

> **Note**
> When defining $Params, you may want to change the default values mentioned here if you have a different Azure SQL Server, database, and credential values.

You should get an output as shown in the following screenshot:

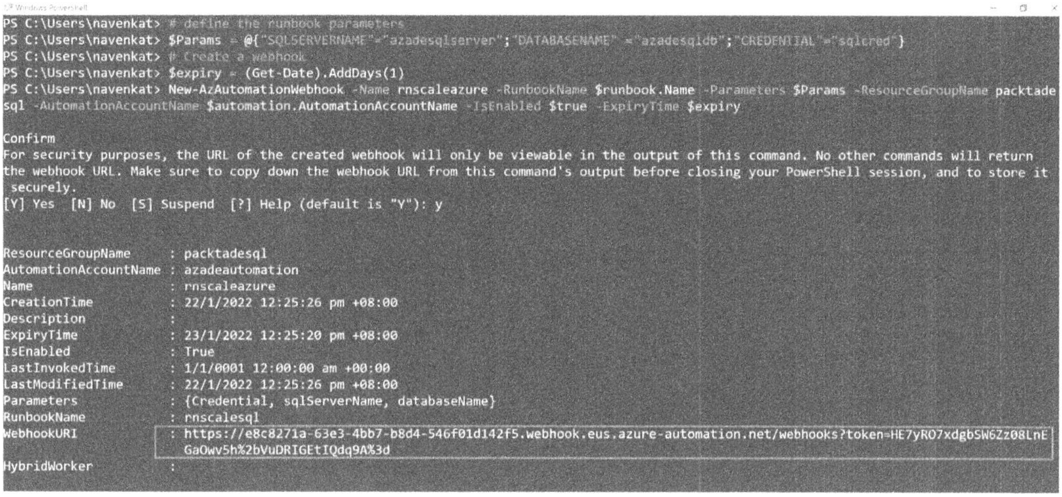

Figure 6.29 – Creating a webhook

Copy and save the `WebhookURI` value for later use.

14. The next step is to create an alert for an Azure SQL database that, when triggered, will call the webhook URI. Execute the following query to create an alert action group receiver:

```
#Create action group reciever

$whr = New-AzActionGroupReceiver -Name agrscalesql
-WebhookReceiver -ServiceUri "https://e8c8271a-63e3-
4bb7-b8d4-546f01d142f5.webhook.eus.azure-automation.net/
webhooks?token=HE7yRO7xdgbSW6Zz08LnEGaOwv5h%2bVuDRIGEtIQd
q9A%3d"
```

> **Note**
> Replace the value of the `ServiceUri` parameter with your webhook URI from the previous step.

15. Execute the following query to create an action group with an action receiver as defined by the preceding command:

```
#Create a new action group

$ag = Set-AzActionGroup -ResourceGroupName packtade -Name
ScaleSQLAzure -ShortName scaleazure -Receiver $whr
```

16. Execute the following query to create an alert condition to trigger the alert:

```
#define the alert trigger condition
$condition = New-AzMetricAlertRuleV2Criteria -MetricName
"cpu_percent" -TimeAggregation maximum -Operator
greaterthan -Threshold 40 -MetricNamespace "Microsoft.
Sql/servers/databases"
```

The condition defines that the alert should trigger when the metric CPU percentage is greater than 40%.

17. Execute the following query to create an alert on the Azure SQL database:

```
#Create the alert with the condition and action defined
in previous steps.
$rid = (Get-AzSqlDatabase -ServerName azadesqlserver
-ResourceGroupName packtadesql -DatabaseName azadesqldb).
Resourceid
Add-AzMetricAlertRuleV2 -Name monitorcpu
-ResourceGroupName packtadesql -WindowSize 00:01:00
-Frequency 00:01:00 -TargetResourceId $rid -Condition
$condition -Severity 1 -ActionGroupId $ag.id
```

You should get an output as shown in the following screenshot:

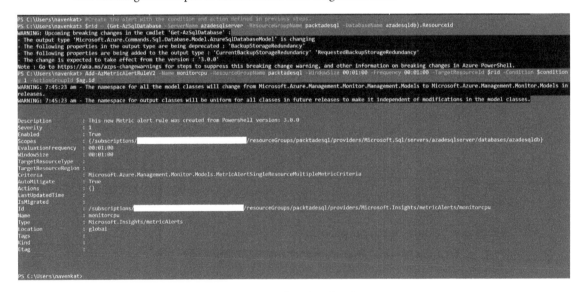

Figure 6.30 – Creating the alert

The preceding command creates an Azure SQL database alert. The alert is triggered when the cpu_percent metric is greater than 40% for more than 1 minute. When the alert is triggered, as defined in the action group, the webhook is called. The webhook in turn runs the runbook. The runbook modifies the service tier of the database to Standard S0.

Go to portal.azure.com, find **azadesqlserver**, and go to the **azadesqldb** database. Click on the **Alerts** option on the left-hand side. Click **Alert rules**:

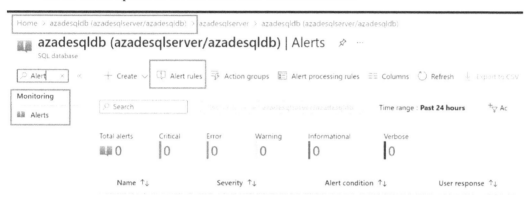

Figure 6.31 – Alert rules

You will notice that an alert condition called **monitorcpu** has been created:

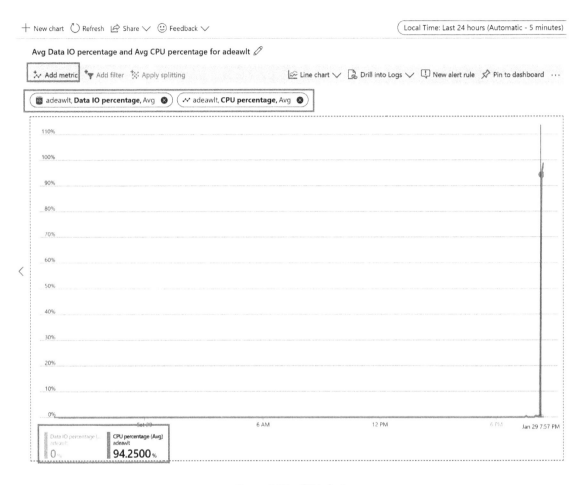

Figure 6.32 – CPU alert

18. To ensure that the runbook is able to connect to Azure SQL Database, execute the following PowerShell command to allow all connections from Azure services to connect to the Azure SQL database server named `azadesqlserver`:

```
New-AzSqlServerFirewallRule -FirewallRuleName
"AllowAllAzureIPs" -StartIpAddress "0.0.0.0"
EndIpAddress "0.0.0.0" -ServerName azadesqlserver
-ResourceGroupName packtadesql
```

19. The PowerShell command will return a result as shown here:

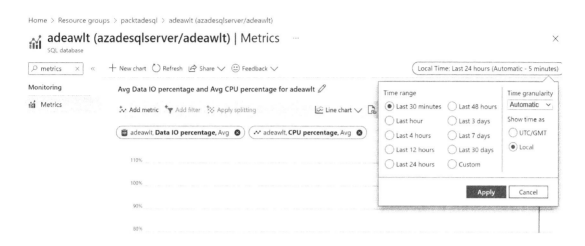

Figure 6.33 – Allow all Azure connections

20. To see the alert in action, connect to the Azure SQL database and execute the following query to simulate high CPU usage:

```
--query to simulate high CPU usage
While(1=1)
Begin
Select cast(a.object_id as nvarchar(max)) from sys.
objects a, sys.objects b,sys.objects c, sys.objects d
End
```

As soon as the alert condition is triggered, the webhook is called and the database service tier is modified to Standard S0. You can check that the alert was triggered by navigating to the **azadesqldb** database on the Azure portal and clicking on the **Alerts** option. **Fired time** confirms that a CPU spike was detected and the alert was fired:

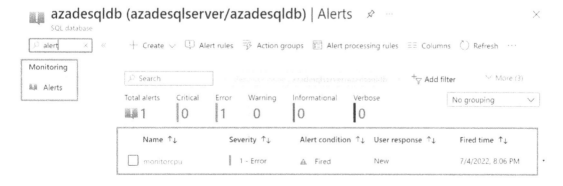

Figure 6.34 – Alert triggered

How it works...

To configure automatic scaling for an Azure SQL database, we create an Azure Automation runbook. The runbook specifies the PowerShell code to modify the service tier of an Azure SQL database.

We create a webhook to trigger the runbook. We create an Azure SQL database alert and define the alert condition as `cpu_percent` metric is greater than 40% for at least 1 minute. In the alert action, we call the webhook defined earlier.

When the alert condition is reached, the webhook is called, which in turn executes the runbook, resulting in the Azure SQL Database service tier change.

Monitoring an Azure SQL database using the Azure portal

Azure SQL Database has built-in monitoring features, such as Query Performance Insights, performance overview, and diagnostic logging. In this recipe, we'll learn how to use the monitoring capabilities using the Azure portal.

Getting ready

We'll use PowerShell to create an Azure SQL database, so open a PowerShell window and log in to your Azure account by executing the `Connect-AzAccount` command.

We'll use the Azure portal to monitor the Azure SQL database. Open `https://portal.azure.com` and log in to your Azure account.

How to do it...

First, let's execute a sample workload.

Creating an Azure SQL database and executing a sample workload

The steps are as follows:

1. Execute the following PowerShell command to create an Azure SQL database using the `AdventureWorksLT` sample database:

    ```
    # create the resource group
    New-AzResourceGroup -Name packtadesql -Location "central
    us" -force
    #create credential object for the Azure SQL Server admin
    credential
    $sqladminpassword = ConvertTo-SecureString 'Sql@
    Server@1234' -AsPlainText -Force
    ```

```
$sqladmincredential = New-Object System.Management.
Automation.PSCredential ('sqladmin', $sqladminpassword)
# create the Azure SQL Server
New-AzSqlServer -ServerName azadesqlserver
-SqlAdministratorCredentials $sqladmincredential
-Location "central us" -ResourceGroupName packtadesql
#Create the SQL Database
New-AzSqlDatabase -DatabaseName adeawlt -Edition basic
-ServerName azadesqlserver -ResourceGroupName packtadesql
-SampleName AdventureWorksLT
```

2. Execute the following command to add the client IP to the Azure SQL Server firewall:

```
$clientip = (Invoke-RestMethod -Uri https://ipinfo.io/
json).ip
New-AzSqlServerFirewallRule -FirewallRuleName "home"
-StartIpAddress $clientip -EndIpAddress $clientip
-ServerName azadesqlserver -ResourceGroupName packtadesql
```

3. Download the workload.sql file from https://github.com/PacktPublishing/
   Azure-Data-Engineering-Cookbook-2nd-edition/blob/main/Chapter06/
   workload.sql and execute the following command to run a workload against the Azure SQL
   Database. Ensure to edit the script to set your correct local path for input and output parameters:

```
sqlcmd -S azadesqlserver.database.windows.net -d adeawlt
-U sqladmin -P Sql@Server@1234 -i "C:\ADECookbook\
Chapter06\workload.sql" > "C:\ADECookbook\Chapter06\
workload_output.txt"
```

It can take 4 to 5 minutes for the workload to complete. You can execute the preceding command multiple times; however, you should run it at least once.

Monitoring Azure SQL Database metrics

The steps are as follows:

1. In the Azure portal, navigate to **All resources** | **azadesqlserver** and find the **adeawlt** database. Search for and open **Metrics**:

Figure 6.35 – Opening the Metrics section in the Azure portal

The **Metrics** page allows you to monitor different available metrics over time.

2.   To select metrics, click **Add metric | CPU percentage | Data IO percentage**:

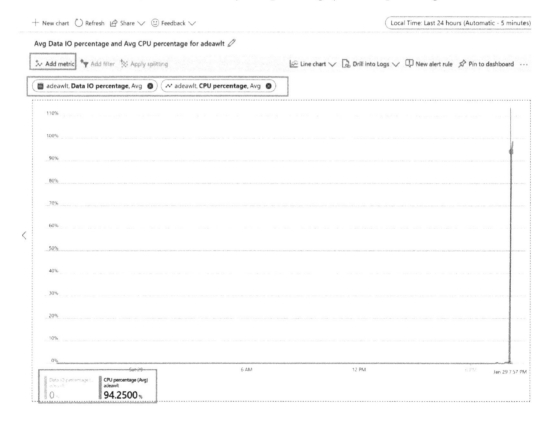

Figure 6.36 – Monitoring the metrics for a SQL database

We can select the metrics we are interested in monitoring and use the **Pin to dashboard** feature to pin the chart to the portal dashboard. We can also create an alert rule from the **Metrics** page by clicking on **New alert rule**. We can select a time range to drill down to specific times or investigate spikes in the chart.

3. To select a time range, select the **Time range** dropdown in the top-right corner of the **Metrics** page and select the desired time range:

Figure 6.37 – Selecting a time range to monitor

Using Query Performance Insight to find resource-consuming queries

Query Performance Insight is an intelligent performance feature that allows us to find any resource-consuming and long-running queries. The steps are as follows:

1. In the Azure portal, navigate to **All resources | azadesqlserver** and find the **adeawlt** database. Find and open **Query Performance Insight**:

Figure 6.38 – Selecting Query Performance Insight for the SQL database

2. On the **Query Performance Insight** page, observe that there are three tabs: **Resource consuming queries**, **Long running queries**, and **Custom**. We can select resource-consuming queries by **CPU**, **Data IO**, and **Log IO**. The table at the bottom of the page lists the query IDs by CPU consumption. Click on the color-coded box in the **QUERY ID** column to get the query text:

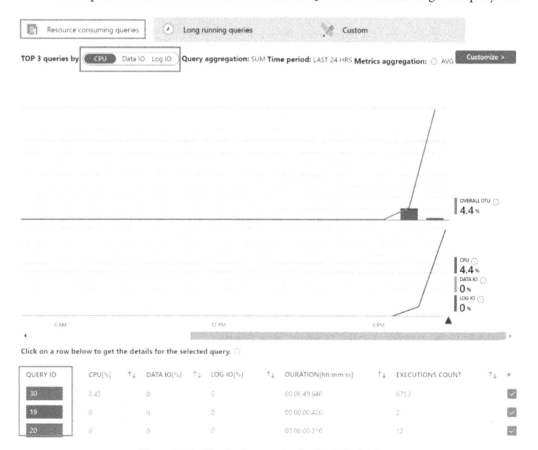

Figure 6.39 – Monitoring queries for the SQL database

3. The query details for query ID **30** are shown in the following screenshot:

Home > Resource groups > packtadesql > adeawlt (azadesqlserver/adeawlt) >

Query details ···
adeawlt - Query ID 30

⚙ Settings   ↻ Refresh   Recommendations   </> Query Text

Query ID 30:

```
1  Select count(*),sod.productid From  [SalesLT].[Product] CROSS JOIN SalesLT.
   SalesOrderDetail sod CROSS JOIN [SalesLT].[SalesOrderDetail] GROUP BY sod.productid
   Order by count(*) desc
```

Details of Query ID 30 (Query aggregation: sum) Last 24 hrs

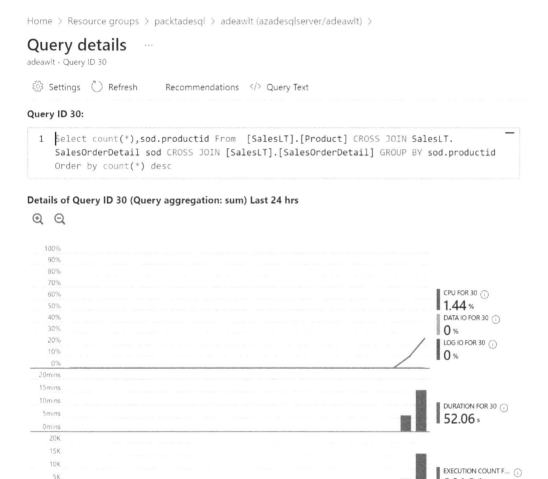

Figure 6.40 – Viewing the query details

We can look at the query text and optimize it for better performance.

4. The **Custom** tab allows us to select resource-consuming queries by duration and execution count. We can also specify a custom time range, the number of queries, and the query and metric aggregation:

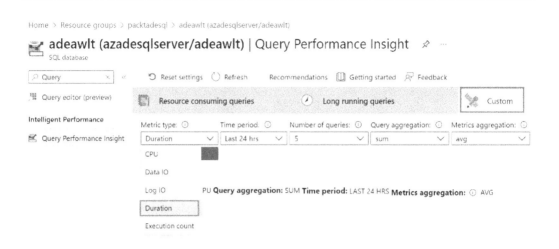

Figure 6.41 – Providing custom monitoring configuration

5. Select the options and click the **Go** button to refresh the chart as per the custom settings. We will get the top three queries by duration:

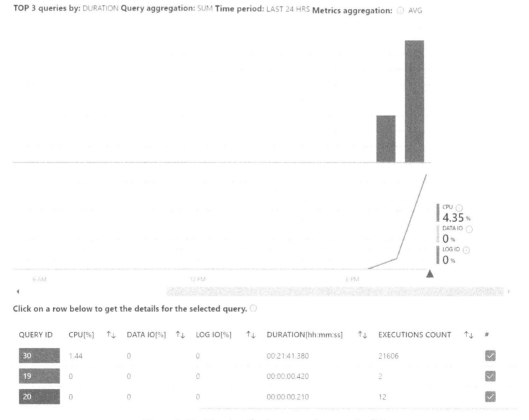

Figure 6.42 – Viewing the long-running queries list

We can further look into the query text and other details by selecting the query ID.

Monitoring an Azure SQL database using diagnostic settings

In addition to metrics and Query Performance Insight, we can also monitor an Azure SQL database by collecting diagnostic logs. The diagnostic logs can be sent to a Log Analytics workspace or Azure Storage, or can be streamed to Azure Event Hubs.

Log Analytics is an Azure service that can store diagnostic data from various Azure services and allows us to analyze the data in one place. Let's create a Log Analytics workspace to store the diagnostics details of the SQL database. The steps to create a Log Analytics workspace and configure the diagnostics for a SQL database are as follows:

1. Execute the following command to create a Log Analytics workspace named `packtadesqllgw`:

    ```
    $ResourceGroup = "packtadesql"
    $WorkspaceName = "packtadesqllgw"
    $Location = "central us"
    # Create the workspace
    New-AzOperationalInsightsWorkspace -Location $Location
    -Name $WorkspaceName -ResourceGroupName $ResourceGroup
    ```

2. To enable diagnostic logging using the Azure portal, navigate to **All resources | azadesqlserver | adeawlt**. Find and open **Diagnostic settings**:

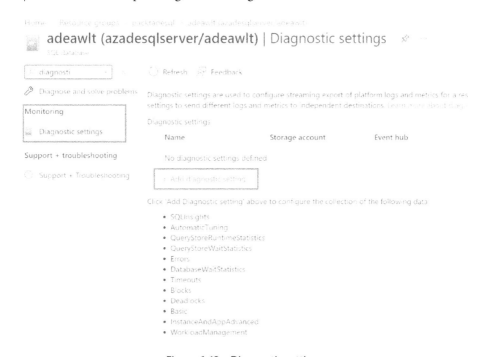

Figure 6.43 – Diagnostic settings

3. Click + **Add diagnostic setting** to add a new diagnostic setting.

4. Select the categories to be included in the logs and their destination. Provide any name for the diagnostic setting. Select **Send to Log Analytics workspace** under **Destination details**. Select **packtadesqllgw** as the Log Analytics workspace to store the diagnostic log. Click **Save** to create the new diagnostic setting.

Home > Resource groups > packtadesql > adeawlt (azadesqlserver/adeawlt) >

Diagnostic setting   ···

[💾 Save]   ✕ Discard   🗑 Delete   ℛ Feedback

A diagnostic setting specifies a list of categories of platform logs and/or metrics that you want to collect from a resource, and one or more destinations that you would stream them to. Normal usage charges for the destination will occur. Learn more about the different log categories and contents of those logs

Diagnostic setting name *   [sqldiagnostics]   ✓

Logs   **Destination details**

  Category groups ⓘ   ✓ Send to Log Analytics workspace

    ☐ allLogs   ☐ audit

  Categories   Subscription
                                               [Visual Studio Ultimate with MSDN   ∨]
    ✓ SQLInsights
                                               Log Analytics workspace
    ✓ AutomaticTuning   [packtadesqllgw (central us)   ∨]

    ✓ QueryStoreRuntimeStatistics   ☐ Archive to a storage account

    ✓ QueryStoreWaitStatistics   ☐ Stream to an event hub

    ☐ Errors   ☐ Send to partner solution

    ✓ DatabaseWaitStatistics

    ☐ Timeouts

    ✓ Blocks

    ✓ Deadlocks

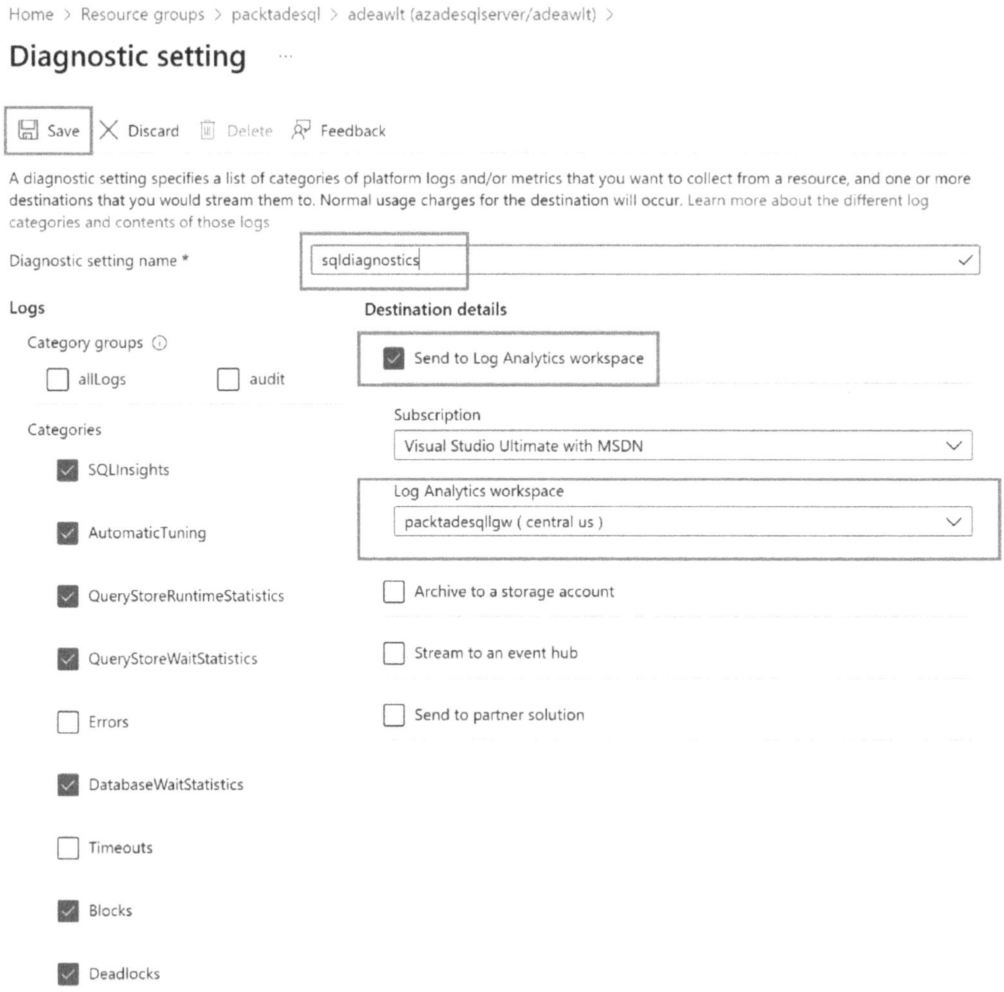

Figure 6.44 – Selecting categories

> **Important Note**
>
> Diagnostic settings add an additional cost to the Azure SQL database. It may also take some time for the logs to be available after creating a diagnostic setting.

Automatic tuning in an Azure SQL database

Automatic tuning provides three features: **FORCE PLAN**, **CREATE INDEX**, and **DROP INDEX**. Automatic tuning can be enabled for an Azure SQL Server, in which case it's applied to all of the databases in that Azure SQL Server. Automatic tuning can be enabled for individual Azure SQL databases as well. The steps are as follows:

1. To enable automatic tuning, in the Azure portal, navigate to **All resources | azadesqlserver | adeawlt**. Find and select **Automatic tuning** under **Intelligent Performance**:

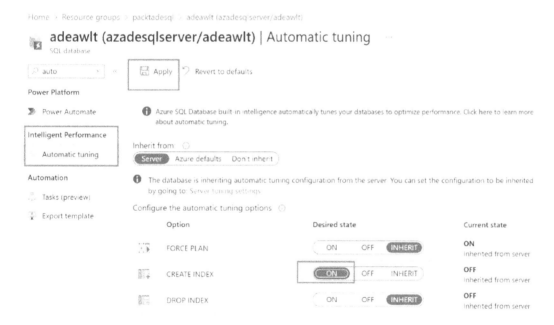

Figure 6.45 – Automatic tuning in the SQL database

2. Enable the **CREATE INDEX** tuning option by clicking **ON** under the **Desired state** option.

3. Click **Apply** to save the configuration.

> **Important Note**
>
> It may take some time for the recommendation to show up.
>
> The recommendations will show up in the performance recommendations in the **Intelligent Performance** section.

Configuring auditing for Azure SQL Database

SQL Auditing is a process in which key activities or changes in the database are recorded not only for monitoring but also for compliance to global security standards. In the following recipe, we will configure an audit for Azure SQL Database to record the database activities, store the details in Azure Log Analytics, and verify the recorded data.

Getting ready

Create an Azure SQL database and a Log Analytics workspace, as mentioned in the *Monitoring an Azure SQL database using the Azure portal* recipe of this chapter.

How to do it...

1. In the Azure portal, navigate to **All resources | azadesqlserver | adeawlt**. Find **Auditing**:

Figure 6.46 – Azure SQL Database Auditing

2. Turn on **Azure SQL Auditing**. Enable **Log Analytics**. Pick the **packtadesqllgw** Log Analytics workspace, which was created in the *Monitoring an Azure SQL Database using the Azure portal* recipe of this chapter. Hit the **Save** button:

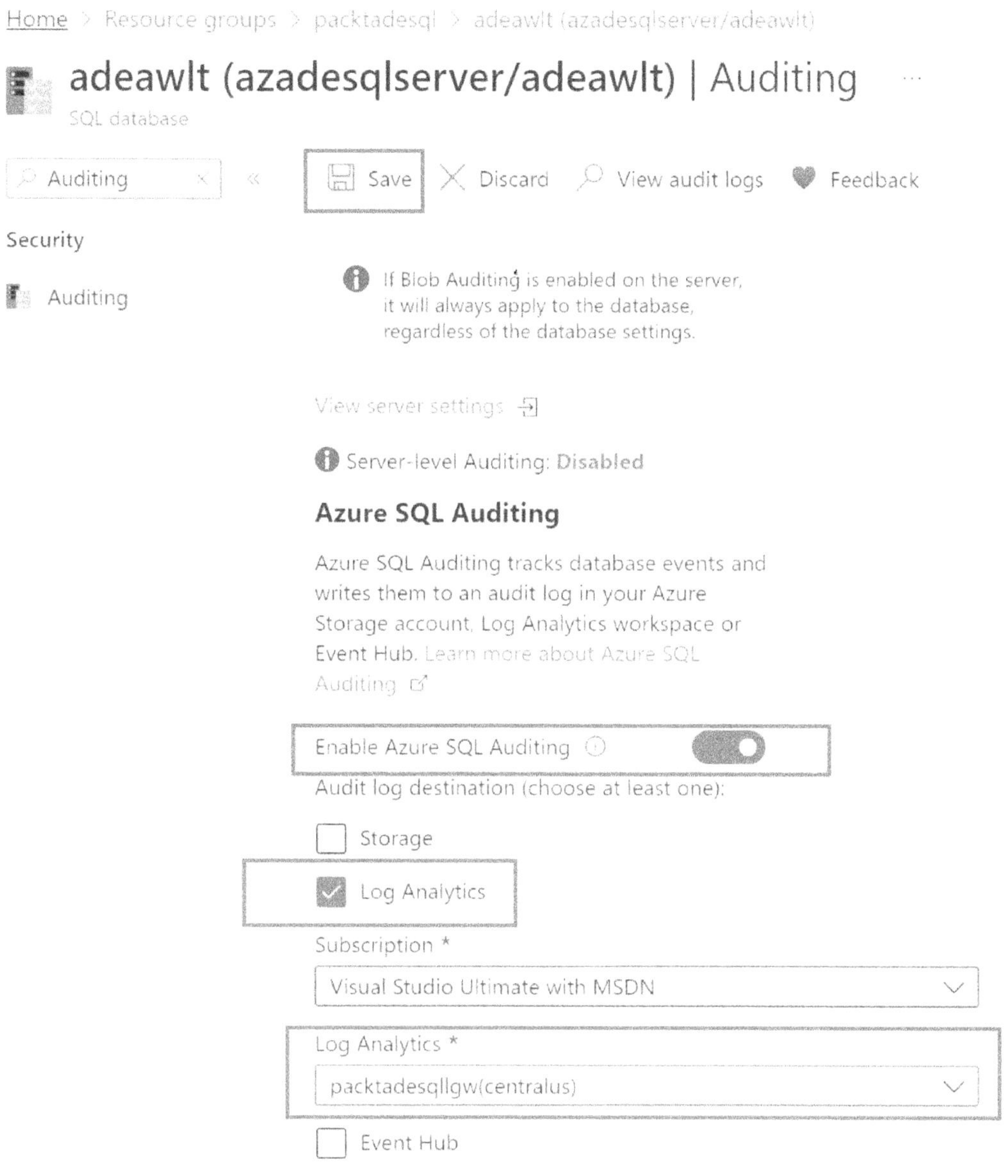

Home > Resource groups > packtadesql > adeawlt (azadesqlserver/adeawlt)

adeawlt (azadesqlserver/adeawlt) | Auditing   ···
SQL database

🔍 Auditing   ×   « | 💾 Save   ✕ Discard   🔍 View audit logs   ♥ Feedback

Security

🔳 Auditing

ℹ️ If Blob Auditing is enabled on the server, it will always apply to the database, regardless of the database settings.

View server settings ⤴

ℹ️ Server-level Auditing: Disabled

Azure SQL Auditing

Azure SQL Auditing tracks database events and writes them to an audit log in your Azure Storage account, Log Analytics workspace or Event Hub. Learn more about Azure SQL Auditing ☑

| Enable Azure SQL Auditing ⓘ   ⬤▬ |

Audit log destination (choose at least one):

☐ Storage

☑ Log Analytics

Subscription *

| Visual Studio Ultimate with MSDN   ⌄ |

Log Analytics *

| packtadesqllgw(centralus)   ⌄ |

☐ Event Hub

Figure 6.47 – Configure Auditing

3. Connect to the database using SSMS and execute the following queries:

    ```
    select ProductID from SalesLT.SalesOrderDetail where
    OrderQty > 2
    GO
    Delete from SalesLT.SalesOrderDetail where ProductID
    between 900 and 1000
    GO
    ```

4. Go to portal.azure.com and then to **All Resources**. Find **packtadesqllgw**. Click on it to open the Log Analytics workspace:

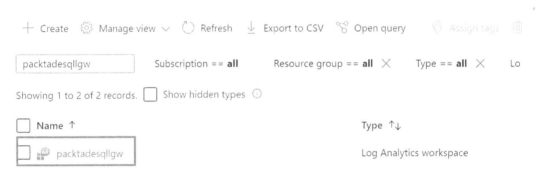

All resources

Figure 6.48 – Open the Log Analytics workspace

5. Click **Logs** in the **General** section. Paste the following **Kusto Query Language** (**KQL**) script in the query window and hit the **RUN** button:

    ```
    AzureDiagnostics
    | where Category == 'SQLSecurityAuditEvents' and action_
    name_s == "BATCH COMPLETED"
    | project event_time_t,ResourceGroup,server_instance_
    name_s,database_name_s,statement_s,action_name_s,server_
    principal_name_s,application_name_s,duration_
    milliseconds_d,response_rows_d,affected_rows_d
    |sort by event_time_t desc
    ```

The following screenshot shows the script execution:

Figure 6.49 – Running the KQL script in the Log Analytics workspace

6. Scroll to the right on the **Results** screen to verify whether the fired queries are being recorded by the audit in the Log Analytics workspace. The fired DELETE and SELECT script should be listed in the **statement_s** column:

```
1  AzureDiagnostics
2  | where Category == 'SQLSecurityAuditEvents' and action_name_s == "BATCH COMPLETED"
3  | project event_time_t,ResourceGroup,server_instance_name_s,database_name_s,statement_s,action_name_s,
   server_principal_name_s,application_name_s,duration_milliseconds_d,response_rows_d,affected_rows_d
4  |sort by event_time_t desc
```

Figure 6.50 – KQL results of the audit data

How it works...

Please refer to the following to understand how auditing works in Azure SQL Database:

1. Auditing is not enabled by default. We enabled a database-level audit for the **adeawlt** database in Azure Portal.

2. Azure SQL Database auditing records login success, login failures, and batch completion events when configured using Azure portal. If you need additional events to be added to the audit policy, you may do so using the PowerShell command, **Set-AzSqlDatabaseAudit**.

3. We configured the audit to store the data to our **packtadesqllgw** Log Analytic workspace.

4. We can use a single Log Analytic workspace (**packtadesqllgw**) to store both the audit data and SQL diagnostics data. How long the Log Analytic workspace will retain the audit logs depends on the workspace's data retention settings.

5. KQL scripts are used to query the Log Analytic workspace. **AzureDiagnostics** is the single table that maintains both SQL audit data and database diagnostics. By filtering for SQL audit events using the `Category == 'SQLSecurityAuditEvents'` condition, we were able to find audit records. By filtering using the `action_name_s == "BATCH COMPLETED"` condition, we were able to find queries or batch completion events and verify whether our actions on the database were recorded or not.

7

Processing Data Using Azure Databricks

Databricks is a data engineering product built on top of Apache Spark that provides a unified, cloud-optimized platform so that you can perform **Extract, Transform, and Load** (ETL), **Machine Learning** (**ML**), and **Artificial Intelligence** (**AI**) tasks on a large quantity of data.

Azure Databricks, as its name suggests, is the Databricks integration with Azure, which also provides fully managed Spark clusters, an interactive workspace for data visualization and exploration, integration with data sources such as Azure Blob Storage, Azure Data Lake Storage, Azure Cosmos DB, and Azure SQL Data Warehouse.

Azure Databricks can process data from multiple and diverse data sources, such as SQL or NoSQL, structured or unstructured data, and streaming data sources, and also scale up as many servers as required to cater to any data growth.

By the end of the chapter, you will have learned how to configure Databricks, work with storage accounts, process data using Scala, store processed data in Delta Lake, and visualize the data in Power BI.

In this chapter, we'll cover the following recipes:

- Configuring the Azure Databricks environment
- Integrate Databricks with Azure Key Vault
- Mounting an Azure Data Lake container in Databricks
- Processing data using notebooks
- Scheduling notebooks using job clusters
- Working with Delta Lake tables
- Connecting a Databricks Delta Lake to Power BI

Technical requirements

For this chapter, you will need the following:

- A Microsoft Azure subscription

- PowerShell 7

- Microsoft Azure PowerShell

- Power BI Desktop

Configuring the Azure Databricks environment

In this recipe, we'll learn how to configure the Azure Databricks environment by creating an Azure Databricks workspace, cluster, and cluster pools.

Getting ready

To get started, log in to `https://portal.azure.com` using your Azure credentials.

How to do it...

An Azure Databricks workspace is the starting point for writing solutions in Azure Databricks. A workspace is where you create clusters, write notebooks, schedule jobs, and manage the Azure Databricks environment.

An Azure Databricks workspace can be created in an Azure-managed virtual network or customer-managed virtual network. In this recipe, we will create a Databricks cluster in an Azure-managed network. Let's get started:

1. Go to `portal.azure.com` and click **Create a resource**. Search for `Azure Databricks`. Click **Create**, as shown in the following screenshot:

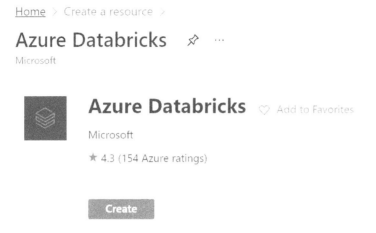

Figure 7.1 – Creating a Databricks resource

2. Provide the resource group name and workspace name, as shown in the following screenshot, and click **Review + Create**:

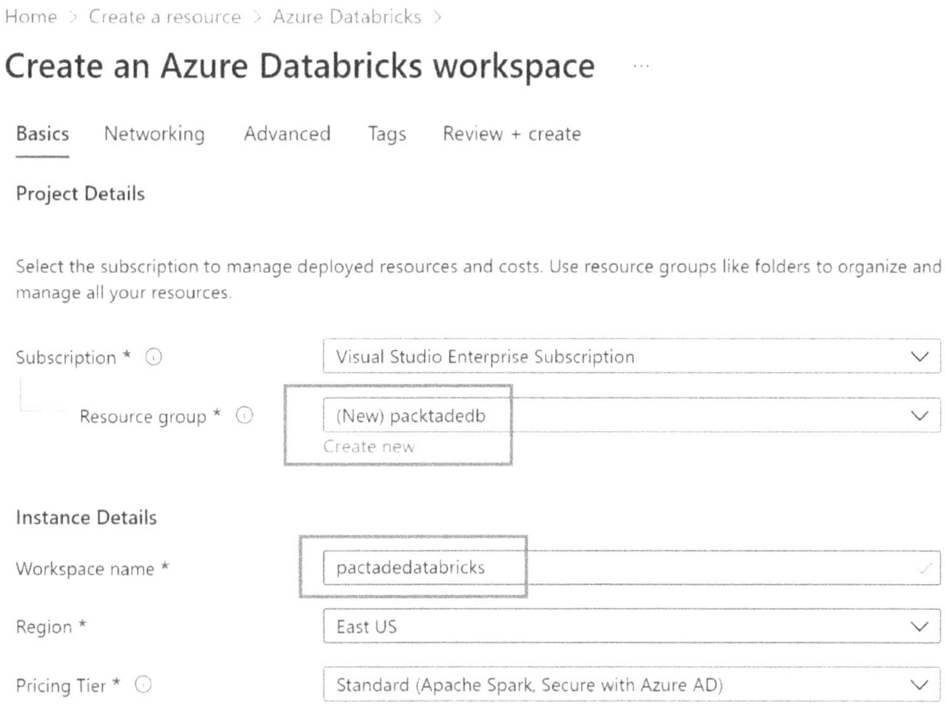

Home > Create a resource > Azure Databricks >

Create an Azure Databricks workspace   ...

Basics   Networking   Advanced   Tags   Review + create

Project Details

Select the subscription to manage deployed resources and costs. Use resource groups like folders to organize and manage all your resources.

Subscription * ⓘ	Visual Studio Enterprise Subscription ∨
Resource group * ⓘ	(New) packtadedb ∨
	Create new

Instance Details

Workspace name *	pactadedatabricks
Region *	East US ∨
Pricing Tier * ⓘ	Standard (Apache Spark, Secure with Azure AD) ∨

Figure 7.2 – Creating a Databricks workspace

Creating Azure Databricks clusters

Once the resource is created, go to the Databricks workspace that we created (go to `portal.azure.com`, click on **All resources** and search for `pactadedatabricks`). Perform the following steps to create a Databricks cluster:

    1.   Click **Launch Workspace**:

         🗑 Delete

         ⌃ Essentials

Status	: Active	Managed Resource Group	: databricks-rg-pactadedatabricks-6qmidqjocusbo
Resource group	: packtadedb	URL	: https://adb-7675839323985314.14.azuredatabricks.net
Location	: East US 🗋	Pricing Tier	: standard
Subscription	: Visual Studio Enterprise Subscription		
Subscription ID	:		
Tags (edit)	: Click here to add tags		

Figure 7.3 – Launching the workspace

    2.   To create a cluster, select **Compute** from the left-hand menu of the Databricks workspace:

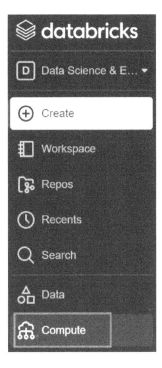

Figure 7.4 – Creating a cluster

There are two types of cluster: **Interactive** and **Automated**. Interactive clusters are created manually by users so that they can interactively analyze the data while working on, or even developing, a data engineering solution. Automated clusters are created automatically when a job starts and are terminated as and when the job completes.

3. Click **Create Cluster** to create a new cluster. On the **New Cluster** page, provide a cluster name of dbcluster01. Then, set **Cluster mode** to **Standard**, **Terminate after** to 10 **minutes of inactivity**, **Min workers** to 1, and **Max workers** to 2, and leave the rest of the options as their defaults:

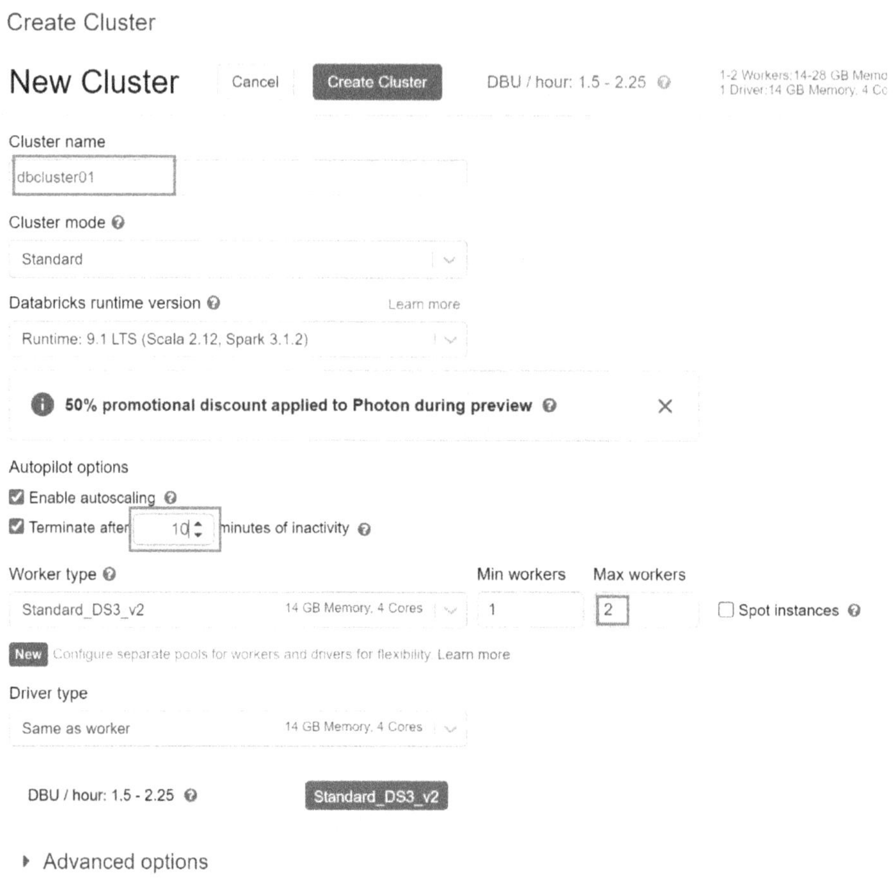

Create Cluster

New Cluster   Cancel   Create Cluster   DBU / hour: 1.5 - 2.25 ⓘ   1-2 Workers:14-28 GB Memo
1 Driver:14 GB Memory, 4 Co

Cluster name

dbcluster01

Cluster mode ⓘ

Standard

Databricks runtime version ⓘ   Learn more

Runtime: 9.1 LTS (Scala 2.12, Spark 3.1.2)

ⓘ 50% promotional discount applied to Photon during preview ⓘ   ✕

Autopilot options
☑ Enable autoscaling ⓘ
☑ Terminate after   10 ⇕   minutes of inactivity ⓘ

Worker type ⓘ   Min workers   Max workers

Standard_DS3_v2   14 GB Memory, 4 Cores   1   2   ☐ Spot instances ⓘ

New Configure separate pools for workers and drivers for flexibility. Learn more

Driver type

Same as worker   14 GB Memory, 4 Cores

DBU / hour: 1.5 - 2.25 ⓘ   Standard_DS3_v2

▸ Advanced options

Figure 7.5 – Creating a cluster configuration

There are two major cluster modes: **Standard** and **High Concurrency**. **Standard** cluster mode uses single-user clusters, optimized to run tasks one at a time. The **High Concurrency** cluster mode is optimized to run multiple tasks in parallel; however, it only supports R, Python, and SQL workloads, and doesn't support Scala.

These autoscaling options allow Databricks to provision as many nodes as required to process a task within the limit, as specified by the **Min workers** and **Max workers** options.

The **Terminate after** option terminates the clusters when there's no activity for a given amount of time. In our case, the cluster will auto-terminate after 10 minutes of inactivity. This option helps save costs.

There are two types of cluster nodes: **Worker type** and **Driver type**. The **Driver type** node is responsible for maintaining a notebook's state information, interpreting the commands being run from a notebook or a library, and co-ordinates with Spark executors. The **Worker type** nodes are the Spark executor nodes, which are responsible for distributed data processing.

The **Advanced options** section can be used to configure Spark configuration parameters, environment variables, and tags, configure **Secure Shell (SSH)** in the clusters, enable logging, and run custom initialization scripts at the time of cluster creation.

4. Click **Create Cluster** to create the cluster. It will take around 5 to 1 0 minutes to create the cluster, and may take more time depending on the number of worker nodes that you have selected.

Creating Azure Databricks pools

Azure Databricks pools reduce cluster startup and autoscaling time by keeping a set of idle, ready-to-use instances without the need for creating instances when required. To create Azure Databricks pools, execute the following steps:

1. In your Azure Databricks workspace, on the **Compute** page, select the **Pools** option, and then select **Create Pool** to create a new pool. Provide the pool's name, then set **Min Idle** to 2, **Max Capacity** to 4, and **Idle Instance Auto Termination** to 10. Leave **Instance Type** as its default of **Standard_DS3_v2** and set **Preloaded Databricks Runtime Version** to **Runtime: 9.1 LTS (Scala 2.12, Spark 3.1.2)**:

Figure 7.6 – Creating a Databricks cluster pool

Min Idle specifies the number of instances that will be kept idle and available without terminating. The **Idle Instance Auto Terminate** setting doesn't apply to these instances. **Max Capacity** limits the maximum number of instances to this number, including idle and running ones. This helps with managing cloud quotas and their costs.

The Azure Databricks runtime is a set of core components or software that runs on your clusters. There are different runtimes, depending on the type of workload you have.

2. Click **Create** to create the pool. We can attach a new or existing cluster to a pool by specifying the pool name under the **Worker type** option and **Driver type** option. In the workspace, navigate to the **Clusters** page and select **dbcluster01**, which we created in *step 2* of the previous section. On the **dbcluster01** page, click **Edit**, and select **dbclusterpool** from the **Worker type** drop-down list and **Driver type** drop-down list:

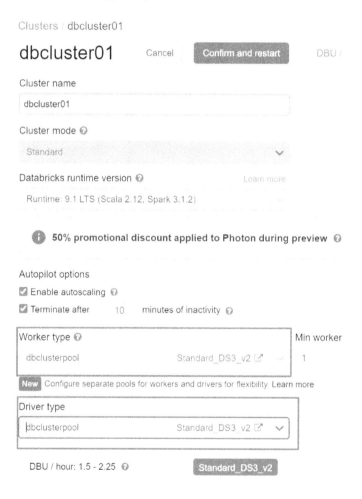

Figure 7.7 – Attaching a Databricks cluster to a pool

3. Click **Confirm** to apply these changes. The cluster will now show up in the **Attached Clusters** list on the pool's page:

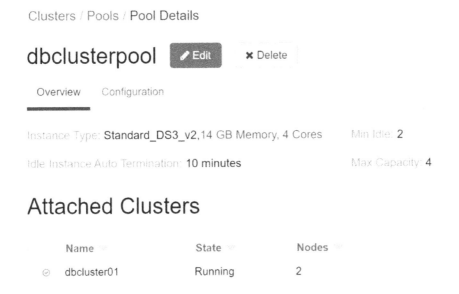

Clusters / Pools / Pool Details

dbclusterpool   ✎ Edit   ✕ Delete

Overview   Configuration

Instance Type: **Standard_DS3_v2**, 14 GB Memory, 4 Cores   Min Idle: **2**

Idle Instance Auto Termination: **10 minutes**   Max Capacity: **4**

Attached Clusters

Name	State	Nodes
⊘ dbcluster01	Running	2

Figure 7.8 – Databricks clusters attached to a pool

We can add multiple clusters to a pool. Whenever an instance, such as **dbcluster01**, requires an instance, it'll attempt to allocate the pool's idle instance. If an idle instance isn't available, the pool expands to get new instances, as long as the number of instances is under the maximum capacity.

Integrating Databricks with Azure Key Vault

Azure Key Vault is a useful service for storing keys and secrets that are used by various other services and applications. It is important to integrate Azure Key Vault with Databricks, as you could store the credentials of objects such as a SQL database or data lake inside the key vault. Once integrated, Databricks can reference the key vault, obtain the credentials, and access the database or data lake account. In this recipe, we will cover how you can integrate Databricks with Azure Key Vault.

Getting ready

Create a Databricks workspace and a cluster as explained in the *Configuring the Azure Databricks environment* recipe of this chapter.

Log in to `portal.azure.com`, click **Create a resource**, search for **Key Vault**, and click **Create**. Provide the key vault details, as shown in the following screenshot, and click **Review + create**:

Home > packtadedb > Create a resource > Key Vault >

Create a key vault ...

Basics   Access policy   Networking   Tags   Review + create

Azure Key Vault is a cloud service used to manage keys, secrets, and certificates. Key Vault eliminates the need for developers to store security information in their code. It allows you to centralize the storage of your application secrets which greatly reduces the chances that secrets may be leaked. Key Vault also allows you to securely store secrets and keys backed by Hardware Security Modules or HSMs. The HSMs used are Federal Information Processing Standards (FIPS) 140-2 Level 2 validated. In addition, key vault provides logs of all access and usage attempts of your secrets so you have a complete audit trail for compliance.

Project details

Select the subscription to manage deployed resources and costs. Use resource groups like folders to organize and manage all your resources.

Subscription *	Visual Studio Enterprise Subscription ⌄
└─ Resource group *	packtadedb ⌄
	Create new

Instance details

Key vault name * ⓘ	packtadedbkv ✓
Region *	East US ⌄
Pricing tier * ⓘ	Standard ⌄

Recovery options

Soft delete protection will automatically be enabled on this key vault. This feature allows you to recover or permanently delete a key vault and secrets for the duration of the retention period. This protection applies to the key vault and the secrets stored within the key vault.

To enforce a mandatory retention period and prevent the permanent deletion of key vaults or secrets prior to the retention period elapsing, you can turn on purge protection. When purge protection is enabled, secrets cannot be purged by users or by Microsoft.

Review + create   < Previous   Next : Access policy >

Figure 7.9 – Creating a key vault

How to do it...

Perform the following steps to integrate Databricks with Azure Key Vault:

1. Open the **pactadedatabricks** Databricks workspace on the Azure portal and click **Launch Workspace**.

2. This will open up the Databricks portal. Copy the URL, which will be set out as `https://adb-xxxxxxxxxxxxxxxx.xx.azuredatabricks.net/?o=` `xxxxxxxxxxxxxxx#`:

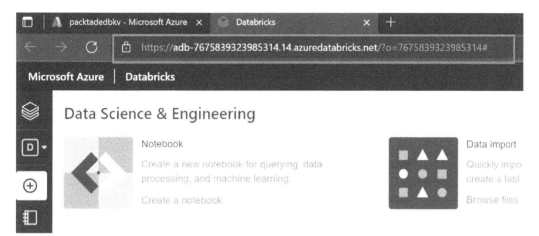

Figure 7.10 – The Databricks URL

3. Add `secrets/createScope` to the end of the URL and go to the URL `https://adb-xxxxxxxxxxxxxxxx.xx.azuredatabricks.net/?o=` `xxxxxxxxxxxxxxx-#secrets/createScope`. It should open a page to create a secret scope. Provide the following details:

    A. **Scope name**: Any name that relates to the key or password that you will access through the key vault. In our case, let's use `datalakekey`.

    B. **DNS Name**: DNS would be the DNS name of Key Vault, which will have the following format: `<key vault name>.vault.azure.net`. In our case, it is, `packtadedbkv.vault.azure.net`.

    C. Set **Manage Principal** to `All Users`.

    D. **Resource ID**: Resource ID of the key vault. To obtain it, go to **Key Vault** in the Azure portal (under **All resources**, search for `packtadedbkv`). Go to **Properties** and copy the **Resource ID**:

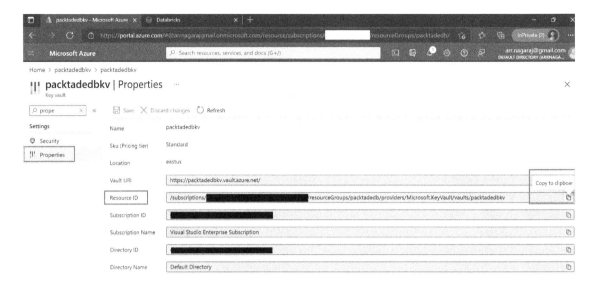

Figure 7.11 – The key vault resource ID

4. Return to the Databricks portal secret scope page and fill in the details, as shown in the following screenshot:

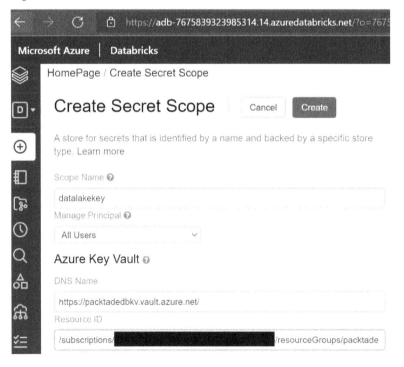

Figure 7.12 – Key vault scope creation

You will receive confirmation that a secret scope called **datalakekey** has been successfully added. This completes the integration between Databricks and Azure Key Vault.

How it works...

Upon completion of the preceding steps, all users with access to the Databricks workspace will be able to extract secrets and keys from the key vault and use them in a Databricks notebook to perform the desired operations.

Behind the scenes, Databricks uses a service principal to access the key vault. As we create the scope in the Databricks portal, Azure will grant the relevant permissions required for the Databricks service principal on the key vault. You can verify as much using the following steps:

1. Go to the **packtadedbkv** key vault on the Azure portal.
2. Click on **Access policies**. You will notice the Azure Databricks service principal being granted **Get** and **List** permissions on secrets. This will allow the Databricks workspace to read secrets from the key vault:

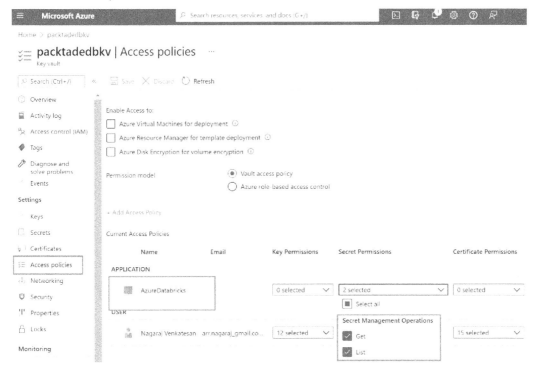

Figure 7.13 – Verifying permissions

Mounting an Azure Data Lake container in Databricks

Accessing data from Azure Data Lake is one of the fundamental steps of performing data processing in Databricks. In this recipe, we will learn how to mount an Azure Data Lake container in Databricks using the Databricks service principal. We will use Azure Key Vault to store the Databricks service principal ID and the Databricks service principal secret that will be used to mount a data lake container in Databricks.

Getting ready

Create a Databricks workspace and a cluster, as explained in the *Configuring the Azure Databricks environment* recipe of this chapter.

Create a key vault in Azure and integrate it with Azure Databricks, as explained in the *Integrating Databricks with Azure Key Vault* recipe.

Create an Azure Data Lake account, as explained in the *Provisioning an Azure Storage account using the Azure portal* recipe of *Chapter 1, Creating and Managing Data in Azure Data Lake*.

Go to the Azure Data Lake Storage account created in the Azure portal (click **All resources**, then search for `packtadestoragev2`). Click **Containers**. Click **+ Container**:

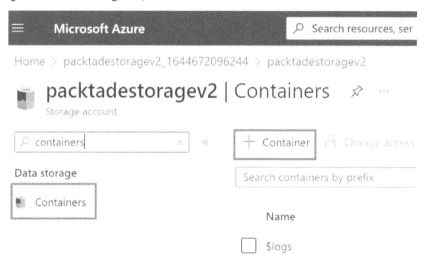

Figure 7.14 – Adding a container

Provide a container name of `databricks` and click **Create**:

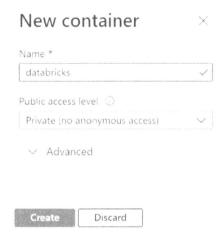

Figure 7.15 – Creating a container

How to do it...

Mounting the container in Databricks will involve the following high-level steps:

1. Registering an application in Azure **Active Directory** (**AD**) and obtaining the service principal secret from Azure AD

2. Storing the Application ID and service principal secret in a key vault

3. Granting permission on the data lake container to the Application ID

4. Mounting the data lake container on Azure Databricks

The detailed steps are as follows:

1. Go to `portal.azure.com` and click **Azure Active Directory**. Click **App registrations** and then hit + **New registration**:

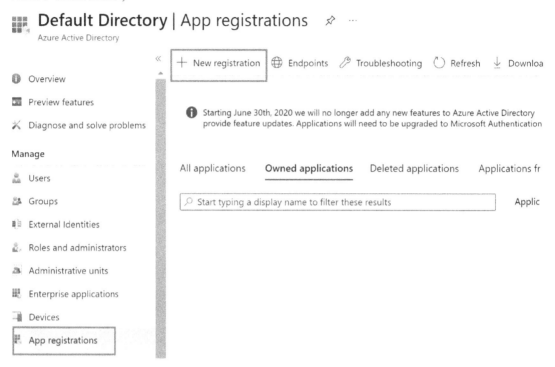

Figure 7.16 – App registration

2.   Provide a name of PacktDatabricks and click on the **Register** button:

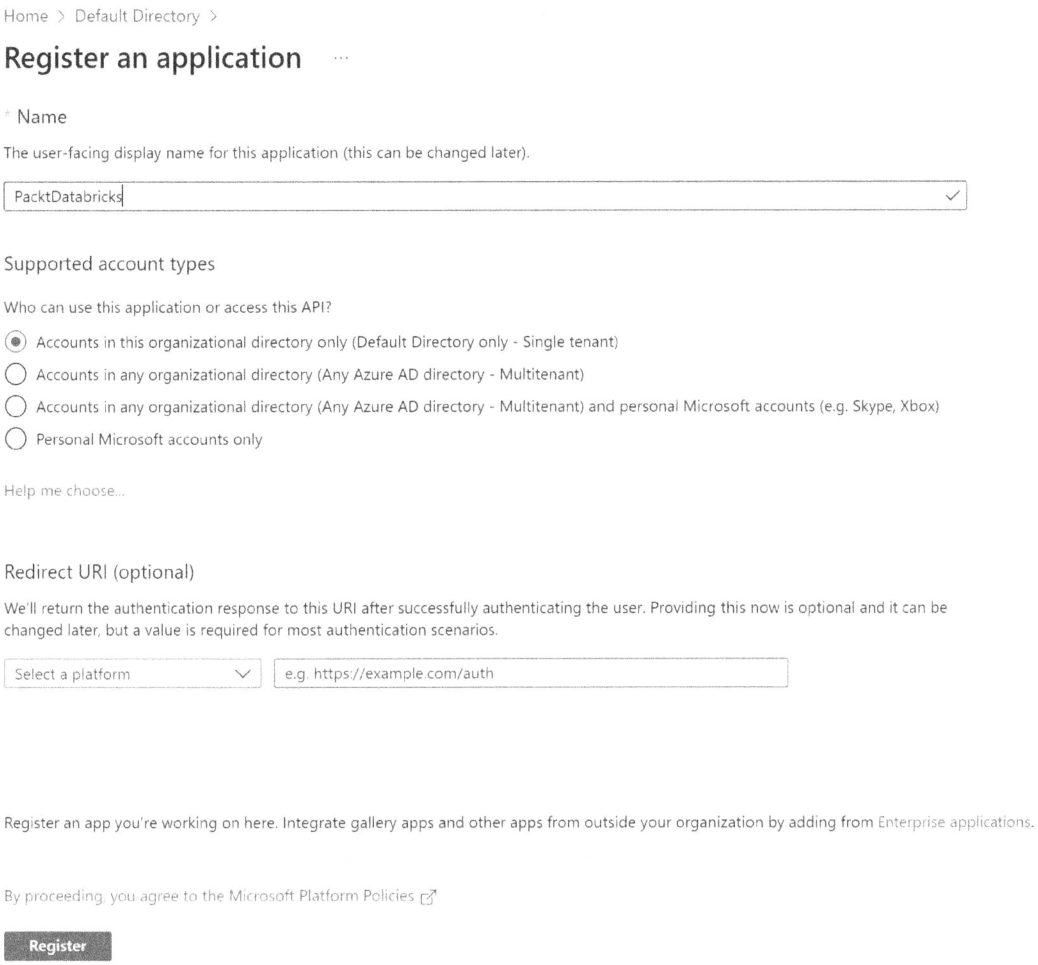

Home > Default Directory >

Register an application   ⋯

* Name

The user-facing display name for this application (this can be changed later).

| PacktDatabricks | ✓ |

Supported account types

Who can use this application or access this API?

◉ Accounts in this organizational directory only (Default Directory only - Single tenant)

○ Accounts in any organizational directory (Any Azure AD directory - Multitenant)

○ Accounts in any organizational directory (Any Azure AD directory - Multitenant) and personal Microsoft accounts (e.g. Skype, Xbox)

○ Personal Microsoft accounts only

Help me choose...

Redirect URI (optional)

We'll return the authentication response to this URI after successfully authenticating the user. Providing this now is optional and it can be changed later, but a value is required for most authentication scenarios.

| Select a platform ∨ | e.g. https://example.com/auth |

Register an app you're working on here. Integrate gallery apps and other apps from outside your organization by adding from Enterprise applications.

By proceeding, you agree to the Microsoft Platform Policies ☑

Register

Figure 7.17 – Registering an application

3.   After the registration is done, copy the **Application (client) ID** and the **Directory (tenant) ID**:

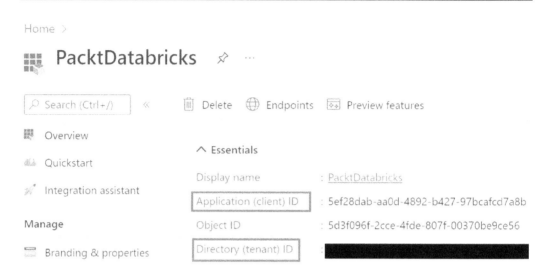

Figure 7.18 – Copying the Application ID

4. Click **Certificates & secrets**. Click + **New client secret**. For **Description**, provide any relevant description and click **Add**:

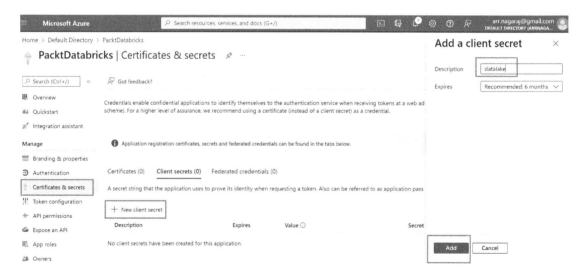

Figure 7.19 – Adding a client secret

5. Copy the generated secret value. Ensure you copy it before closing the screen, as you only get to see it once:

Figure 7.20 – The secret value

6. Go to the **packtadedbkv** Azure Key Vault. Click **Secrets**. Click + **Generate/Import**:

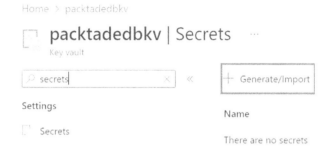

Figure 7.21 – Creating secrets

7. Set **Name** as appsecret and **Value** as the secret value copied in *step 5*. Hit the **Create** button:

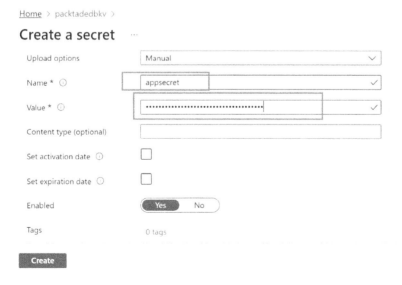

Figure 7.22 – Storing secrets

8. Repeat *step 6* and *step 7*, and add the application ID and directory ID values saved in *step 3* as secrets inside the key vault. Provide the name for the secrets as `ApplicationID` and `DirectoryID`. Once done, the key vault should have three secrets, as shown in the following screenshot:

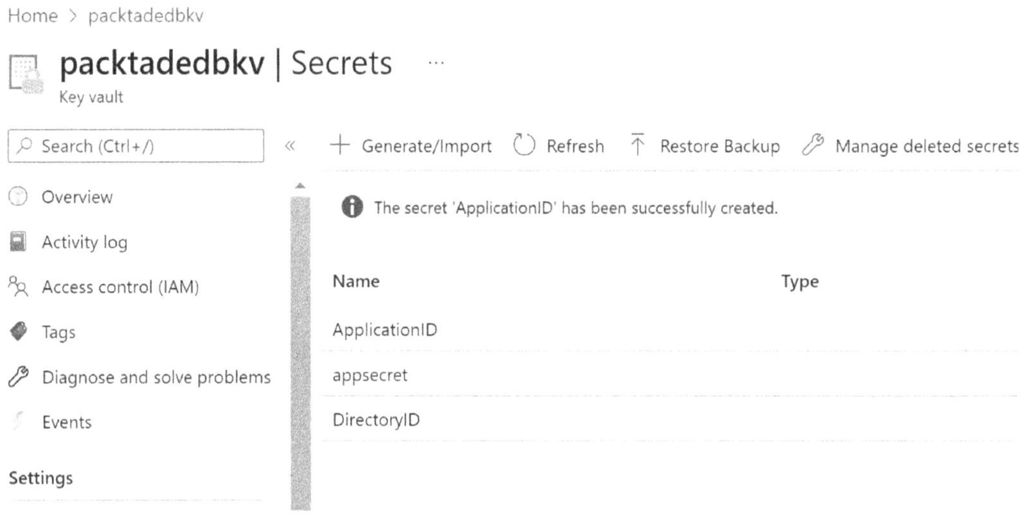

Figure 7.23 – secrets added

9. Go to the **packtadestoragev2** Data Lake account. Click **Containers** and open the **databricks** container. Click **Access Control (IAM)** and then click **Add role assignment**:

Home > packtadestoragev2 > databricks

Figure 7.24 – Adding role assignment

10. Search for the `Storage Blob Data Contributor` role, select it, and click **Next**:

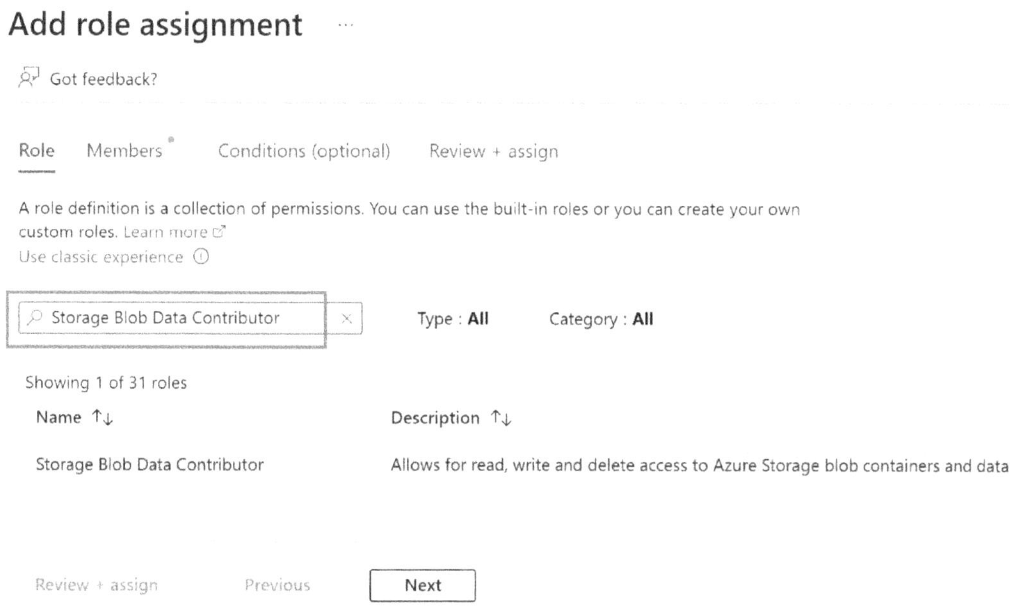

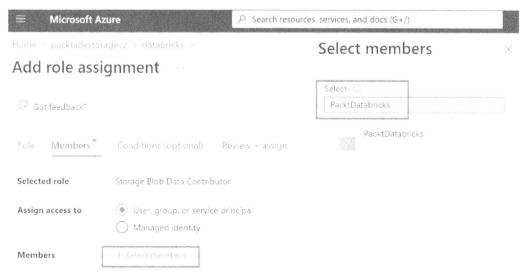

Figure 7.25 – Storage Blob Data Contributor

11. Click + **Select members** and, in the **Select members** box, type the app name registered in *step 2*, which is `PacktDatabricks`:

Figure 7.26 – Adding members to the Storage Blob Data Contributor role

12. Select **PacktDatabricks**, hit the **Select** button, and then click **Review + assign** to assign permission:

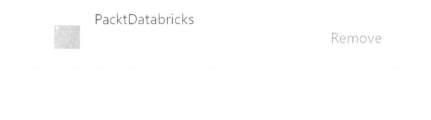

Figure 7.27 – Selecting members for the Storage Blob Data Contributor role

13. Launch the Databricks workspace if you need to and go back to the Databricks portal. From the **Create** menu, select **Notebook**:

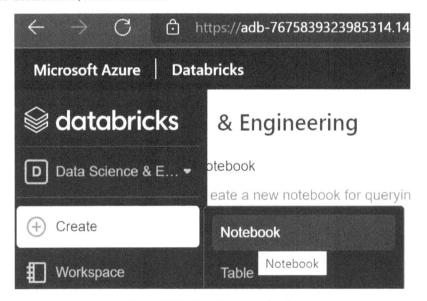

Figure 7.28 – Creating a notebook

14. Provide any notebook name. Set **Default Language** to **Scala**:

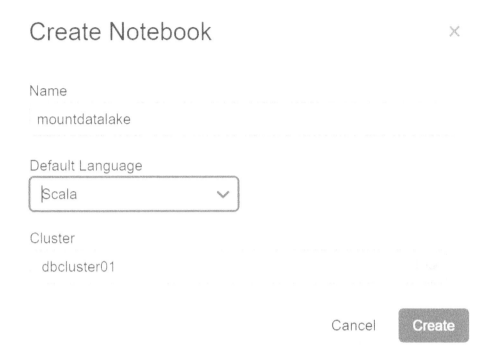

Figure 7.29 – The notebook name

15. Use the following Scala code to mount the data lake container in Databricks. The code extracts the application ID, application secret, and directory ID from the key vault using the dbutils. secrets.get function available in Scala. The dbutils.secret.get function takes the scope name (provided in the *Integrating Databricks with Azure Key Vault* recipe), and the secret names provided in *step 7* and *step 8*. The dbutils.fs.mount command has a parameter called source, which takes the URL of the data lake container to be mounted. The data lake container URL format is abfss://<containername>@<storageaccountname>. dfs.core.windows.net/ and, in our case, the URL would be abfss://databricks@ packtadestoragev2.dfs.core.windows.net/:

```
val appsecret = dbutils.secrets.
get(scope="datalakekey",key="appsecret")
val ApplicationID = dbutils.secrets.
get(scope="datalakekey",key="ApplicationID")
val DirectoryID = dbutils.secrets.
get(scope="datalakekey",key="DirectoryID")
val endpoint = "https://login.microsoftonline.com/" +
DirectoryID + "/oauth2/token"
```

```
val configs = Map(
  "fs.azure.account.auth.type" -> "OAuth",
  "fs.azure.account.oauth.provider.type" -> "org.apache.
hadoop.fs.azurebfs.oauth2.ClientCredsTokenProvider",
  "fs.azure.account.oauth2.client.id" -> ApplicationID,
  "fs.azure.account.oauth2.client.secret" -> appsecret,
  "fs.azure.account.oauth2.client.endpoint" -> endpoint)
// Optionally, you can add <directory-name> to the source
URI of your mount point.
dbutils.fs.mount(
  source = "abfss://databricks@packtadestoragev2.dfs.
core.windows.net/",
  mountPoint = "/mnt/datalakestorage",
  extraConfigs = configs)
```

Upon running the script, the data lake container will be successfully mounted:

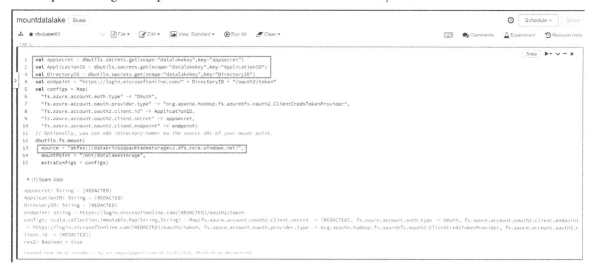

Figure 7.30 – Mounting the data lake container notebook

How it works...

On Azure AD, we registered an application that created a service principal. Service principals are entities that applications can use to authenticate themselves to Azure services. We provided permissions for the application ID on the container to be accessed, which grants permission to the service principal created. We stored the credentials of the service principal (the application ID and secret) in Azure Key

Vault to ensure secure access to credentials. Databricks obtains the service principal credentials (the application ID and secret) from the key vault and uses the security context of the service principal to access the Azure Data Lake account. Databricks, while mounting the data lake account, retrieves the application ID, directory ID, and secret from the key vault and uses the service principal context to access the Azure Data Lake account. This process makes for a very secure method of accessing a data lake account for the following reasons:

- Sensitive information such as an application secret or directory ID is accessed programmatically and not used in plain text inside the notebook. This ensures sensitive information is not exposed to anyone who accesses the notebook.

- Developers who access the notebook needn't know the password or key of the data lake account. Using a key vault ensures that they can continue their development work, even without having direct access to the data lake account or database.

- Data lake accounts can be accessed through account keys too. However, we used service principals for authentication, as using service principals restricts access to the data lake account via applications, while accessing them using an account key or **Shared Access Signature** (**SAS**) token would provide direct login access to the data lake account via tools such as Azure Storage Explorer.

Processing data using notebooks

Databricks notebooks are the fundamental component in Databricks for performing data processing tasks. In this recipe, we will perform operations such as reading, filtering, cleaning a **Comma-Separated Value** (**CSV**) file, and gaining insights from it using a Databricks notebook written in Scala code.

Getting ready

Create a Databricks workspace and a cluster, as explained in the *Configuring the Azure Databricks environment* recipe.

Download the `covid-data.csv` file from the path at `https://github.com/PacktPublishing/Azure-Data-Engineering-Cookbook-2nd-edition/blob/main/chapter07/covid-data.csv`.

How to do it...

Let's process some data using Scala in a Databricks notebook by following the steps provided here:

1. Log in to `portal.azure.com`. Go to **All resources** and find **pactadedatabricks**, the Databricks workspace created in the *Configuring the Azure Databricks environment* recipe. Click **Launch Workspace** to log in to the Databricks portal.

2.   From the **Create** menu, select **Notebook**:

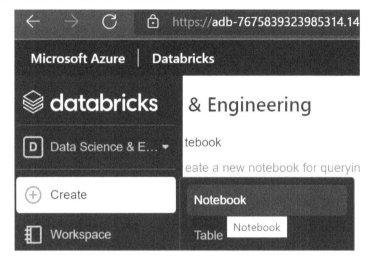

Figure 7.31 – Creating a notebook

3.   Set the notebook name as processdata and **Default Language** to **Scala**:

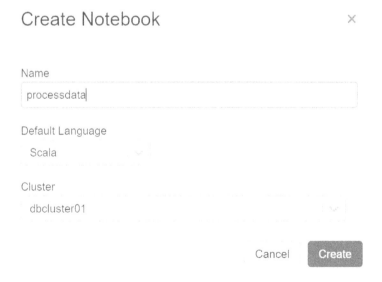

Figure 7.32 – Creating the processdata notebook

4.   In the **File** menu, click **Upload Data**:

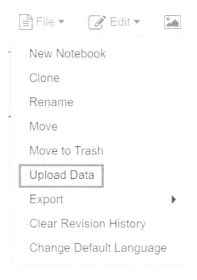

Figure 7.33 – Uploading data

5. Click **Drop files to upload, or click to browse**. Make sure to also note down the default path in your environment. Usually, it will follow the following format – `/FileStore/shared_uploads/<loginname>`:

Upload Data

DBFS Target Directory ❓

| /FileStore/ | shared_uploads/arr.nagaraj@gmail.com | Select |

Files uploaded to DBFS are accessible by everyone who has access to this workspace. Learn more

Files ❓

Drop files to upload, or click to browse

Figure 7.34 – Uploading data

6. Upload the `covid-data.csv` file downloaded at the beginning of this recipe:

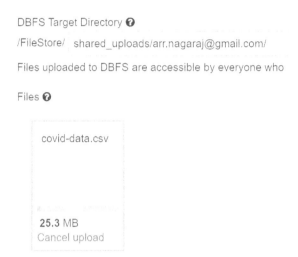

Upload Data

DBFS Target Directory ❷

/FileStore/ shared_uploads/arr.nagaraj@gmail.com/

Files uploaded to DBFS are accessible by everyone who

Files ❷

covid-data.csv

25.3 MB
Cancel upload

Figure 7.35 – Uploading the covid-data.csv file

7. Execute the following command to read the data to a DataFrame. Ensure to replace the file path noted in *step 6*. Notice from the output message that the DataFrame contains over 60 fields:

```
val covid_raw_data = spark.read.format("csv")
.option("header", "true")
.option("inferSchema", "true")
.load("/FileStore/shared_uploads/arr.nagaraj@gmail.com/
covid_data.csv")
```

The result of the command is provided here:

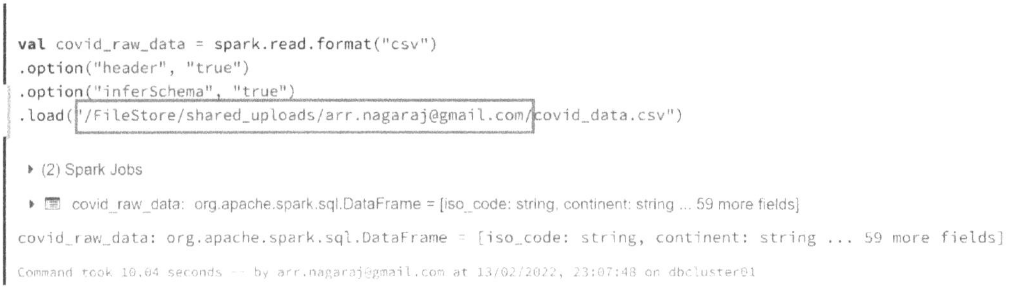

```
val covid_raw_data = spark.read.format("csv")
.option("header", "true")
.option("inferSchema", "true")
.load("/FileStore/shared_uploads/arr.nagaraj@gmail.com/covid_data.csv")

 ▶ (2) Spark Jobs

 ▶ 🖩  covid_raw_data: org.apache.spark.sql.DataFrame = [iso_code: string, continent: string ... 59 more fields]

covid_raw_data: org.apache.spark.sql.DataFrame = [iso_code: string, continent: string ... 59 more fields]

Command took 10.04 seconds -- by arr.nagaraj@gmail.com at 13/02/2022, 23:07:48 on dbcluster01
```

Figure 7.36 – Reading the CSV data

8. At the right-hand corner of the cell, click on the drop-down button, and click **Add Cell Below**. Provide the following command to display the DataFrame:

```
display(covid_raw_data)
```

The result of the command is provided here:

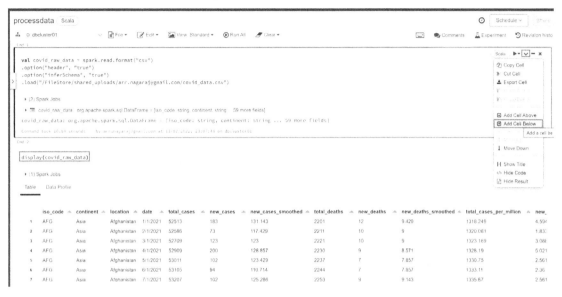

Figure 7.37 – Displaying the CSV data

9. Execute the following command to get the row count in the CSV file or DataFrame. There are over 94,000 rows:

```
covid_raw_data.count()
```

The result of the command is provided here:

```
Cmd 3

covid_raw_data.count()

▶ (2) Spark Jobs

res6: Long = 94342

Command took 1.74 seconds -- by arr.nagaraj@gmail.com at 13/02/2022, 12:22:56 on dbcluster01
```

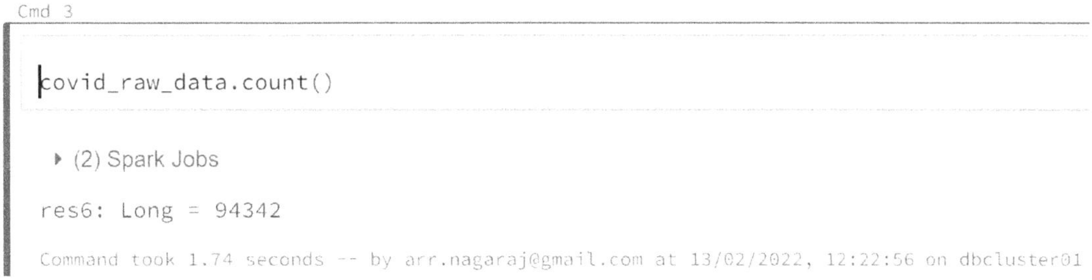

Figure 7.38 – Displaying the row count

10. Let's remove any duplicates. The `dropDuplicates()` function removes duplicates and stores the result in a new DataFrame:

```
val covid_remove_duplicates = covid_raw_data.
dropDuplicates()
```

The result of the command is provided here:

```
Cmd 4

val covid_remove_duplicates = covid_raw_data.dropDuplicates()

▶ 🖻  covid_remove_duplicates: org.apache.spark.sql.Dataset[org.apache.spark.sql.Row] = [iso_code: string, continent: string ... 59 more fields]

covid_remove_duplicates: org.apache.spark.sql.Dataset[org.apache.spark.sql.Row] = [iso_code: string, continent

Command took 0.41 seconds -- by arr.nagaraj@gmail.com at 13/02/2022, 12:24:32 on dbcluster01
```

Figure 7.39 – Removing duplicates

11. Let's look at the fields and their data type in the DataFrame. The `printSchema()` function provides the DataFrame structure:

```
covid_remove_duplicates.printSchema()
```

The result of the preceding command is provided in the following screenshot:

```
Cmd 5

covid_remove_duplicates.printSchema()

root
 |-- iso_code: string (nullable = true)
 |-- continent: string (nullable = true)
 |-- location: string (nullable = true)
 |-- date: string (nullable = true)
 |-- total_cases: integer (nullable = true)
 |-- new_cases: integer (nullable = true)
 |-- new_cases_smoothed: double (nullable = true)
 |-- total_deaths: integer (nullable = true)
 |-- new_deaths: integer (nullable = true)
 |-- new_deaths_smoothed: double (nullable = true)
 |-- total_cases_per_million: double (nullable = true)
 |-- new_cases_per_million: double (nullable = true)
 |-- new_cases_smoothed_per_million: double (nullable = true)
 |-- total_deaths_per_million: double (nullable = true)
 |-- new_deaths_per_million: double (nullable = true)
 |-- new_deaths_smoothed_per_million: double (nullable = true)
 |-- reproduction_rate: double (nullable = true)
 |-- icu_patients: integer (nullable = true)
 |-- icu_patients_per_million: double (nullable = true)
 |-- hosp_patients: integer (nullable = true)

Command took 0.21 seconds -- by arr.nagaraj@gmail.com at 13/02/2022, 12:24:45 on dbcluster01
```

Figure 7.40 – Removing duplicates

12. The data is about the impact of COVID across the globe. Let's focus on a handful of columns that are required, instead of all the columns provided. The `select` function can help us to specify the columns that we need out of a DataFrame. Execute the following command to load the selected columns to another DataFrame:

```
val covid_selected_columns = covid_remove_duplicates.
select("iso_code","location","continent","date","new_
deaths_per_million","people_fully_
vaccinated","population")
```

The result of the preceding command is provided in the following screenshot:

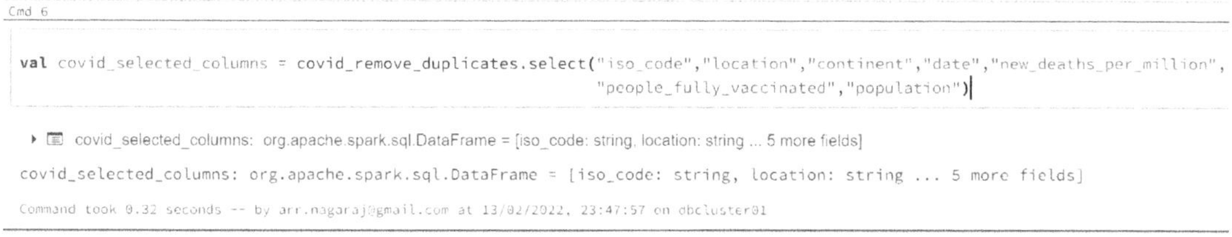

```
Cmd 6

val covid_selected_columns = covid_remove_duplicates.select("iso_code","location","continent","date","new_deaths_per_million",
                                                "people_fully_vaccinated","population")

  ▸ ▦ covid_selected_columns: org.apache.spark.sql.DataFrame = [iso_code: string, location: string ... 5 more fields]

covid_selected_columns: org.apache.spark.sql.DataFrame = [iso_code: string, location: string ... 5 more fields]

Command took 0.32 seconds -- by arr.nagaraj@gmail.com at 13/02/2022, 23:47:57 on dbcluster01
```

Figure 7.41 – Loading selected columns

13. For our analysis, let's remove any rows that contain NULL values in any of the columns. The `na.drop()` function can help to achieve this. Execute the following command. The `covid_clean_data` DataFrame will only contain rows without any NULL value in them once the command has been executed. The `covid_clean_data.count()` command shows that only 32,000+ rows were without any NULL values:

```
val covid_clean_data = covid_selected_columns.na.drop()
covid_clean_data.count()
```

The result of the preceding command is provided in the following screenshot:

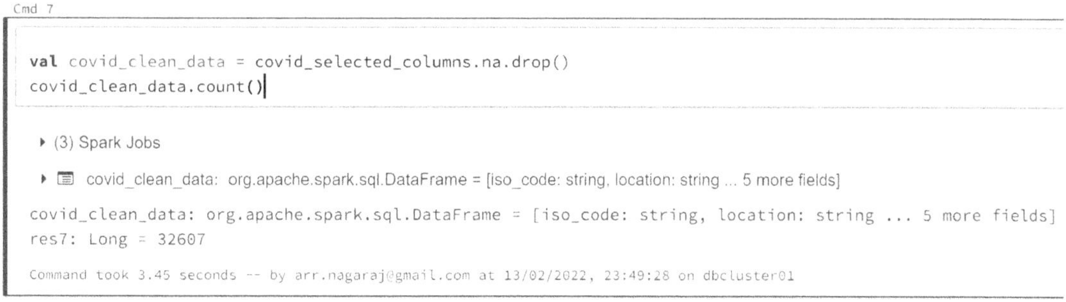

```
Cmd 7

val covid_clean_data = covid_selected_columns.na.drop()
covid_clean_data.count()

  ▸ (3) Spark Jobs

  ▸ ▦ covid_clean_data: org.apache.spark.sql.DataFrame = [iso_code: string, location: string ... 5 more fields]

covid_clean_data: org.apache.spark.sql.DataFrame = [iso_code: string, location: string ... 5 more fields]
res7: Long = 32607

Command took 3.45 seconds -- by arr.nagaraj@gmail.com at 13/02/2022, 23:49:28 on dbcluster01
```

Figure 7.42 – Removing NULL values

14. Data analysis is easier to perform using Spark SQL commands. To use SQL commands to analyze a DataFrame, we need to create a temporary view. The following command will create a temporary view called `covid_view`:

```
covid_clean_data.createOrReplaceTempView("covid_view")
```

The result of the preceding command is provided in the following screenshot:

```
Cmd 8

covid_clean_data.createOrReplaceTempView("covid_view")|


Command took 0.17 seconds -- by arr.nagaraj@gmail.com at 13/02/2022, 23:55:46 on dbcluster01
```

Figure 7.43 – Creating a temporary view

15. Execute the following command to get some insights out of the data. `%sql`, on the first line, lets us switch from **Scala** to **SQL**. The SQL query provides the number of deaths, and the percentage of people vaccinated in each country with a population of over 1 million, ordered by countries with the highest deaths per million people:

```
%sql
SELECT iso_code, location, continent,
SUM(new_deaths_per_million) as death_sum,
MAX(people_fully_vaccinated * 100 / population) as
percentage_vaccinated FROM covid_view
WHERE population > 1000000
GROUP BY iso_code,location,continent
ORDER BY death_sum desc
```

The result of the command is provided here:

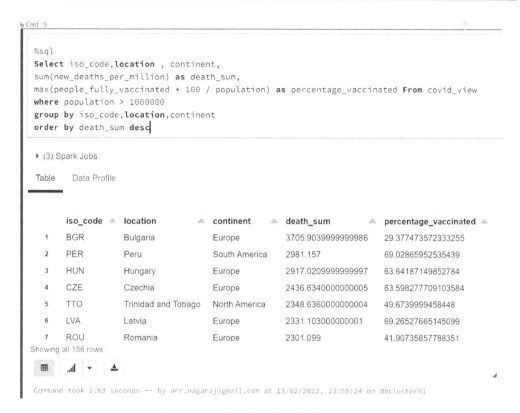

Figure 7.44 – Insights using the SQL query

16. To visualize the output from the previous SQL query, click on the chart icon and select **Bar**:

Figure 7.45 – Generating a bar graph

17. Click on **Plot Options...**:

Command took 1.69 seconds -- by arr.nagaraj@gmail.com

Figure 7.46 – Plot options

18. Set **Plot Options...** as follows. Set **location** in the **Keys** section, **continent** in the **Series groupings** section, and **death_sum** in the **Values** section, as shown in the following screenshot. This will provide a bar graph with a line for each country, with the color of the line based on the continent that the country belongs to:

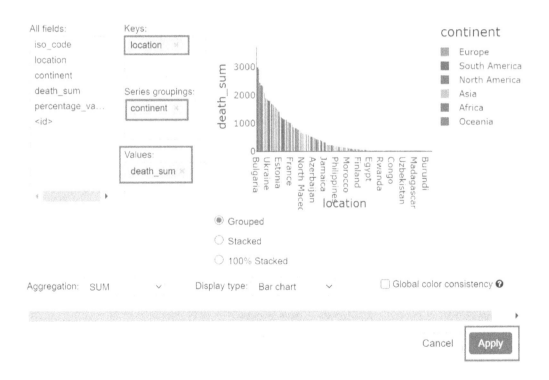

Figure 7.47 – Customizing the plot

19. You will get the following output. High spikes with green lines indicate that European countries were heavily affected by COVID:

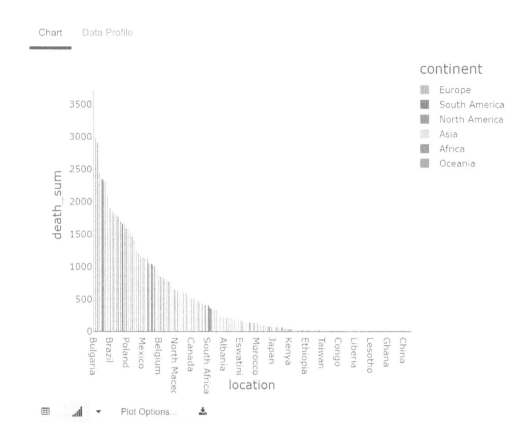

Figure 7.48 – Visual insights

How it works...

DataFrames are the fundamental objects used to store runtime data during data processing in Databricks. DataFrames are in-memory objects and extremely well-optimized for performing advanced analytics operations.

A CSV file was loaded to the **Databricks File System** (**DBFS**) storage, which is the default local storage available when a Databricks workspace is created. We can perform the same activities in a data lake account too, by uploading the CSV file to the data lake container and mounting the data lake container, as explained in the *Mounting an Azure Data Lake container in Databricks* recipe.

After loading the data to a DataFrame, we were able to cleanse the data by performing operations such as removing unwanted columns, dropping duplicates, and deleting rows with NULL values easily using Spark functions. Finally, by creating a temporary view out of the DataFrame, we were able to analyze the DataFrame's data using SQL queries and get visual insights using Databricks' visualization capabilities.

Scheduling notebooks using job clusters

Data processing can be performed using notebooks, but to operationalize it, we need to execute it at a specific scheduled time, depending upon the demands of the use case or problem statement. After a notebook has been created, you can schedule a notebook to be executed at a preferred frequency using job clusters. This recipe will demonstrate how you could schedule a notebook using job clusters.

Getting ready

Create a Databricks workspace, as explained in the *Configuring the Azure Databricks environment* recipe.

How to do it...

In the following steps, we will import the `SampleJob.dbc` notebook file into the Databricks workspace and schedule it to be run daily:

1. Log in to `portal.azure.com`. Go to **All resources** and find **pactadedatabricks**, the Databricks workspace created in the *Configuring the Azure Databricks environment* recipe. Click **Launch Workspace** to log in to the Databricks portal.

2. Navigate to **Workspace | Create | Folder**, as shown in the following screenshot:

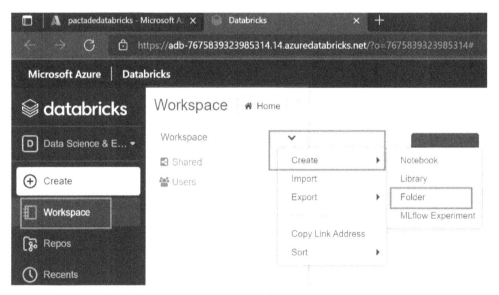

Figure 7.49 – Creating folder insights

3. Create a folder called `Job`:

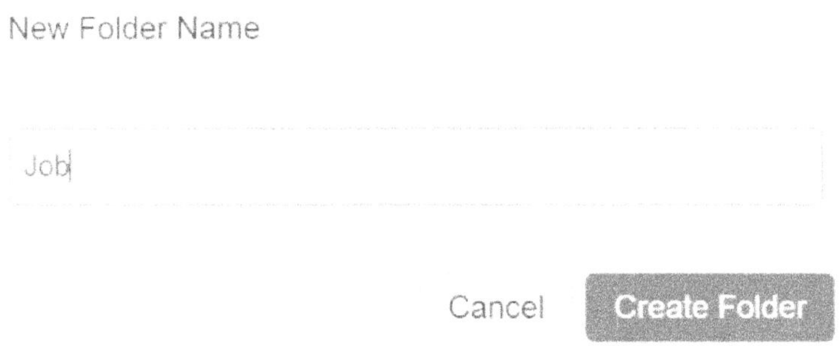

New Folder Name

Job

Cancel   **Create Folder**

Figure 7.50 – Create a Job folder

4. Click **Workspace**, click on the **Job** folder, and pick the **Import** option:

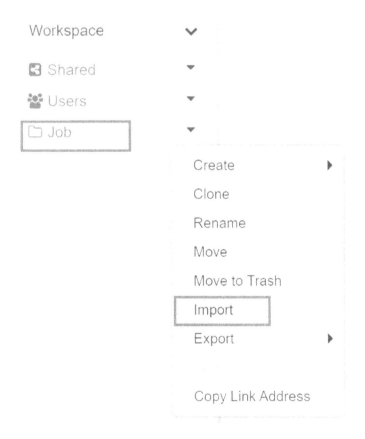

Figure 7.51 – Importing a notebook

5. Pick the **URL** option to import the notebook. Paste the `https://github.com/` `PacktPublishing/Azure-Data-Engineering-Cookbook-2nd-edition/` `blob/main/chapter07/SampleJob.dbc` path to import a notebook called `SampleJob`:

Figure 7.52 – Importing a notebook

6. In the **Job** folder, you will have a notebook called `SampleJob`. The `SampleJob` notebook will read a sample CSV file and provide insights from it:

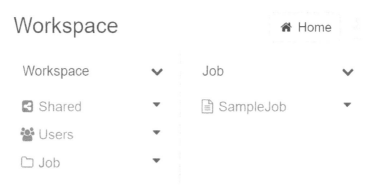

Figure 7.53 – The imported notebook

7. From the menu on the left, click **Jobs**:

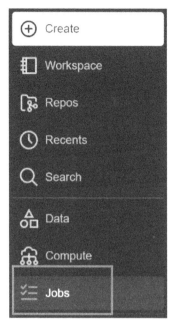

Figure 7.54 – Jobs in the Create menu

8. Click **Create Job**:

Figure 7.55 – Create Job

9. Provide a job name of SampleJob. Select the imported **SampleJob** notebook. On the **New Cluster** configuration, click on the edit icon to set the configuration options. The job will create a cluster each time it runs based on the configuration and delete the cluster once the job is completed:

Figure 7.56 – Editing the cluster configuration

10. Reduce the total number of cores to **2** and hit the **Confirm** button to save the cluster configuration. Hit the **Create** button on the job creation screen:

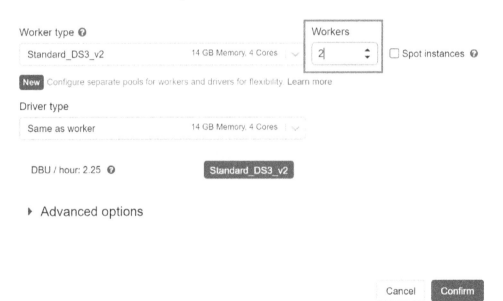

Figure 7.57 – Save the cluster configuration

11. Hit the **Edit schedule** button to edit the schedule:

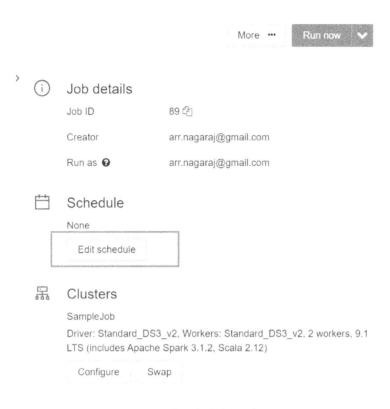

Figure 7.58 – Edit schedule configuration

12. Set **Schedule Type** to **Scheduled** and set a frequency as per your needs. Hit the **Save** button:

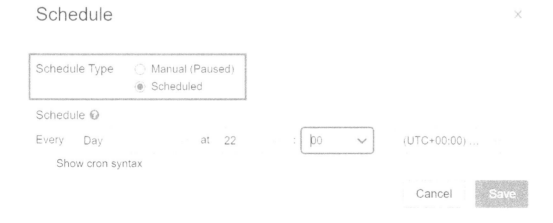

Figure 7.59 – Setting the schedule configuration

13. Hit **Run now** to trigger the job:

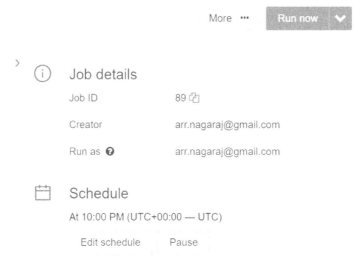

Figure 7.60 – Run now

14. Click on **Runs** on the left tab to view the job run result:

Figure 7.61 – Job runs

15. The **Active runs** section will provide the currently active jobs. After a few minutes, the job will move to the **Running** state from the initial **Pending** state:

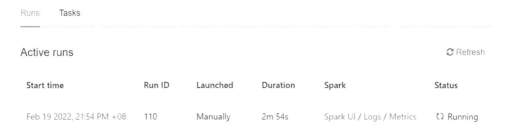

Figure 7.62 – Job runs

16. Upon completion, the job result will be listed as completed in the **Completed runs (past 60 days)** section. Click **Start time** to view the result of the completed notebook:

Completed runs (past 60 days)   ⟳ Refresh

Latest successful run (refreshes automatically)

Start time	Run ID	Launched	Duration	Spark	Status
Feb 19 2022, 21:54 PM +08	110	Manually	3m 22s	Spark UI / Logs / Metrics	⊘ Succeeded

Figure 7.63 – A completed run

17. The job execution result is shown in the following screenshot:

Jobs > SampleJob > Run 110

SampleJob run

Output

```
val diamonds = spark.read.format("csv")
  .option("header", "true")
  .option("inferSchema", "true")
  .load("/databricks-datasets/Rdatasets/data-001/csv/ggplot2/diamonds.csv")
```

▸ 🖼 diamonds: org.apache.spark.sql.DataFrame = [_c0: integer, carat: double ... 9 more fields]

```
diamonds: org.apache.spark.sql.DataFrame = [_c0: int, carat: double ... 9 more fields]
```

Command took 21.47 seconds

```
diamonds.createOrReplaceTempView("diamonds_view")
```

Command took 0.30 seconds

```
%sql
Select cut, color, avg(price) as avg_price, max(price) as max_price
From diamonds_view
Group by cut,color
order by avg_price desc, cut, color
```

	cut	color	avg_price	max_price
1	Premium	J	6294.591584158416	18710
2	Premium	I	5946.180672268908	18823
3	Very Good	I	5255.879568106312	18500
4	Premium	H	5216.706779661017	18795
5	Fair	H	5135.683168316832	18565
6	Very Good	J	5103.513274336283	18430
7	Good	I	5078.532567049809	18707

Showing all 35 rows.

⬇

Command took 2.88 seconds

Figure 7.64 – The notebook output

18. If we need to execute another notebook at the same schedule, we could create an additional task within the same job. Go back to the job menu and click on **SampleJob**. Switch to the **Tasks** section. Click on the + button to add a new task:

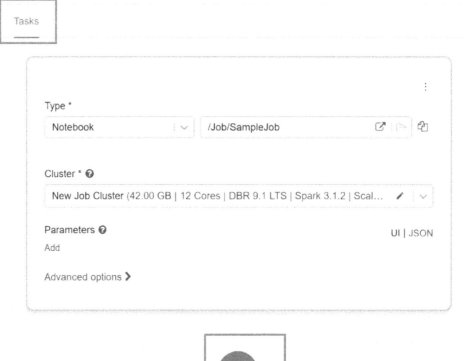

Figure 7.65 – Adding a new task

19. Provide a task name. You may add any notebook to the task. Notice that the cluster name is **SampleJob_cluster**, which implies that it will use the cluster created for the previous task. The **Depends on** option controls whether the job needs to run after the previous **SampleJob** task is completed:

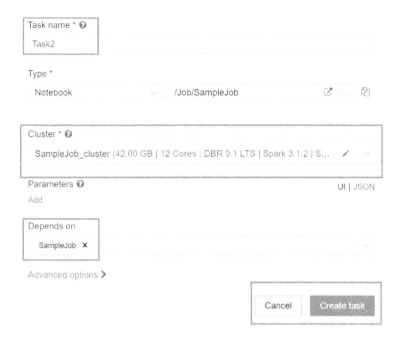

Figure 7.66 – Adding a new task

20. Upon adding it, the two tasks will appear, as shown here:

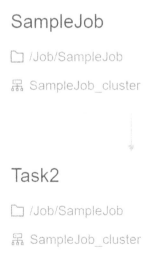

SampleJob

📁 /Job/SampleJob

🔜 SampleJob_cluster

Task2

📁 /Job/SampleJob

🔜 SampleJob_cluster

Figure 7.67 – Multiple tasks

How it works...

The imported notebook was set to run at a specific schedule using the Databricks job scheduling capabilities. While scheduling the jobs, **New Cluster** was selected, instead of picking any cluster available in the workspace. Picking **New Cluster** implies a cluster will be created each time the job runs and will be destroyed once the job completes. This also means the jobs need to wait for an additional 2 minutes for the cluster to be created for each run.

Adding multiple notebooks to the same job via additional tasks allows us to reuse the job cluster created for the first notebook execution, and the second task needn't wait for another cluster to be created. Usage of multiple tasks and the dependency option allows us to orchestrate complex data processing flows using Databricks notebooks.

Working with Delta Lake tables

Delta Lake databases are **Atomicity, Consistency, Isolation, and Durability (ACID)** property-compliant databases available in Databricks. Delta Lake tables are tables in Delta Lake databases that use Parquet files to store data and are highly optimized for performing analytic operations. Delta Lake tables can be used in a data processing notebook for storing preprocessed or processed data. The data stored in Delta Lake tables can be easily consumed in visualization tools such as Power BI.

In this recipe, we will create a Delta Lake database and Delta Lake table, load data from a CSV file, and perform additional operations such as UPDATE, DELETE, and MERGE on the table.

Getting ready

Create a Databricks workspace and a cluster, as explained in the *Configuring the Azure Databricks environment* recipe of this chapter.

Download the `covid-data.csv` file from this link: `https://github.com/PacktPublishing/Azure-Data-Engineering-Cookbook-2nd-edition/blob/main/chapter07/covid-data.csv`.

Upload the `covid-data.csv` file to the workspace, as explained in *step 1* to *step 6* of the *How to do it...* section of the *Processing data using notebooks* recipe.

How to do it...

In this recipe, let's create a Delta Lake database and tables, and process data using the following steps:

1. From the Databricks menu, click **Create** and create a new notebook:

Figure 7.68 – A new notebook

2. Create a notebook called `Covid-DeltaTables` with **Default Language** set to **SQL**:

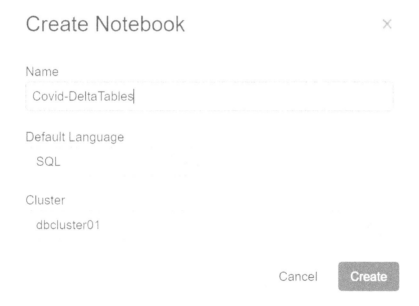

Figure 7.69 – Delta notebook creation

3. Add a cell to the notebook and execute the following command to create a Delta database called `covid`:

```
CREATE DATABASE covid
```

The output is displayed in the following screenshot:

Figure 7.70 – Delta database creation

4. Execute the following command to read the `covid-data.csv` file to a temporary view. Please note that the path depends on the location where `covid-data.csv` was uploaded and ensure to provide the correct path for your environment:

```
CREATE TEMPORARY VIEW covid_data
USING CSV
OPTIONS (path "/FileStore/shared_uploads/arr.nagaraj@
gmail.com/covid_data.csv", header "true", mode
"FAILFAST")
```

The output is displayed in the following screenshot:

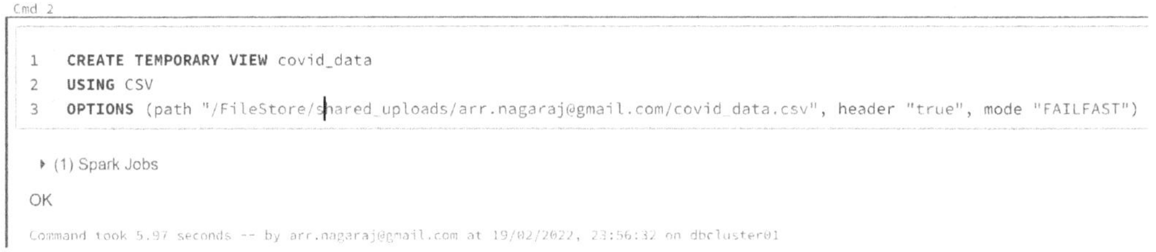

```
Cmd 2

1    CREATE TEMPORARY VIEW covid_data
2    USING CSV
3    OPTIONS (path "/FileStore/shared_uploads/arr.nagaraj@gmail.com/covid_data.csv", header "true", mode "FAILFAST")

  ▶ (1) Spark Jobs

OK

Command took 5.97 seconds -- by arr.nagaraj@gmail.com at 19/02/2022, 23:56:32 on dbcluster01
```

Figure 7.71 – Temporary view creation

5. Execute the following command to create a Delta table using the temporary view. `USING DELTA` in the `CREATE TABLE` command indicates that it's a Delta table being created. The location specifies where the table is stored. If you have mounted a data lake container (as we did in the *Mounting an Azure Data Lake container in Databricks* recipe), you can use the data lake mount point to store the Delta Lake table in your Azure Data Lake account. In this example, we use the default storage provided by Databricks, which comes with each Databricks workspace.

The table name is provided in `<database name>.<table name>` to ensure it belongs to the Delta Lake database created. To insert the data from the view into the new table, the table is created using the `CREATE TABLE` command, followed by the `AS` command, followed by the `SELECT` statement against the view:

```
CREATE OR REPLACE TABLE covid.covid_data_delta
USING DELTA
LOCATION '/FileStore/shared_uploads/arr.nagaraj@gmail.
com/covid_data_delta'
AS
SELECT iso_code,location,continent,date,new_deaths_per_
million,people_fully_vaccinated,population FROM covid_
data
```

The output is displayed in the following screenshot:

Figure 7.72 – Creating the Delta table

6. Go to the Databricks menu, click **Data**, and then click on the **covid** database. You will notice that a **covid_data_delta** table has been created:

Figure 7.73 – The created Delta table

7. Go back to the notebook we were working on. Add a new cell and delete some rows using the following command. The `DELETE` command will delete around 57,000 rows. We can delete, select, and insert data as well as we would in any other commercial database:

```
DELETE FROM covid.covid_data_delta where population is
null or people_fully_vaccinated is null or new_deaths_
per_million is null or location is null
```

The output is displayed in the following screenshot:

Figure 7.74 – Deleting a few rows in a Delta table

8. Add a new cell and execute the following command to delete all the rows from the table. Let's run a `select count(*)` query against the table, which will return **0** if all the rows have been deleted:

```
delete from covid.covid_data_delta;
Select count(*) from covid.covid_data_delta;
```

The output is displayed in the following screenshot:

Figure 7.75 – Deleting all rows

9. Delta tables have the ability to time travel, which allows us to read older versions of the table. Using the `as of version <version number>` keyword, we can read the older version of the table. Version 0 gives the most recent version behind the current version of the table. Add a new cell and execute the following command to read the data before deletion:

```
select * from covid.covid_data_delta version as of 0;
```

The output is displayed in the following screenshot:

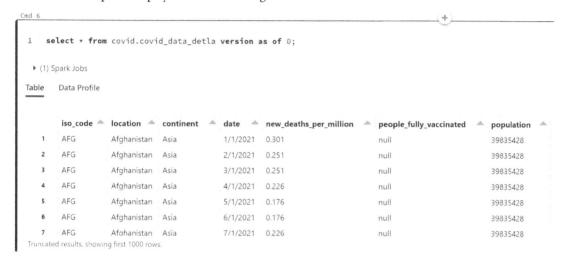

Figure 7.76 – Check out an older version of the Delta table

10. `RESTORE TABLE` can restore the table to the older version. Add a cell and execute the following command to recover all the rows before deletion:

```
RESTORE TABLE covid_data_delta TO VERSION AS OF 0
```

The output is displayed in the following screenshot:

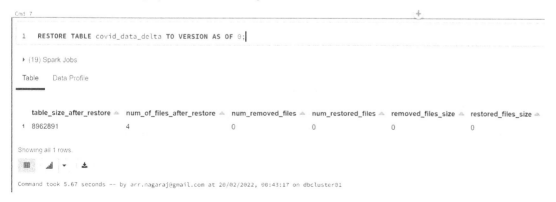

Figure 7.77 – Restoring the table to an older version

11. Let's perform an UPDATE statement, followed by a DELETE statement. Add two cells. Paste the following commands into these cells, as shown in the following screenshot. Execute them in sequence:

```
UPDATE covid_data_delta SET population = population * 1.2
WHERE continent = 'Asia';
DELETE FROM covid_data_delta  WHERE continent = 'Europe';
```

The output is displayed in the following screenshot:

Figure 7.78 – Update and delete

12. If we want to revert these two operations (UPDATE and DELETE), we can use the MERGE statement and the Delta table's row versioning capabilities to achieve this. Using a single MERGE statement, we can perform the following:

- Compare the older version of the table with the current version of the table

- If the rows match, update all the columns with values from the older version

- If the row doesn't match, insert the row from the older version into the current table

This can be achieved using the following code:

```
MERGE INTO covid_data_delta source
  USING covid_data_delta TIMESTAMP AS OF "2022-02-19
16:45:00" target
  ON source.location = target.location and source.date =
target.date
  WHEN MATCHED THEN UPDATE SET *
  WHEN NOT MATCHED
  THEN INSERT *
```

The MERGE command takes the covid_data_delta table as the source table to be updated or inserted. Instead of specifying version numbers, we can also use timestamps to obtain older versions of the table. covid_data_delta TIMESTAMP AS OF "2022-02-19 16:45:00" takes the version of the table as of February 19, 2022, 16:45:00 – UTC time. WHEN MATCHED THEN UPDATE SET * updates all the columns in the table when the condition specified in the ON clause matches. When the condition doesn't match, the rows are inserted from the older version to the current version of the table. The output, as expected, shows that the rows that were deleted in *step 11* were successfully reinserted:

Figure 7.79 – The merge statement

How it works...

Delta tables offer advanced capabilities for processing data, such as support for UPDATE, DELETE, and MERGE statements. MERGE statements and the versioning capabilities of Delta tables are very powerful in ETL scenarios, where we need to perform UPSERT (update if it matches, insert if it doesn't) operations against various tables.

These capabilities for supporting data modifications and row versioning are made possible because Delta tables maintain the changes to the table via a transaction log file stored in JSON format. The transaction files are located in the same location where the table was created but in a subfolder called _delta_log. By default, the log files are retained for 30 days and can be controlled using the delta. logRetentionDuration table property. The ability to read older versions is also controlled by the delta.logRetentionDuration property.

Connecting a Databricks Delta Lake table to Power BI

Delta Lake databases are commonly used to store processed data in Delta Lake tables, which is then ready to be consumed by reporting-layer applications such as Power BI. Delta Lake tables are best suited for handling analytic workloads from Power BI, as Delta Lake tables use Parquet files as storage, which offer optimal performance for analytic workloads.

In this recipe, we will use Power BI Desktop, connect to a Delta Lake table, and build a simple report in Power BI.

Getting ready

Create a Databricks workspace and a cluster as explained in the *Configuring the Azure Databricks environment* recipe.

Download the latest version of Power BI Desktop from https://powerbi.microsoft.com/en-us/downloads/ and install Power BI Desktop on your machine.

How to do it...

Perform the following steps to connect a Delta Lake table to a Power BI report and create visualizations:

1. Log in to the Databricks portal and click on the **Compute** button in the menu:

Figure 7.80 – Compute

2.   Click on the cluster and start it if the cluster is terminated. Click on **Advanced options**:

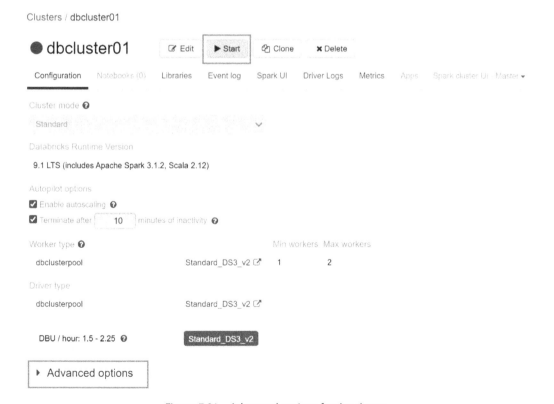

Figure 7.81 – Advanced options for the cluster

3. Click on the **JDBC/ODBC** section. Copy the values from the **Server Hostname** and **HTTP Path** sections. We will need them when connecting to the Delta Lake table using Power BI Desktop:

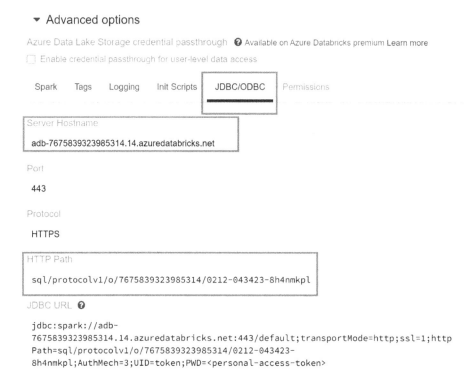

Figure 7.82 – The JDBC/ODBC connection string

4. Click on the **Workspace** button in the Databricks menu. Click on any folder (by default under your username) and click on the **Import** button:

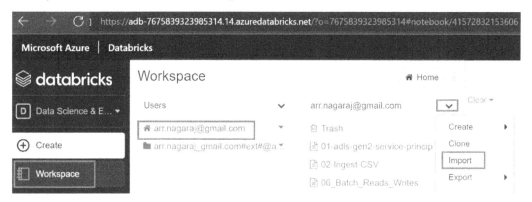

Figure 7.83 – Importing the notebook

5. For the **Import from** option, pick **URL**. Provide https://github.com/PacktPublishing/ Azure-Data-Engineering-Cookbook-2nd-edition/blob/main/chapter07/ Delta_PowerBI.dbc as the input and click the **Import** button:

Figure 7.84 – Import notebook

6. This notebook will read a CSV file, create a database called PowerBI, and create a Delta table called Diamond_Insights_PowerBI. Click on the **Run All** button in the top-right corner of the notebook:

Figure 7.85 – Run all cells

7. Open the installed **Power BI Desktop**:

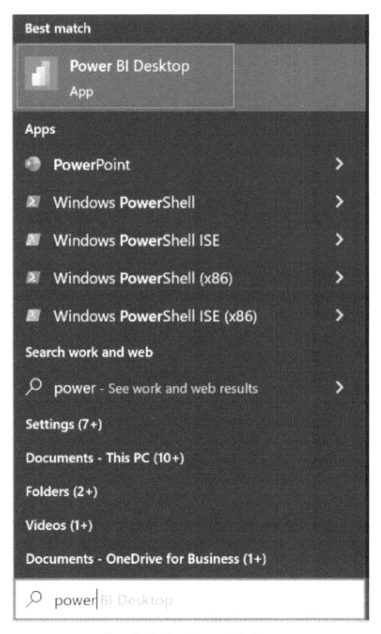

Figure 7.86 – Start Power BI Desktop

8.   Click **Get data** and then **More…**:

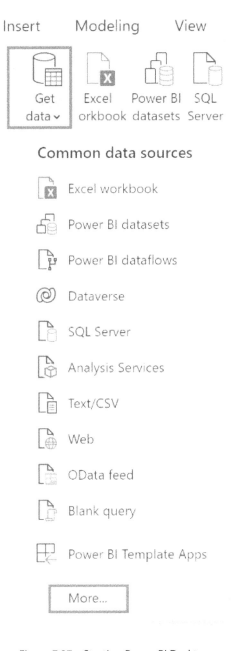

Figure 7.87 – Starting Power BI Desktop

9. Select **Azure Databricks** and click on the **Connect** button:

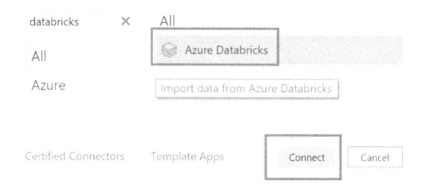

Figure 7.88 – Connecting to Databricks

10. Provide the **Server Hostname** value and the **HTTP Path** value noted in *step 3*, and click on the **Ok** button:

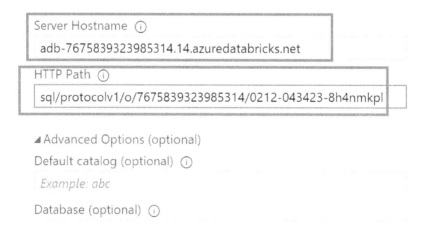

Figure 7.89 – Connecting to the Delta database

11. Sign in to Azure using the Azure account that you used to create the Databricks workspace:

Figure 7.90 – Connecting to Azure

12. Once you're signed in, press the **Connect** button:

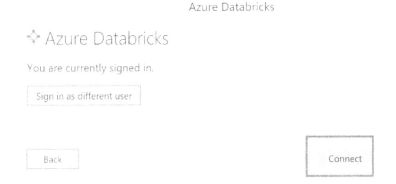

Figure 7.91 – Connecting to the Delta table

13. Expand **hive_metastore** and then the **powerbi** database. Select the **diamond_insights_powerbi** table and click on the **Load** button:

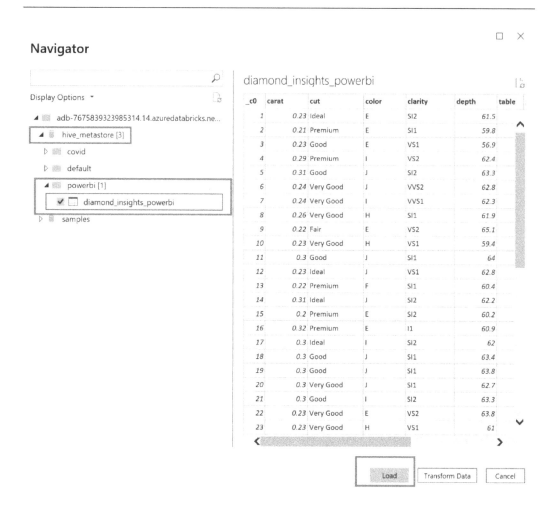

Figure 7.92 – Loading the Delta table

14. Select the `color` and `price` columns on the right-hand side. Set the **Visualizations** type to **Clustered column chart**. This will show the price of diamonds by color code. This will indicate that **color code G** has the highest price:

Figure 7.93 – Adding a clustered column chart

15. Click on the white space anywhere outside of the clustered column chart visual. Click on the cut and price columns and set the **Visualizations** type to **Clustered column chart**. Now, you will have added another visual:

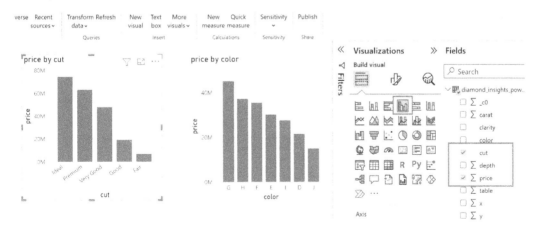

Figure 7.94 – Adding another visual

16. Click on the bar against the **Ideal** value on the **price by cut** visual. The **price by color** visual will be automatically filtered and will show that **color G** contributes close to 20 million in price when the cut is **Ideal**:

Figure 7.95 – Get insights using visuals

17. Open the **File** menu and hit the **Save** button to save the Power BI report.

How it works...

We used the Azure AD credentials to sign in to Azure Databricks and extracted the connection string from the Databricks cluster. When the connection request comes from Power BI, Databricks authenticates using Azure AD credentials. For the authentication to succeed, the Databricks cluster needs to be up and running. Once authenticated, Delta Lake tables are accessible, just as with any other database tables using Power BI. We added two simple visuals to explore the data visually. Clicking on one of the visuals automatically filters the other visual, which allows us to get insights out of the data easily.

Processing Data Using Azure Synapse Analytics

Azure Synapse Analytics workspaces generation 2, formally released in December 2020, is the industry-leading big data solution for processing and consolidating data of business value. Azure Synapse Analytics has three important components:

- SQL pools and Spark pools for performing data exploration and processing
- Integration pipelines for performing data ingestion and data transformations
- Power BI integration for data visualization

Having data ingestion, processing, and visualization capabilities in a single service with seamless integration with all other services makes Azure Synapse Analytics a very powerful tool in big data engineering projects. This chapter will introduce you to Synapse workspaces and cover the following recipes:

- Provisioning an Azure Synapse Analytics workspace
- Analyzing data using serverless SQL pool
- Provisioning and configuring Spark pools
- Processing data using Spark pools and a lake database
- Querying the data in a lake database from serverless SQL pool
- Scheduling notebooks to process data incrementally
- Visualizing data using Power BI by connecting to serverless SQL pool

By the end of the chapter, you will have learned how to provision a Synapse workspace and Spark pools, explore and analyze data using serverless SQL pool and Spark pools, create a lake database and query a lake database from serverless SQL pool and a Spark pool, and finally, visualize the lake database data in Power BI.

Technical requirements

For this chapter, you will need the following:

- A Microsoft Azure subscription
- PowerShell 7
- Power BI Desktop

Provisioning an Azure Synapse Analytics workspace

In this recipe, we'll learn how to provision a Synapse Analytics workspace. A Synapse Analytics workspace is the logical container that will hold the Spark pools, SQL pool, and integration pipelines that are required for the data engineering tasks.

Getting ready

To get started, log into `https://portal.azure.com` using your Azure credentials.

How to do it...

Follow these steps to create a Synapse Analytics workspace:

1. Go to the Azure portal home page at `portal.azure.com` and click on **Create a resource**. Search for `Synapse` and select **Azure Synapse Analytics**. Click the **Create** button:

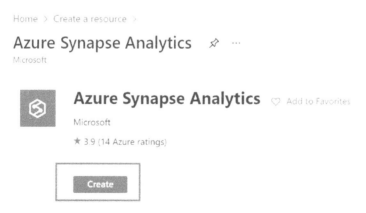

Figure 8.1 – Creating a Synapse Analytics workspace

2. Provide the following details:

   I.   Create a new resource group called `PacktADESynapse` by clicking the **Create new** link.

II.   Provide a unique workspace name. For this example, we are using `packtadesynapse`. You may pick any location.

III.   Provisioning a Synapse Analytics workspace requires an Azure Data Lake Storage Gen2 account. You may use an existing account or create a new account using the **Create new** link. Let's create a new Azure Data Lake Storage Gen2 account called `packatadesynapse` by clicking the **Create new** link.

IV.   A Synapse Analytics workspace requires a container on the new data lake account provisioned. Let's create a new container named `synapse` by clicking the **Create new** link. After the Synapse workspace is provisioned, the Synapse workspace service account is given permissions on the data lake account and the container by default. We will be able to connect to the data stored in the storage account from the Synapse workspace seamlessly. The data lake account will be used to store a lake database (a component of Synapse Spark pools) and other artifacts of the Synapse workspace. Click the **Next: Security >** button at the bottom:

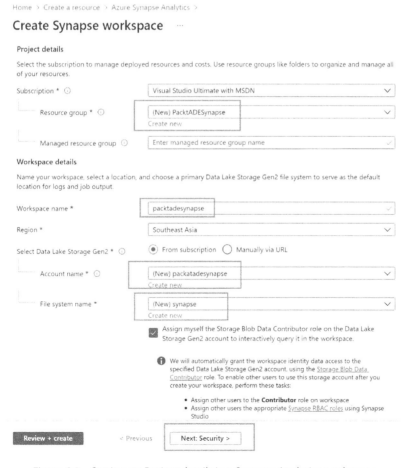

Figure 8.2 – Setting up Project details in a Synapse Analytics workspace

V.   On occasion, Azure subscriptions may not be able to create a Synapse workspace because of a **Resource provider not registered for the subscription** error. To resolve the error, go to the Azure portal and open your subscription. Click the **Resource providers** section, search for the Synapse resource, and register **Microsoft.Synapse** on the subscription. Please refer to https://docs.microsoft.com/en-us/azure/azure-resource-manager/management/resource-providers-and-types#azure-portal for detailed steps on how to register a resource on a subscription.

3.   On the **Security** page, we will be filling in the **user ID** and **password** details for SQL pool. Provide the user ID and password as sqladminuser and PacktAdeSynapse123. Click **Review + Create** to create the Synapse Analytics workspace:

Home > Create a resource > Azure Synapse Analytics >

Create Synapse workspace   ...

* Basics   * **Security**   Networking   Tags   Review + create

Configure security options for your workspace.

Authentication

Choose the authentication method for access to workspace resources such as SQL pools. The authentication method can be changed later on. Learn more

Authentication method ⓘ   (●) Use both local and Azure Active Directory (Azure AD) authentication
                              ( ) Use only Azure Active Directory (Azure AD) authentication

SQL Server admin login * ⓘ   sqladminuser

SQL Password ⓘ   ·················

Confirm password   ·················   ✓ Pass

System assigned managed identity permission

Select to grant the workspace network access to the Data Lake Storage Gen2 account using the workspace system identity. Learn more

☐ Allow network access to Data Lake Storage Gen2 account. ⓘ

ⓘ The selected Data Lake Storage Gen2 account does not restrict network access using any network access rules, or you selected a storage account manually via URL under Basics tab. Learn more

Workspace encryption

⚠ Double encryption configuration cannot be changed after opting into using a customer-managed key at the time of workspace creation.

Choose to encrypt all data at rest in the workspace with a key managed by you (customer-managed key). This will provide double encryption with encryption at the infrastructure layer that uses platform-managed keys. Learn more

Double encryption using a customer-   ( ) Enable (●) Disable

[Review + create]   [< Previous]   [Next: Networking >]

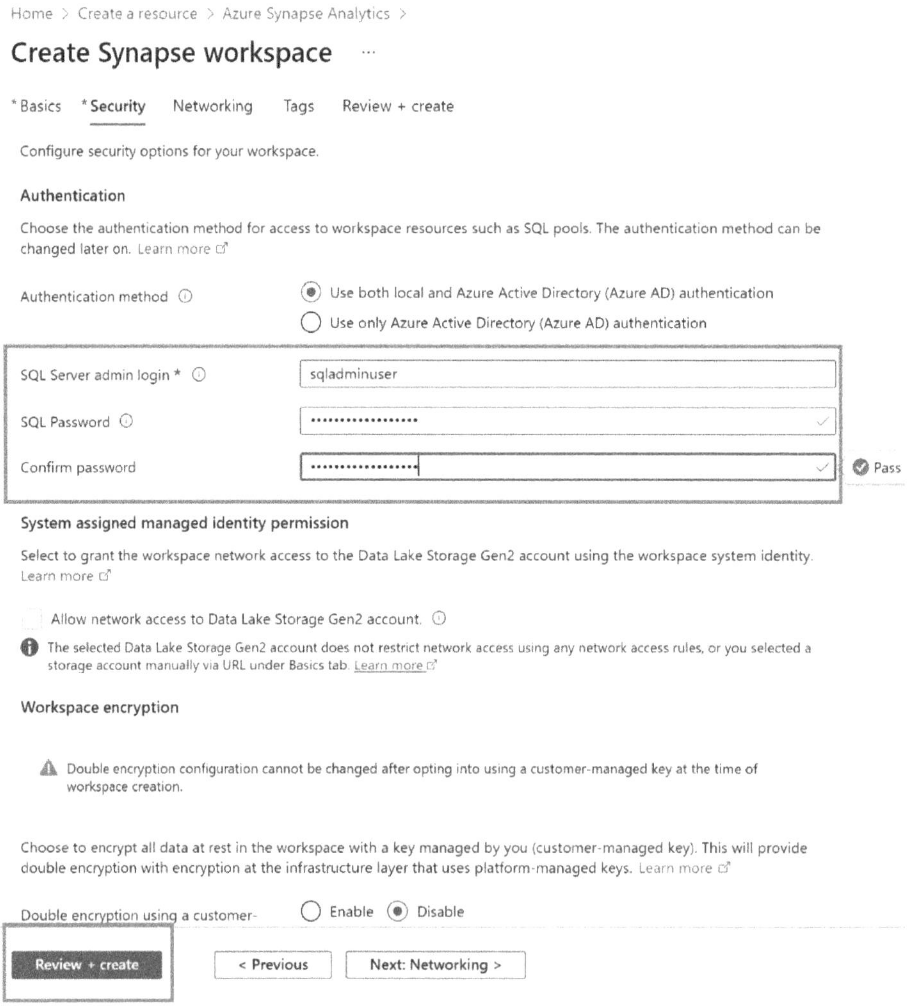

Figure 8.3 – Creating SQL pool in a Synapse Analytics workspace

The preceding steps will create a Synapse Analytic workspace and a storage account. In the subsequent recipes of this chapter, we will use the workspace to explore and process data using SQL pool and Spark pool.

Analyzing data using serverless SQL pool

Serverless SQL pool allows us to explore data using T-SQL commands in a Synapse Analytics workspace. The key advantage of serverless SQL pool is that it is available by default once a Synapse Analytics workspace is provisioned with no cluster or additional resources to be created. In serverless SQL pool, you will be charged only for the data processed by the queries as it is designed as a pure pay-per-use model.

Getting ready

Create a Synapse Analytics workspace, as explained in the *Provisioning an Azure Synapse Analytics workspace* recipe of this chapter.

How to do it...

In this recipe, we will perform the following:

- Uploading sample data into a Synapse Analytics workspace data lake account and querying it using serverless SQL pool

- Creating a serverless SQL database and defining a view to read the data stored in the data lake account

The detailed steps to perform these tasks are as follows:

1. Log in to `portal.azure.com`, go to **All resources**, and search for `packtadesynapse`, the Synapse Analytics workspace we created in the *Provisioning an Azure Synapse Analytics workspace* recipe. Click the workspace. Click **Open Synapse Studio**:

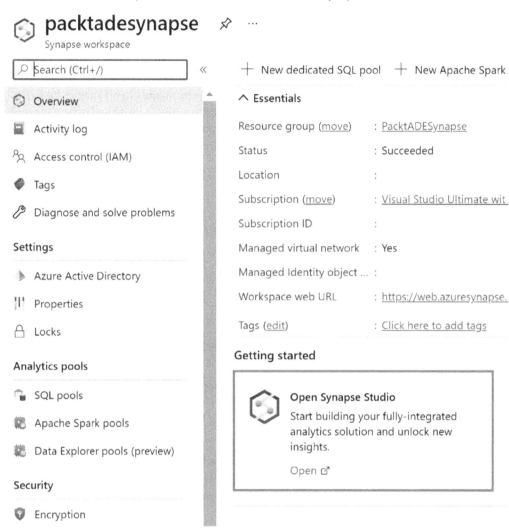

Figure 8.4 – Open Synapse Studio

2.   Click on the blue cylinder (the data symbol) on the left, which will take you to the **Data** section. Click the **Linked** tab. Expand the data lake account of the Synapse workspace (packtadesynapse for this example) and click on the **synapse (Primary)** container. Click the + **New folder** button and create a folder called CSV:

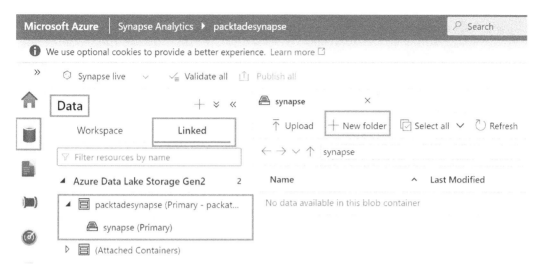

Figure 8.5 – Synapse studio folder creation

3. Download the `covid-data.csv` file from `https://github.com/PacktPublishing/ Azure-Data-Engineering-Cookbook-2nd-edition/blob/main/chapter07/ covid-data.csv` to your local machine.

4. Double click on the CSV folder in Synapse Studio. Click on the **Upload** button and upload the `covid-data.csv` file from your local machine into the data lake account of the Synapse workspace:

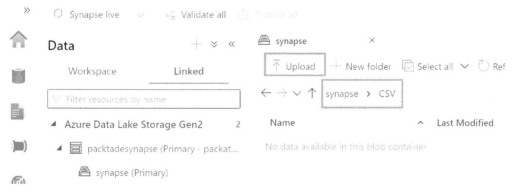

Figure 8.6 – Synapse Studio folder creation

5. After the file has uploaded, right-click the `covid-data.csv` file in the CSV folder, select **New SQL script**, and click the **Select TOP 100 rows** option:

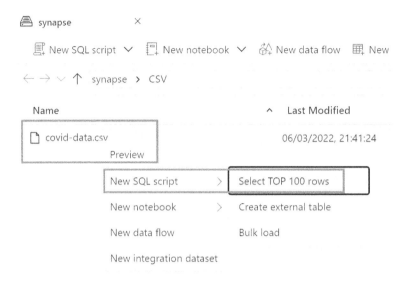

Figure 8.7 – Synapse Studio folder creation

6. A new query window will open with a script using the OPENROWSET command. Click the
    Run button to execute the query and preview the data:

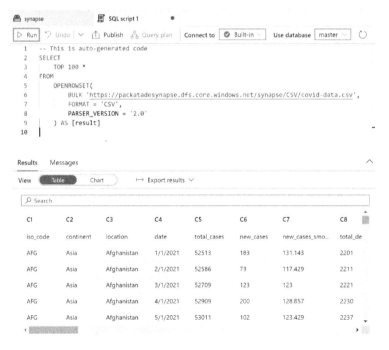

Figure 8.8 – Querying data using OPENROWSET

7. Let's perform the following. Let's create a **serverless database** and a **view** referencing the OPENROWSET command to read the covid-data.csv file. We notice that the actual column names (iso_code, continent, location, and date) are listed in the first row. These columns (iso_code, continent, location, and date) need to move up to become the table's column names and we need to remove the existing column names (C1, C2, C3, C4, and so on). We can fix that by adding the HEADER_ROW = TRUE option after the PARSER_VERSION option in the OPENROWSET command. Use the following command to create a database, create a view, and fix the header:

```
CREATE DATABASE serverless
GO
USE serverless
GO
CREATE VIEW covid AS
SELECT
    *
FROM
    OPENROWSET(
        BULK 'https://packatadesynapse.dfs.core.windows.
net/synapse/CSV/covid-data.csv',
        FORMAT = 'CSV',
    PARSER_VERSION = '2.0'
, HEADER_ROW = TRUE
        ) AS [result]
```

The result of the query execution is demonstrated here:

Figure 8.9 – Create view in serverless SQL

8. Click on the icon that looks like a notebook on the left-hand side of the screen. This will take you to the **Develop** section. Click the + button on top and click **SQL script**:

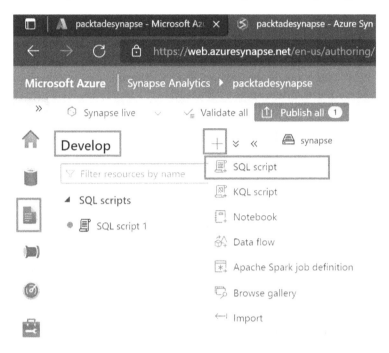

Figure 8.10 – A new SQL script

9. Use the following script, referencing the serverless database and the view created, to find a list of countries that have the maximum number of deaths per million people on a given day:

```
use serverless
GO
Select iso_code,location , continent,
max(isnull(new_deaths_per_million,0)) as death_sum,
max(isnull(people_fully_vaccinated,0) / isnull(popula-
tion,0)) * 100 as percentage_vaccinated From covid
where isnull(population,0) > 1000000
group by iso_code,location,continent
order by death_sum desc
```

The output of the preceding query is shown in the following screenshot:

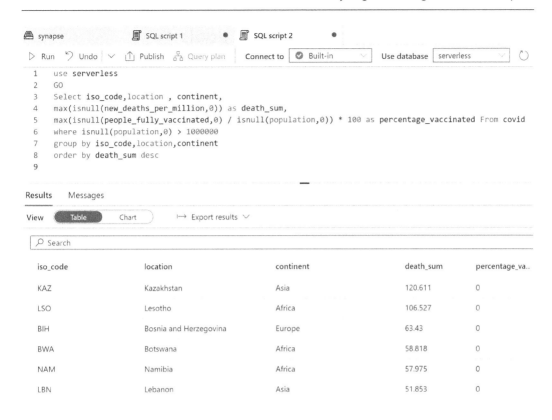

```
     synapse              SQL script 1    •        SQL script 2    •

   ▷ Run   Undo  ∨   ⏱ Publish   Query plan    Connect to   ✓ Built-in   ∨    Use database   serverless   ∨    ↻

   1    use serverless
   2    GO
   3    Select iso_code,location , continent,
   4    max(isnull(new_deaths_per_million,0)) as death_sum,
   5    max(isnull(people_fully_vaccinated,0) / isnull(population,0)) * 100 as percentage_vaccinated From covid
   6    where isnull(population,0) > 1000000
   7    group by iso_code,location,continent
   8    order by death_sum desc
   9
```

Results   Messages

View   (Table   Chart)   ⟼ Export results ∨

🔍 Search

iso_code	location	continent	death_sum	percentage_va...
KAZ	Kazakhstan	Asia	120.611	0
LSO	Lesotho	Africa	106.527	0
BIH	Bosnia and Herzegovina	Europe	63.43	0
BWA	Botswana	Africa	58.818	0
NAM	Namibia	Africa	57.975	0
LBN	Lebanon	Asia	51.853	0

Figure 8.11 – Querying the serverless view

How it works...

After we uploaded the covid-data.csv file to the Azure Data Lake Storage account associated with the Synapse workspace, we were able to query the data at the click of a button, without providing any credentials or provisioning any other resources. Serverless SQL pool, which is available by default in a Synapse workspace, allows us to interact with the data with minimal effort.

We used the OPENROWSET function to read the data from a CSV file and we encapsulated it inside a view for easier access in subsequent scripts. The serverless database and the view can also be accessed from other services such as Power BI and Data Factory. Serverless SQL pool, with its ability to define views, can be used to create a logical data warehouse on top of a data lake storage account, which will serve as a powerful tool for data analysis and exploration.

Provisioning and configuring Spark pools

A Spark pool is an important component of Azure Synapse Analytics that allows us to perform data exploration and processing using the Apache Spark engine. Spark pools in Azure Synapse Analytics allow us to process data using programming languages such as PySpark, Scala, C#, and Spark SQL. In this recipe, we will learn how to provision and configure Spark pools in Synapse Analytics.

Getting ready

Create a Synapse Analytics workspace, as explained in the *Provisioning an Azure Synapse Analytics workspace* recipe.

How to do it...

Let's perform the following steps to provision a Spark pool in an Azure Synapse Analytics workspace:

1. Log in to portal.azure.com and click **All Resources**. Search for packtadesynapse, the Synapse Analytics workspace created in the *Provisioning an Azure Synapse Analytics workspace* recipe. Click on the workspace. Search for **Apache Spark pools** under **Analytics pools**. Click + **New**:

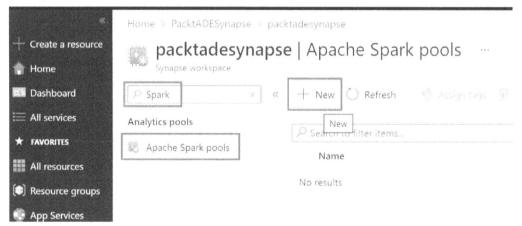

Figure 8.12 – Creating Spark pools

2. Fill in the details as follows:

    A. Name the Spark pool packtsparkpool.

    B. The **Node size family** property indicates the type of virtual machines that will be running to process the big data workload. Let's pick **Memory Optimized**, as it's typically good enough for general purpose data processing tasks. The **Hardware accelerated** type provides GPU-powered machines meant for performing heavy-duty data science and big data workloads.

C.   The **Node size** property indicates the compute and memory of virtual machines in a Spark pool. Let's pick **Small (4 vCores / 32 GB)**, as it's the cheapest option.

D.   **Autoscale** allows the Spark pool to allocate additional nodes or machines depending upon the workload. Let's leave it as **Enabled**.

E.   Set the minimum and maximum **Number of nodes** to **3** and **10**. A Spark pool can autoscale up to a maximum of 10 machines/nodes. Click on **Next: Additional settings >**:

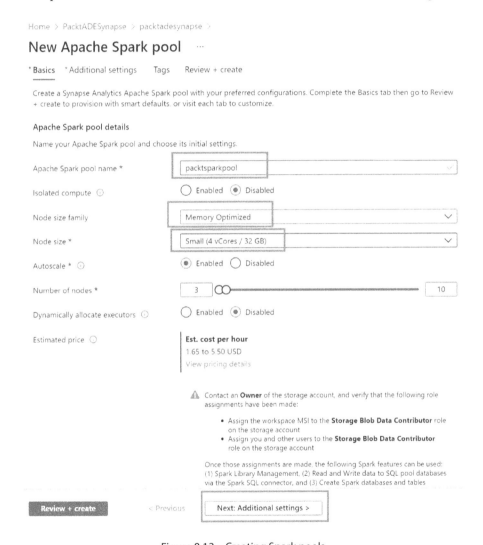

Figure 8.13 – Creating Spark pools

F.   Leave **Automatic pausing** as **Enabled**. Automatic pausing stops the Spark pool when there are no jobs being processed.

G. Set **Number of minutes idle** to **10**. This ensures that the pool is paused if no job is running for 10 minutes.

H. Set the **Apache Spark** version to the latest one (**3.1** here, as of March 2022), as it ensures that we get the latest Java, Scala, .NET, and Delta Lake versions. Click **Review + create**:

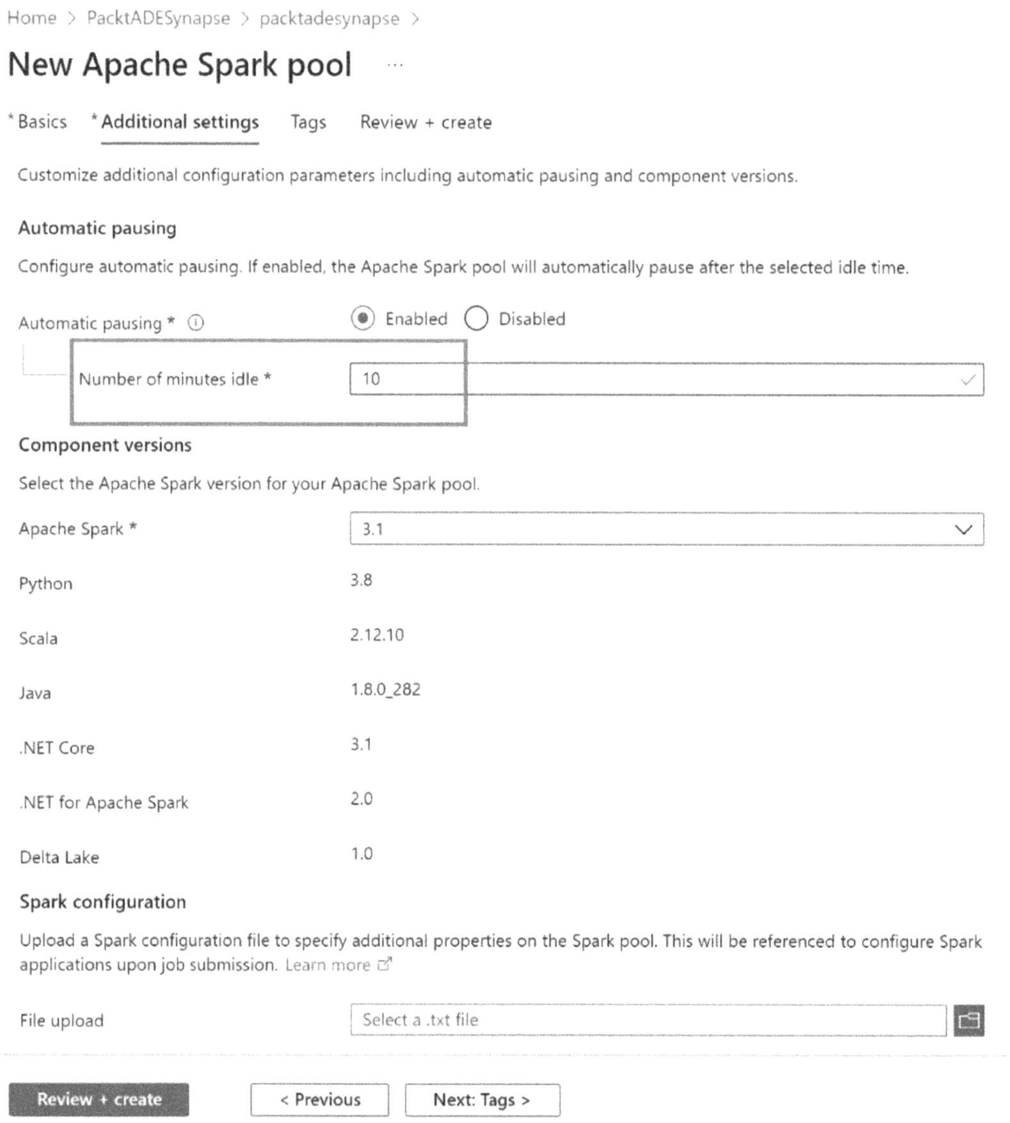

Home > PacktADESynapse > packtadesynapse >

New Apache Spark pool   ···

*Basics   *Additional settings   Tags   Review + create

Customize additional configuration parameters including automatic pausing and component versions.

Automatic pausing

Configure automatic pausing. If enabled, the Apache Spark pool will automatically pause after the selected idle time.

Automatic pausing * ⓘ   ◉ Enabled   ○ Disabled

Number of minutes idle *   10   ✓

Component versions

Select the Apache Spark version for your Apache Spark pool.

Apache Spark *   3.1   ⌄

Python   3.8

Scala   2.12.10

Java   1.8.0_282

.NET Core   3.1

.NET for Apache Spark   2.0

Delta Lake   1.0

Spark configuration

Upload a Spark configuration file to specify additional properties on the Spark pool. This will be referenced to configure Spark applications upon job submission. Learn more ☐

File upload   Select a .txt file   ☐

Review + create   < Previous   Next: Tags >

Figure 8.14 – Configuring Spark pools

How it works...

Creating a Spark pool defines the configuration for the nodes/virtual machines, which will be processing the big data workload as it arrives. Each time a new user logs in and submits a Spark job/ Spark notebook to process data, an instance of Spark pool is created. An instance of Spark pool is basically a bunch of virtual machines/nodes configured as defined in the Spark pool configuration. A single user can use up to the maximum number of nodes defined in the pool. If there are multiple users connecting to the Spark pool, Synapse will create as many instances of Spark pool as the number of users. You will be billed for the number of active instances and for the time period during which they were active. Billing will stop for a particular instance if it remains idle longer than the idle time period defined in the Spark pool configuration.

Processing data using Spark pools and a lake database

Spark pools in a Synapse workspace allow us to process data and store them as tables inside a lake database. A lake database allows us to create tables using CSV files, Parquet files, or as Delta tables stored in the data lake account. Delta tables use Parquet files for storage and support `insert`, `update`, `delete`, and `merge` operations. Delta tables are stored in a columnar format, which is compressed, ideal for storing processed data and supporting analytic workloads. In this recipe, we will read a CSV file, perform basic processing, and load the data into a Delta table in a lake database.

Getting ready

Create a Synapse Analytics workspace, as explained in the *Provisioning an Azure Synapse Analytics workspace* recipe.

Create a Spark pool cluster, as explained in the *Provisioning and configuring Spark pools* recipe.

We need to upload the `covid-data.csv` file from `https://github.com/PacktPublishing/Azure-Data-Engineering-Cookbook-2nd-edition/blob/main/chapter07/covid-data.csv` to a folder named CSV in the data lake account attached to the Synapse Analytics workspace. To do so, follow *step 1* to *step 4* in the *How to do it...* section from the *Analyzing data using serverless SQL pool* recipe.

How to do it...

After uploading the `covid-data.csv` file to the CSV folder in the data lake account, let's perform the following steps to process the data in the CSV file and load it into a Delta Lake table in a lake database:

1. Log in to `portal.azure.com`, click **All resources**, search for `packtadesynapse`, the Synapse Analytics workspace we created, and click on it. Click **Open Synapse Studio**.

2. Click on the data icon on the left, click the **Linked** tab, and expand the **Azure Data Lake Storage Gen2** | **packtadesynapse** | **synapse (Primary)** container. Navigate to the **CSV** folder inside the **synapse** container, where the `covid-data.csv` file has been uploaded. Right-click on the `covid-data.csv` file, select the option **New notebook** | **Load to DataFrame**:

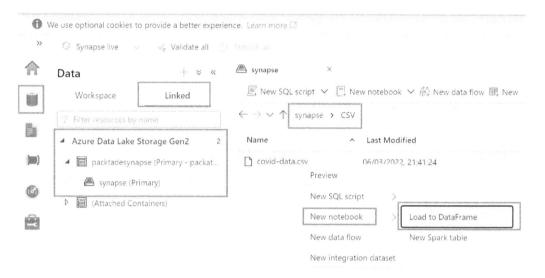

Figure 8.15 – Load to DataFrame

3. A new notebook will be created. Notebooks are used by data engineers to develop the code that will be used to process data using Synapse pools. Attach the notebook to **packtsparkpool**, the Synapse Spark pool created in the *Provisioning and configuring Spark pools* recipe in this chapter, using the **Attach to** drop-down menu at the top of the notebook. Uncomment the fourth line in the first cell by removing ##, as our file contains a header. Hit **Run Cell** (which looks like a play button) on the left:

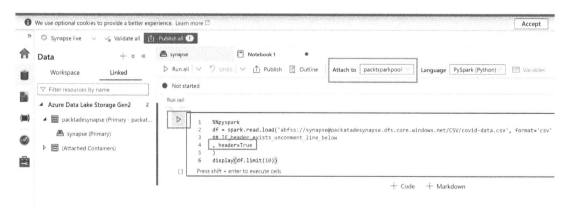

Figure 8.16 – The data loaded to the DataFrame

4. Data will be loaded to a DataFrame called `df` using the automatic PySpark code generated. Let's use the Spark SQL language to understand the data. To interact with the data using Spark SQL, we need to create a temporary view. The `createOrReplaceTempView` command helps to create a temporary view that will be visible only within the notebook. Add a new cell by hitting the **+ Code** button, paste the following command, and run the new cell:

```
df.createOrReplaceTempView("v1")
```

The output is displayed in the following screenshot:

Figure 8.17 – Creating a temporary view

5. To check out which columns are present in the `covid-data.csv` file, let's use the `Describe` Spark SQL command to list the columns in the view. Add a new cell and paste the following command. The `%%sql` command switches the programming language from PySpark to Spark SQL. We will notice that the view contains several columns (use the scroll bar to check out all of them). All the columns are also of the `string` data type:

```
%%sql
Describe v1;
```

The output is as follows:

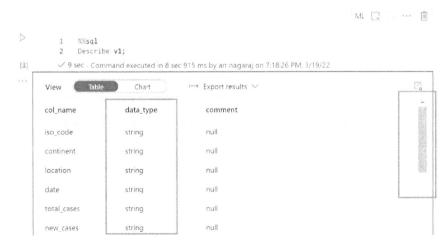

Figure 8.18 – The list columns

6. Let's focus on the following key columns – date, continent, location, new_cases, and new_deaths. Let's also change the data type of new_cases and new_deaths to integer and load it into a Delta table. To load it into a Delta table, we need to create a lake database first, create the new Delta table, and then load the data. The Create database command creates the database, and Create table <tablename> using Delta as <Select statement> creates the table and loads the data. Copy the following command to a new cell and run the new cell:

```
%%sql
Create database sparksqldb;
Create or replace table sparksqldb.covid
USING Delta
AS
Select date, continent,location, CAST(new_cases as int)
as new_cases,
CAST(new_deaths as int) as new_deaths from v1
```

The output of the preceding query is as follows: 

```
1    %%sql
2    Create database sparksqldb;
3    Create or replace table sparksqldb.covid
4    USING Delta
5    AS
6    Select date, continent,location, CAST(new_cases as int) as new_cases,
7    CAST(new_deaths as int) as new_deaths from v1
```

[4]   ✓ 24 sec - Command executed in 1 ms by arr.nagaraj on 7:34:07 PM, 3/19/22

⋯   No data available

Figure 8.19 – Creating a Delta table

7. In the previous step, we created a lake database called sparksqldb and a Delta table inside it called covid. Using the delta option in the CREATE or REPLACE TABLE command ensured that the table was created as a Delta table. The CAST function in the SELECT statement changed the column data type to INTEGER. Verify the data-type change using the DESCRIBE command:

```
%%sql
Describe table sparksqldb.covid;
```

The result of the query execution is demonstrated in the following screenshot:

Figure 8.20 – The Delta table structure

8. Let's delete the rows that have NULL values in the continent column. Add a new cell and copy-paste the following command:

```
%%sql
Delete from sparksqldb.covid where continent is NULL
```

The output of the preceding query is as follows:

Figure 8.21 – Using a Delete statement on a Delta table

9. Delta tables have a feature called time travel, which lets us explore the previous versions of the table. We will use time travel to query the deleted rows (rows with NULL values in the continent column). To perform that as a first step, we need to find the location where the Delta table is stored. The Describe detail command will provide a column called location, which will contain the location of the Delta table. Add a new cell, copy the following command, and run the cell. Copy the contents of the location column. Ensure to expand the column by dragging the slider to your right and copying the full path of the Delta table:

```sql
%%sql
DESCRIBE DETAIL sparksqldb.covid
```

The output of the preceding query is as follows:

Figure 8.22 – Get the Delta table location

10. On the copied location, remove the text that starts with abfss and goes up to windows. net. We only need the path that starts from the container name (synapse), not the storage account or protocol details. For example, if your copied location is abfss://synapse@ packatadesynapse.dfs.core.windows.net/synapse/workspaces/ packtadesynapse/warehouse/sparksqldb.db/covid, remove abfss:// synapse@packatadesynapse.dfs.core.windows.net/ and retain /synapse/ workspaces/packtadesynapse/warehouse/sparksqldb.db/covid.

11. Add a new cell and copy the following Spark command. Paste the edited location path to the load function. The Option("versionAsOf",0) function makes the command read the older version of the table. The second parameter in the option function indicates the version number to be read. Version number 0 would be the most recent previous version of the table, version number 1 would be the next version older than version 0, and so on. The command reads the older data to a DataFrame, which we load to a view called old_Data:

```
df2 = spark.read.format("delta").option("versionAsOf",
0).load("/synapse/workspaces/packtadesynapse/warehouse/
sparksqldb.db/covid")
df2.createOrReplaceTempView("old_Data")
```

The result of the query execution is demonstrated in the following screenshot:

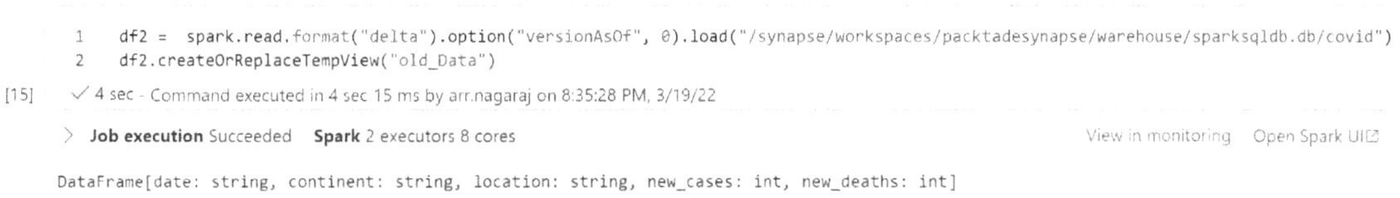

1 df2 = spark.read.format("delta").option("versionAsOf", 0).load("/synapse/workspaces/packtadesynapse/warehouse/sparksqldb.db/covid")
2 df2.createOrReplaceTempView("old_Data")
[15] ✓ 4 sec - Command executed in 4 sec 15 ms by arr.nagaraj on 8:35:28 PM, 3/19/22
Job execution Succeeded Spark 2 executors 8 cores
View in monitoring Open Spark UI
DataFrame[date: string, continent: string, location: string, new_cases: int, new_deaths: int]

Figure 8.23 – Reading an old version of the Delta table

12. Execute a `SELECT` statement against the `old_Data` view to check out the rows that were deleted. Add a new cell, copy the following command, and execute the new cell. We will be able to read the deleted rows using the time travel feature on the Delta table:

```
%%sql
SELECT * FROM old_Data WHERE continent IS NULL
```

The result of the query execution is demonstrated in the following screenshot:

1 %%sql
2 Select * from old_Data where continent is NULL
✓ 6 sec - Command executed in 5 sec 801 ms by arr.nagaraj on 9:25:55 PM, 3/19/22
Job execution Succeeded Spark 2 executors 8 cores
View in monitoring Open Spark UI
View Table Chart Export results
date continent location new_cases new_deaths
1/1/2021 null South America 45902 1080
2/1/2021 null South America 38823 875
3/1/2021 null South America 39551 892
4/1/2021 null South America 50539 1199

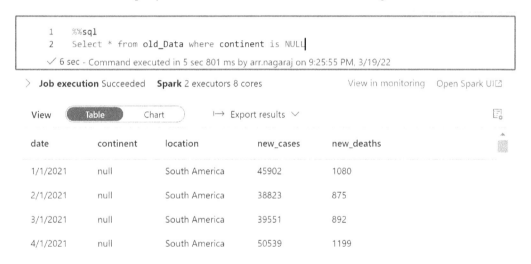

Figure 8.24 – SELECT statement on Delta table

How it works...

The Spark pools in a Synapse workspace allow us to seamlessly load CSV files to Delta tables using notebooks. Notebooks allow us to effortlessly switch between PySpark and SQL. Delta tables support data manipulation commands such as `update`, `delete`, and `merge`, and capabilities such as time travel make it very efficient for data processing tasks in data engineering projects.

Querying the data in a lake database from serverless SQL pool

Lake databases are created from Synapse Spark pools and typically consist of Delta tables. The following recipe will showcase how we could read the data stored in Delta tables from serverless SQL pool.

Getting ready

Create a Synapse Analytics workspace, as explained in the *Provisioning an Azure Synapse Analytics workspace* recipe in this chapter.

Create a Spark pool, as explained in the *Provisioning and configuring Spark pools* recipe in this chapter.

Create a lake database and Delta table, as explained in the *Processing data using Spark pools and lake database* recipe in this chapter.

How to do it...

Perform the following steps to query the data:

1. Log in to portal.azure.com, click **All Resources**, search for **packtadesynapse**, the Synapse Analytics workspace that we created, and click on it. Click **Open Synapse Studio**. Click on the data icon on the left, click the **Linked** tab, and expand the **Azure Data Lake Storage Gen2 | packtadesynapse | synapse (Primary)** container. Navigate to the following folder path: /synapse/ workspaces/packtadesynapse/warehouse/sparksqldb.db. The usual path structure is <SynapseContainerName>/synapse/workspaces/<WorkspaceName>/ warehouse/<lakedatabasename.db>. So, if you have named your lake database or table name differently, then it will vary from mine here. You will find a folder with the Delta table name (**covid**). Right-click on it and select **New SQL script | Create external table**:

Figure 8.25 – Create external table

2. Synapse will detect the file type and schema. Hit **Continue** to generate the external table creation script:

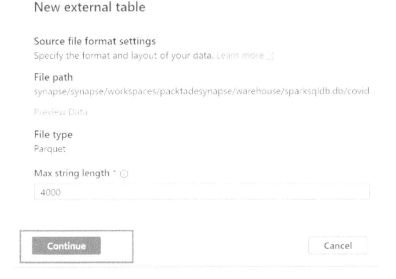

Figure 8.26 – Generation of the external table script

3. Leave the **Select SQL pool** option as **Built-in**. **Built-in** is the in-built serverless SQL pool. Select **+ New** to create a new serverless database:

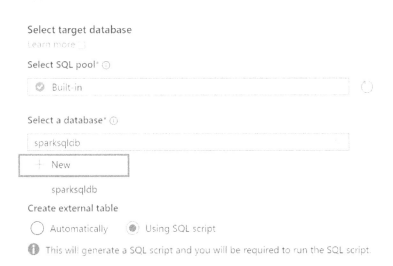

Figure 8.27 – A new serverless database

4. Name the database `ServerlessSQLdb` and click the **Create** button:

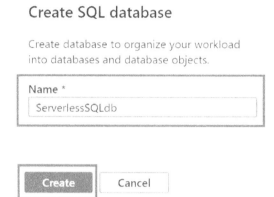

Figure 8.28 – Create a serverless database

5. Name the external table `dbo.covid_ext`. Click **Open script**:

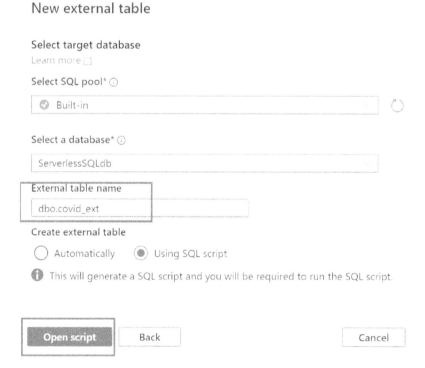

Figure 8.29 – Creating an external table

6. To create an external table in Synapse SQL pool, we need to create the following objects: an **external file format** and an **external data source** first. Using Synapse Studio, we can generate the script to create the **external file format**, **external data source**, and **external table**. Select **ServerlessSQLdb** from the **Use database** dropdown and click the **Run** button to create the external table:

```
1   IF NOT EXISTS (SELECT * FROM sys.external_file_formats WHERE name = 'SynapseParquetFormat')
2       CREATE EXTERNAL FILE FORMAT [SynapseParquetFormat]
3       WITH ( FORMAT_TYPE = PARQUET)
4   GO
5
6   IF NOT EXISTS (SELECT * FROM sys.external_data_sources WHERE name = 'synapse_packatadesynapse_dfs_core_windows_net')
7       CREATE EXTERNAL DATA SOURCE [synapse_packatadesynapse_dfs_core_windows_net]
8       WITH (
9           LOCATION = 'abfss://synapse@packatadesynapse.dfs.core.windows.net'
10      )
11  GO
12
13  CREATE EXTERNAL TABLE covid_ext (
14      [date] nvarchar(4000),
15      [continent] nvarchar(4000),
16      [location] nvarchar(4000),
17      [new_cases] int,
18      [new_deaths] int
19      )
20      WITH (
21      LOCATION = 'synapse/workspaces/packtadesynapse/warehouse/sparksqldb.db/covid',
22      DATA_SOURCE = [synapse_packatadesynapse_dfs_core_windows_net],
23      FILE_FORMAT = [SynapseParquetFormat]
24      )
25  GO
26
27
28  SELECT TOP 100 * FROM dbo.covid_ext
29  GO
```

Figure 8.30 – Creation of the external table

7. Upon clicking the **Run** button, we will be able to see the data read from the external table successfully:

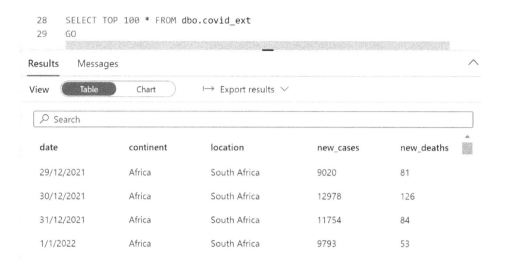

```
28   SELECT TOP 100 * FROM dbo.covid_ext
29   GO
```

date	continent	location	new_cases	new_deaths
29/12/2021	Africa	South Africa	9020	81
30/12/2021	Africa	South Africa	12978	126
31/12/2021	Africa	South Africa	11754	84
1/1/2022	Africa	South Africa	9793	53

Figure 8.31 – Reading the external table

How it works...

To access the Delta table in a lake database, we need to create an external table in serverless SQL pool. We identified the folder where the Delta table was stored and we created an external table against it in serverless SQL pool. The external table acts as a link to the files stored in the Delta table. While the files reside in the Delta table, files appear as a table to end users in Serverless SQL pool. So, when a user queries the external table using a T-SQL script in serverless SQL pool, it will seamlessly read from the Delta table's files and present it in a tabular format.

The lake database created a folder for each Delta table. So, it made it easier for us to create an external table against the folder of the Delta Lake table, which implies that we can seamlessly query the data from lake database Delta table in a serverless SQL pool database via external table. Changes to the Delta table are handled by adding or removing Parquet files inside the Delta Lake table folder using the Apache Spark engine. As we have created the external table against the table's folder (not against any specific file), all the changes happening in the Delta Lake table will immediately be reflected in the serverless SQL pool's external table without any additional effort.

Scheduling notebooks to process data incrementally

Consider the following scenario. Data is loaded daily into the data lake in the form of CSV files. The task is to create a scheduled batch job that processes the files loaded daily, performs basic checks, and loads the data into the Delta table in the lake database. This recipe addresses this scenario by covering the following tasks:

1. Only reading the new CSV files that are loaded to the data lake daily using Spark pools and notebooks

2. Processing and performing `upserts` (update if the row exists, insert if it doesn't), and loading data into the Delta lake table using notebooks

3. Scheduling the notebook to operationalize the solution

Getting ready

Create a Synapse Analytics workspace, as explained in the *Provisioning an Azure Synapse Analytics workspace* recipe in this chapter.

Create a Spark pool, as explained in the *Provisioning and configuring Spark pools* recipe in this chapter.

Download the `TransDtls-2022-03-20.csv` file from `https://github.com/PacktPublishing/Azure-Data-Engineering-Cookbook-2nd-edition/blob/main/chapter08/TransDtls-2022-03-20.csv`. Create a folder called `transaction-data` inside the **synapse** container in the **packtadesynapse** data lake account. You can use Synapse Studio's data pane to do the same. For detailed instructions on creating a folder and manually uploading files to Synapse Studio, refer to *step 1* to *step 4* in the *How to do it…* section of the *Analyzing data using serverless SQL Pool* recipe. Upload the file, as shown in the following picture:

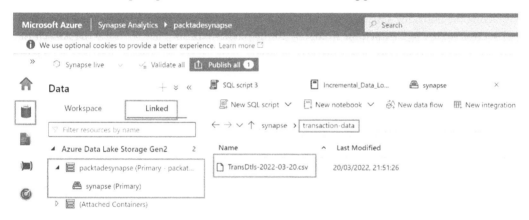

Figure 8.32 – Uploading the file

How to do it...

In this scenario, the `TransDtls-2022-03-20.csv` file contains the data about transactions that have occurred in a store. Let's assume that the file has the loading date suffixed to it. So, a file that was loaded to the data lake on March 20th will be named `TransDtls-2022-03-20.csv`, a load from March 21st will be named `TransDtls-2022-03-21.csv`, and so on. Our notebook and scheduled task should read only the latest file (and not all the files in the `transaction-data` folder), so that it can process the data incrementally. To process the data incrementally and load the data to a Delta Lake, let's perform the following steps:

1. Click the **Develop** icon on the left-hand side of Synapse Studio. In the **Notebooks** section, click on the three dots and select **New notebook**:

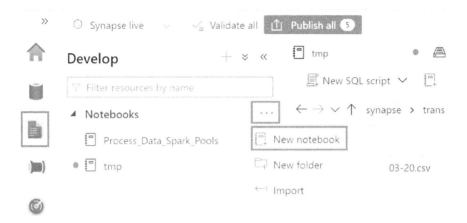

Figure 8.33 – Select New notebook

2. Name the notebook `Incremental_Data_Load` by typing the name into the **Properties** section on the left. Attach the notebook to the **packtsparkpool** cluster using the **Attach to** drop-down option at the top. Copy the following Scala script and paste it into the first cell in the notebook to only read the latest file from the **transaction-data** folder into a DataFrame. The `Java.time.localDate.now` command gets the current date and we use the current date to construct the name of the file to be read. This way, even if there are hundreds of files in the folder, the notebook will only read the latest file. Hit the **Run** button (which looks like a play button) on the left. The latest file is loaded to the DataFrame and named `transaction_today`:

```
%%spark
val date = java.time.LocalDate.now
val transaction_today = spark.read.format("csv").
option("header", "true").option("inferSchema", "true").
load("/transaction-data/TransDtls-" + date +".csv")
display(transaction_today)
```

The output of the preceding query is displayed in the following screenshot:

Figure 8.34 – Reading the latest file

3. Add a new cell using the + **Code** button and copy-paste the command to create a temporary view using the DataFrame. Hit the **Run** button:

```
%%spark
transaction_today.createOrReplaceTempView("transaction_
today")
```

The output of the query execution is demonstrated in the following screenshot:

Figure 8.35 – Reading the DataFrame

4. Add a new cell and use the following SQL script to create a lake database called `DataLoad` and a Delta table called `transaction_Data`. The `If not exists` clause in the `create` statements ensures that the database and the table are only created the first time that the notebook is run. Hit the **Run** button to create the database and table:

```
%%sql
CREATE DATABASE IF NOT EXISTS DataLoad;
CREATE TABLE IF NOT EXISTS DataLoad.transaction_
data(transaction_id int, order_id int, Order_
dt Date,customer_id varchar(100),product_id
varchar(100),quantity int,cost int)
USING DELTA
```

The output is as follows:

```
+ Code    + Markdown                                        M↓  ⌧  ⌣  ···  🗑

▷    1   %%sql
     2   CREATE DATABASE IF NOT EXISTS DataLoad;
     3   CREATE TABLE IF NOT EXISTS DataLoad.transaction_data(transaction_id int, order_id int, Order_dt Date,custom
     4   USING DELTA
[4]  ✓ 23 sec · Command executed in 1 ms by arr.nagaraj on 11:21:54 PM, 3/20/22
···  No data available
```

Figure 8.36 – Table creation

5. Add a new cell and copy the following script to **upsert** data into the table. If the latest file in the transaction-data folder contains information about transactions that already exist in the Delta table, then the Delta table needs to be updated with the values from the latest file, and if the latest file contains transactions that don't exist in the Delta table, then they need to be inserted. The script uses the merge command, which performs the following tasks to achieve the update/insert commands:

A. Compares transaction_id on the latest file in the transaction-data folder (using the transaction_today view's transaction_id column) with the transaction_data table's (as in, the Delta table's) transaction_id column to see whether the file contains data about older transactions or new transactions. The comparison is carried out using the merge statement's ON clause.

B. If transaction_id from the file already exists in the transaction_data table, it implies that the file contains rows about older transactions, and hence, it updates all the columns of the transaction_data table with the latest data from the file. The WHEN Matched clause in the merge statement helps to achieve this.

C. If transaction_id doesn't exist in the Delta table but it does exist in the file, it is a new transaction and it is therefore inserted into the table. The WHEN NOT Matched clause in the merge statement helps to achieve this.

D. On the WHEN Matched clause, additional NULL checks are added, using the is not null clause to ensure that invalid rows are not inserted into the table:

```
%%sql
Merge into DataLoad.transaction_data source
Using transaction_today target on source.transaction_id =
target.transaction_id
WHEN MATCHED THEN UPDATE SET *
WHEN NOT MATCHED AND (target.transaction_id is not null
or target.order_id is not null or target.customer_id is
not null)
THEN INSERT *
```

The output of the query is as follows:

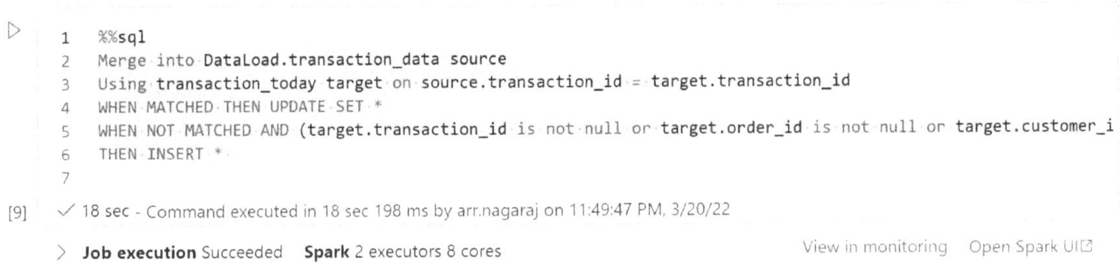

```sql
1    %%sql
2    Merge into DataLoad.transaction_data source
3    Using transaction_today target on source.transaction_id = target.transaction_id
4    WHEN MATCHED THEN UPDATE SET *
5    WHEN NOT MATCHED AND (target.transaction_id is not null or target.order_id is not null or target.customer_i
6    THEN INSERT *
7
```

[9]   ✓ 18 sec - Command executed in 18 sec 198 ms by arr.nagaraj on 11:49:47 PM, 3/20/22

> **Job execution** Succeeded   **Spark** 2 executors 8 cores   View in monitoring   Open Spark UI

Figure 8.37 – A merge statement to upsert data

6. Hit the **Publish** button at the top to save the notebook:

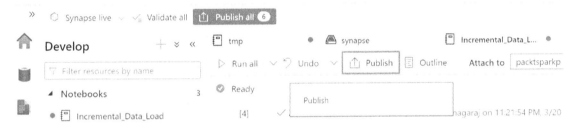

Figure 8.38 – Publishing the notebook

7. To schedule the notebook to run daily, we need to add the notebook to a pipeline. Hit the add to pipeline button in the top-right corner and select **New pipeline**:

Figure 8.39 – Adding a notebook to a pipeline

8. Name the pipeline `Incremental_Data_Load`. Publish it by clicking the **Publish** button. Click **Add trigger** and select **New/Edit**:

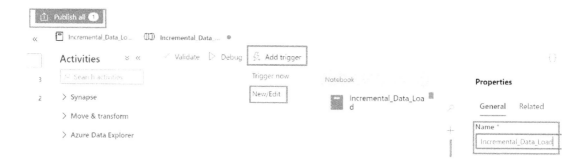

Figure 8.40 – Adding a schedule to the notebook

9. Click **+ New** under **Add triggers**:

Figure 8.41 – Adding a new trigger

Name the trigger Incremental_Data_Load. Under **Recurrence**, set the schedule to run every 1 day. Under **Execute at these times**, type in 9 for **Hours** and 0 for **Minutes**. Click **OK** to schedule

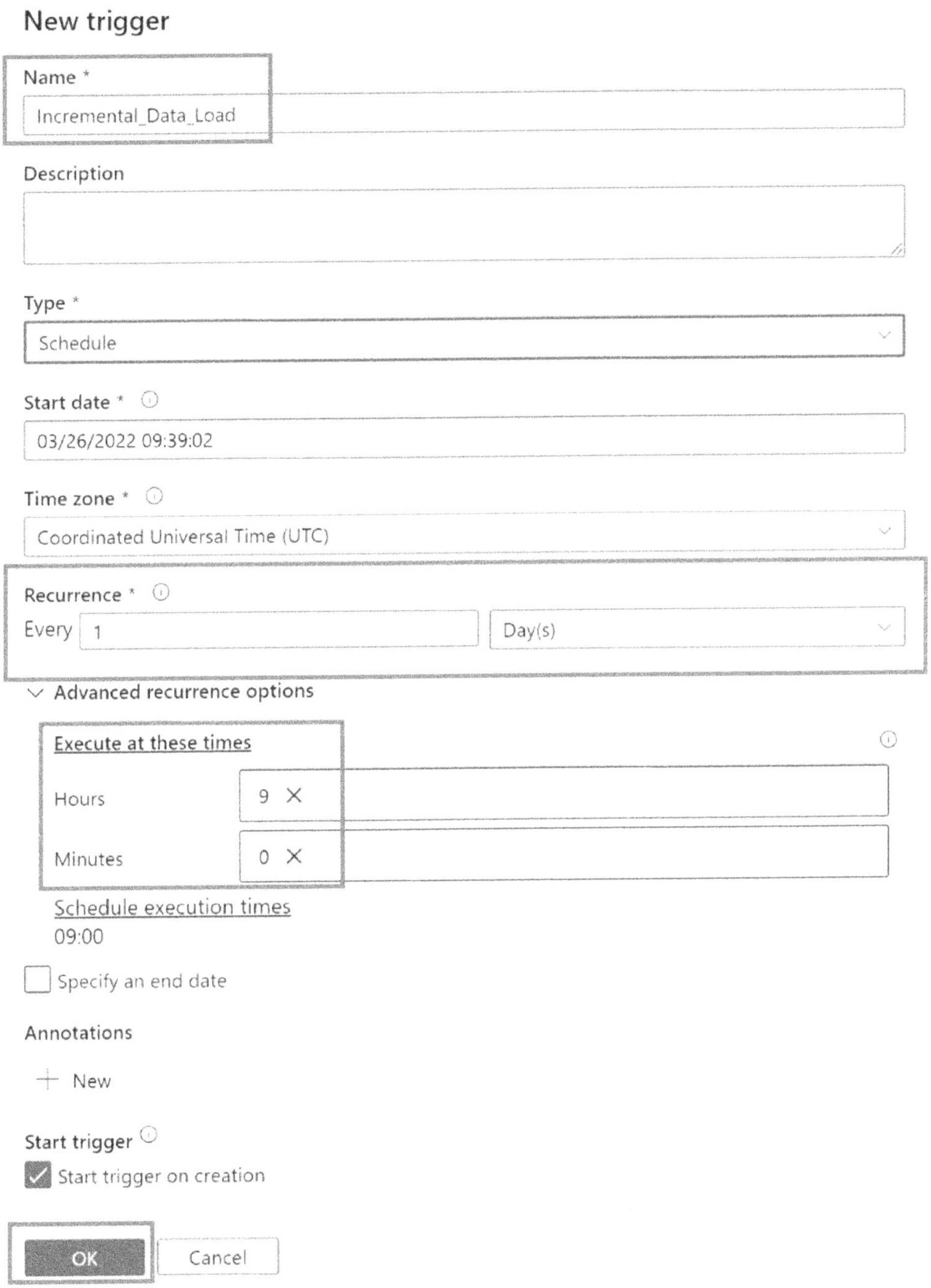

New trigger

Name *

Incremental_Data_Load

Description

Type *

Schedule

Start date * ⓘ

03/26/2022 09:39:02

Time zone * ⓘ

Coordinated Universal Time (UTC)

Recurrence * ⓘ

Every   1   Day(s)

∨ Advanced recurrence options

Execute at these times   ⓘ

Hours   9 ✕

Minutes   0 ✕

Schedule execution times
09:00

☐ Specify an end date

Annotations

+ New

Start trigger ⓘ

☑ Start trigger on creation

OK   Cancel

Figure 8.42 – Adding a trigger

10. There are no parameters to be passed. Hit **OK** and proceed:

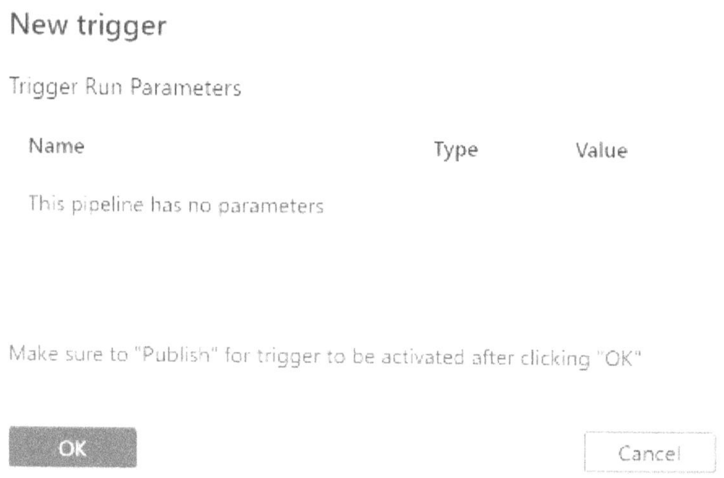

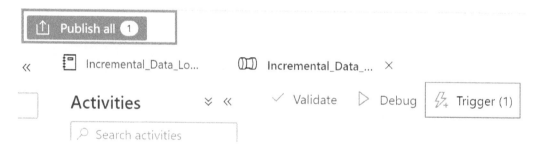

New trigger

Trigger Run Parameters

Name		Type	Value
This pipeline has no parameters			

Make sure to "Publish" for trigger to be activated after clicking "OK"

OK   Cancel

Figure 8.43 – Adding trigger run parameters

11. Hit the **Publish all** button to finish scheduling the notebook via a pipeline for a daily run:

Figure 8.44 – Publishing the trigger

How it works...

Processing data incrementally is a common scenario within data engineering projects. Synapse notebooks are extremely powerful and can be used to identify the new files to be loaded, process them alone, and load them into a Delta table. The MERGE statement is very effective at identifying the new or old transaction records and performing insert/update on the Delta table accordingly. The notebook was added to a pipeline at the click of a button and the pipeline was scheduled to run daily to process the files that are loaded every day. The processed Delta table, which is the outcome of this recipe, is typically consumed by a reporting application such as Power BI to get insights out of the processed data.

Visualizing data using Power BI by connecting to serverless SQL pool

Power BI is an excellent data visualization tool and is often used to consume the processed data in Synapse. Power BI can connect to objects (views or external tables, for example) in serverless SQL pool. In this recipe, we will create a Power BI report that will connect to an external table defined in serverless SQL pool.

Getting ready

Create a Synapse Analytics workspace, as explained in the *Provisioning an Azure Synapse Analytics workspace* recipe of this chapter.

Create a Spark pool, as explained in the *Provisioning and configuring Spark pools* recipe of this chapter.

Create a lake database and Delta table, as explained in the *Processing data using Spark pools and lake database* recipe in this chapter.

Create an external table in serverless SQL pool, as described in the *Querying the data in a lake database from serverless SQL pool* recipe.

Download the latest version of Power BI Desktop from `https://powerbi.microsoft.com/ en-us/downloads/` and install Power BI Desktop on your machine.

How to do it...

Perform the following steps to connect a Delta Lake table to a Power BI report via serverless SQL pool:

1. Open Power BI Desktop and click the **Get data** button:

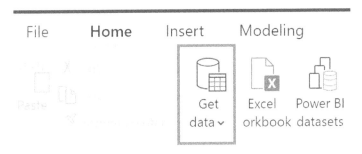

Figure 8.45 – Get data

2. Search for `Synapse`, select **Azure Synapse Analytics SQL**, and click the **Connect** button:

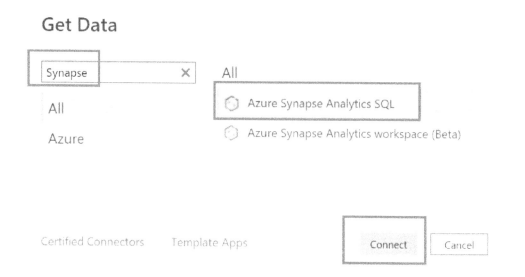

Figure 8.46 – Connecting data

3. Log in to `portal.azure.com`, click **All Resources**, search for `packtadesynapse` (the Synapse Analytics workspace that we created), and click on it. On the **Overview** page, copy the **Serverless SQL endpoint**:

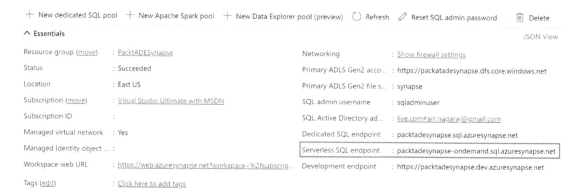

Figure 8.47 – The Serverless SQL endpoint

4. Paste the copied serverless SQL endpoint into the Power BI connection details prompt and click **OK**:

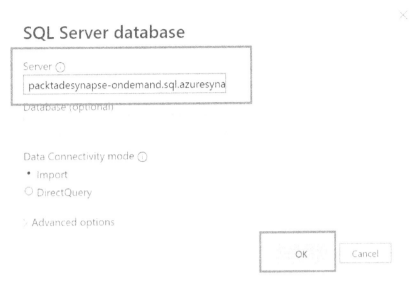

Figure 8.48 – Connect to Power BI

5. Select **Microsoft account** as the connection option. Sign in using the same account that you
    used to connect to `portal.azure.com`:

Figure 8.49 – Connect to Synapse

6. After signing in, click the **Connect** button:

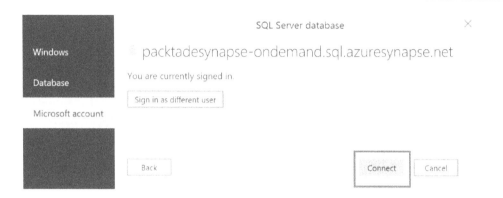

Figure 8.50 – Connect to Synapse

7. Expand **ServerlessSQLdb** and select **covid_ext**, the external table that we created in the *Querying the data in a lake database from serverless SQL pool* recipe. Click **Load**:

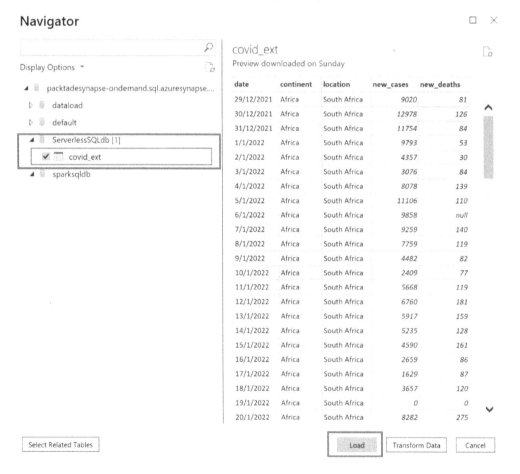

Figure 8.51 – Loading the Power BI data

8. Select the `location` column and the `new_deaths` column from the **covid_ext** table. Select the map visual icon. Place the `location` column under the **Location** property of the map visual and `new_deaths` as the **Size** property of the map visual. We are able to visualize the data effectively as follows:

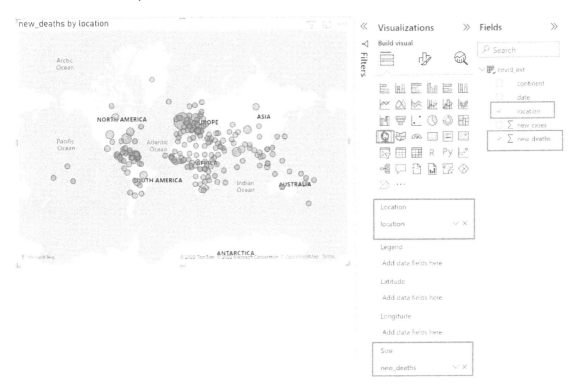

Figure 8.52 – Visualizing the data from Power BI

9. Go to the **File** menu and select the **Save** option to save the report.

How it works...

The Power BI report uses the Azure Active Directory account's context to connect to the Synapse serverless SQL pool. The serverless SQL pool's external table, which the Power BI report reads, is connected to a Delta table created in a lake database, and hence, we are able to visualize the data processed in the lake database in Power BI via a serverless SQL pool.

Transforming Data Using Azure Synapse Dataflows

As introduced in *Chapter 8*, *Processing Data Using Azure Synapse Analytics*, Azure Synapse Analytics comprises three key components – a Synapse integration pipeline to ingest and transform the data, a SQL pool and a Spark pool to process and serve the data, and Power BI integration to visualize the data. This chapter will focus on performing data transformations via Synapse data flows, a critical component of Synapse integration pipelines. Azure Synapse data flows allow us to perform transformations on the data such as converting data format, filtering, merging, and sorting while moving the data from the source to the destination. The best part about Synapse data flows is the user-friendly interface that allows us to perform complex data transformations in a few clicks.

In this recipe, we will be performing the following:

- Copying data using a Synapse data flow
- Performing data transformation using activities such as join, sort, and filter
- Monitoring data flows and pipelines
- Configuring partitions to optimize data flows
- Parameterizing mapping data flows
- Handling schema changes dynamically in data flows using schema drift

By the end of the chapter, you will have learned how to copy data to Parquet files using data flows, perform data transformation using data flows, build dynamic and resilient data flows using parameterization and schema drifting, and monitor data flows and pipelines.

Technical requirements

For this chapter, you will need the following:

- A Microsoft Azure subscription

Copying data using a Synapse data flow

In this recipe, we will convert a CSV file into the Parquet format using a Synapse data flow. We will be performing the following tasks to achieve this:

- Provisioning a Synapse Analytics workspace and a Synapse integration pipeline.

- Creating a data flow activity in the Synapse integration pipeline.

- Building the data flow activity to copy and convert the CSV file to the Parquet format.

Getting ready

To get started, do the following:

1. Log in to `https://portal.azure.com` using your Azure credentials.

2. Create a Synapse Analytics workspace, as explained in the *Provisioning an Azure Synapse Analytics workspace* recipe of *Chapter 8, Processing Data Using Azure Synapse Analytics*.

3. Download the files – `transaction_table-t1.zip` and `transaction_table-t2.zip` – from `https://github.com/PacktPublishing/Azure-Data-Engineering-Cookbook-2nd-edition/upload/main/chapter9`:

4. Unzip them. In the Synapse Analytics workspace, create a folder named `csv` (if it doesn't already exist) in the data lake account attached to the Synapse Analytics workspace.

5. Upload the files, `transaction_table-t1.csv` and `transaction_table-t2.csv`, to the `csv` folder.

    To see detailed example screenshots for a similar task, follow *steps 1 to 4* in the *How to do it…* section of the *Analyzing data using Serverless SQL pool* recipe from *Chapter 8, Processing Data Using Azure Synapse Analytics*.

How to do it...

Perform the following steps to copy data using a Synapse data flow:

1. Log in to `portal.azure.com`, go to **All resources,** and search for **packtadesynapse**, the Synapse Analytics workspace created in the *Provisioning an Azure Synapse Analytics workspace* recipe of *Chapter 8, Processing Data Using Azure Synapse Analytics*. Click on the workspace. Click on **Open Synapse Studio**:

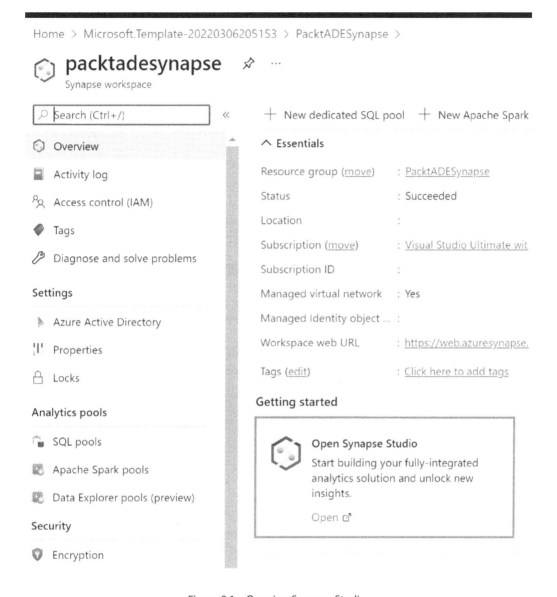

Figure 9.1 – Opening Synapse Studio

2. Click on the blue cylinder (the data symbol) on the left-hand side, which will take you to the **Data** section. Click on the **Linked** tab. Expand the data lake account of the Synapse workspace (**packtadesynapse** for this example). Click on the **synapse (Primary)** container. Click on the + **New folder** button and create a folder called `transaction_table-t1-parquet`. We will use a Parquet folder as the destination to copy the Parquet file converted from the CSV file.

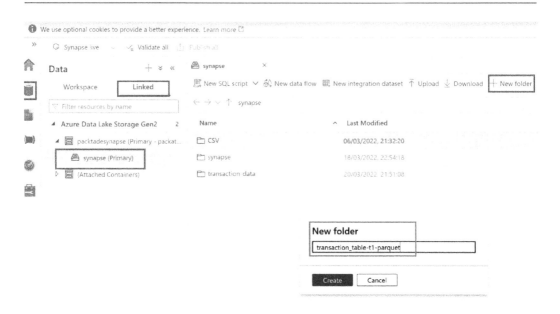

Figure 9.2 – Creating the Parquet folder

3. Click on the pipe-like icon on the left-hand side of Synapse Studio. It will open the integration pipeline development area. Hit the + button and click on **Pipeline**:

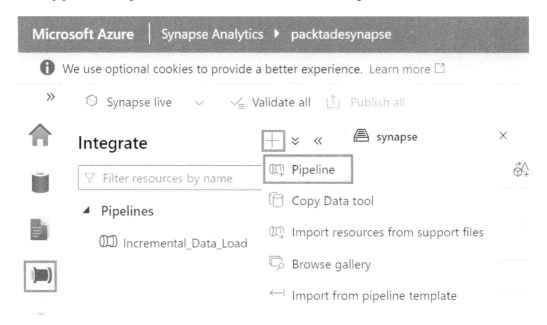

Figure 9.3 – Creating a pipeline

4. Name the pipeline `Copy_Data_Flow` by editing the **Name** property field on the right. Search for **Data flow** under **Activities**. Drag and drop the **Data flow** activity into the pipeline development area:

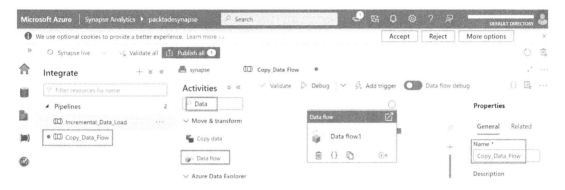

Figure 9.4 – Adding a data flow

5. Scroll down and under **Settings**, click on the + **New** button to create a data flow:

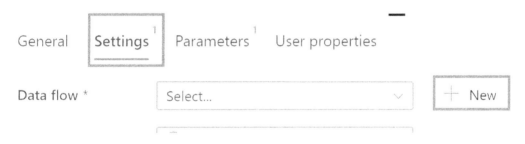

Figure 9.5 – Creating a data flow

6. Change the data flow name to `Copy_CSV_to_Parquet` on the left. Click on the **Add Source** button:

Figure 9.6 – Adding a source

7. Scroll down and name the output stream csv. Click on the **+ New** button to create a new dataset:

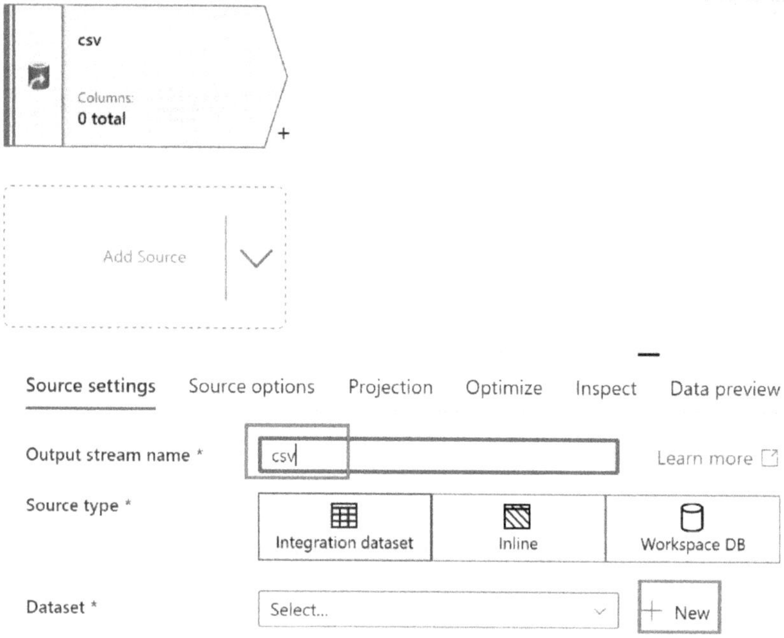

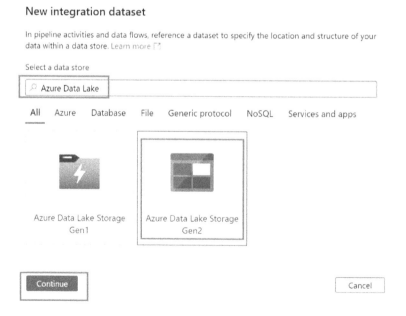

csv

Columns:
0 total

+

Add Source ∨

—

Source settings   Source options   Projection   Optimize   Inspect   Data preview

Output stream name *   csv   Learn more ⬀

Source type *   [⊞ Integration dataset] [⬚ Inline] [🛢 Workspace DB]

Dataset *   Select...   ∨   + New

Figure 9.7 – Adding a dataset

8. Search for Azure Data Lake, select **Azure Data Lake Storage Gen2**, and click on **Continue**:

New integration dataset

In pipeline activities and data flows, reference a dataset to specify the location and structure of your data within a data store. Learn more ⬀

Select a data store

🔍 Azure Data Lake

All   Azure   Database   File   Generic protocol   NoSQL   Services and apps

Azure Data Lake Storage Gen1   Azure Data Lake Storage Gen2

Continue   Cancel

Figure 9.8 – Connecting to Azure Data Lake Gen 2

9. Select CSV as the data format, as the input file is CSV:

Select format

Choose the format type of your data

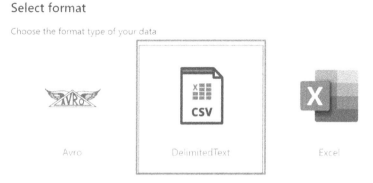

| Avro | DelimitedText | Excel |

Figure 9.9 – Connecting to Azure Data Lake Gen 2

10. Provide the dataset name, transactiontablet1. Set **Linked service** to **packtadesynapse-WorkspaceDefaultStorage**. Set **File path** to synapse/csv/transaction_table-t1. csv, the location to which we uploaded the CSV file. Check the **First row as header** checkbox:

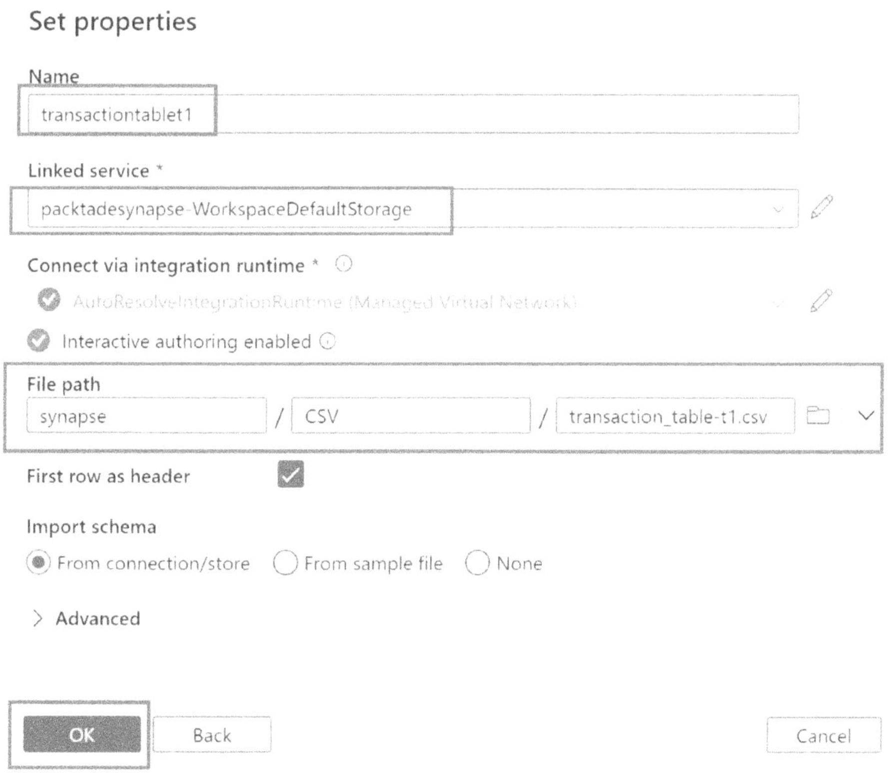

Set properties

Name

    transactiontablet1

Linked service *

    packtadesynapse-WorkspaceDefaultStorage   ∨   ✎

Connect via integration runtime * ⓘ

    ✅ AutoResolveIntegrationRuntime (Managed Virtual Network)   ∨   ✎

    ✅ Interactive authoring enabled ⓘ

File path

    synapse   /   CSV   /   transaction_table-t1.csv   🗁   ∨

First row as header   ✅

Import schema

◉ From connection/store   ○ From sample file   ○ None

> Advanced

    OK   Back   Cancel

Figure 9.10 – Connecting to Azure Data Lake Gen 2

11. Click on the + button, search for `sink`, and select **Sink**. **Sink** is the destination component (transformation) that we will be copying the data to:

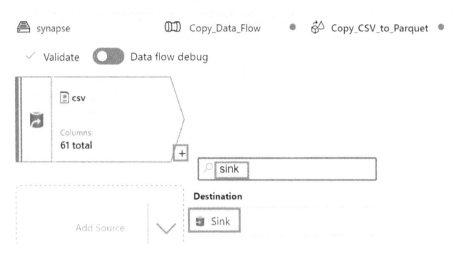

Figure 9.11 – Adding a sink

12. Set **Output stream name** to `Parquet`. Click + **New**:

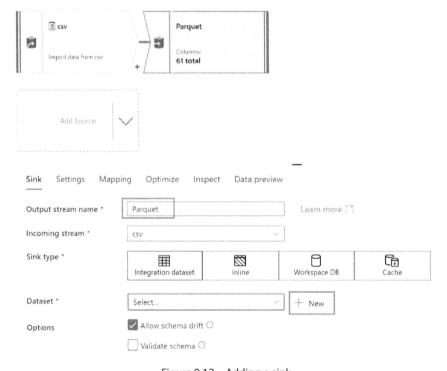

Figure 9.12 – Adding a sink

13. We need to define the destination dataset. Search for Azure Data Lake, select **Azure Data Lake Storage Gen2**, and click **Continue**:

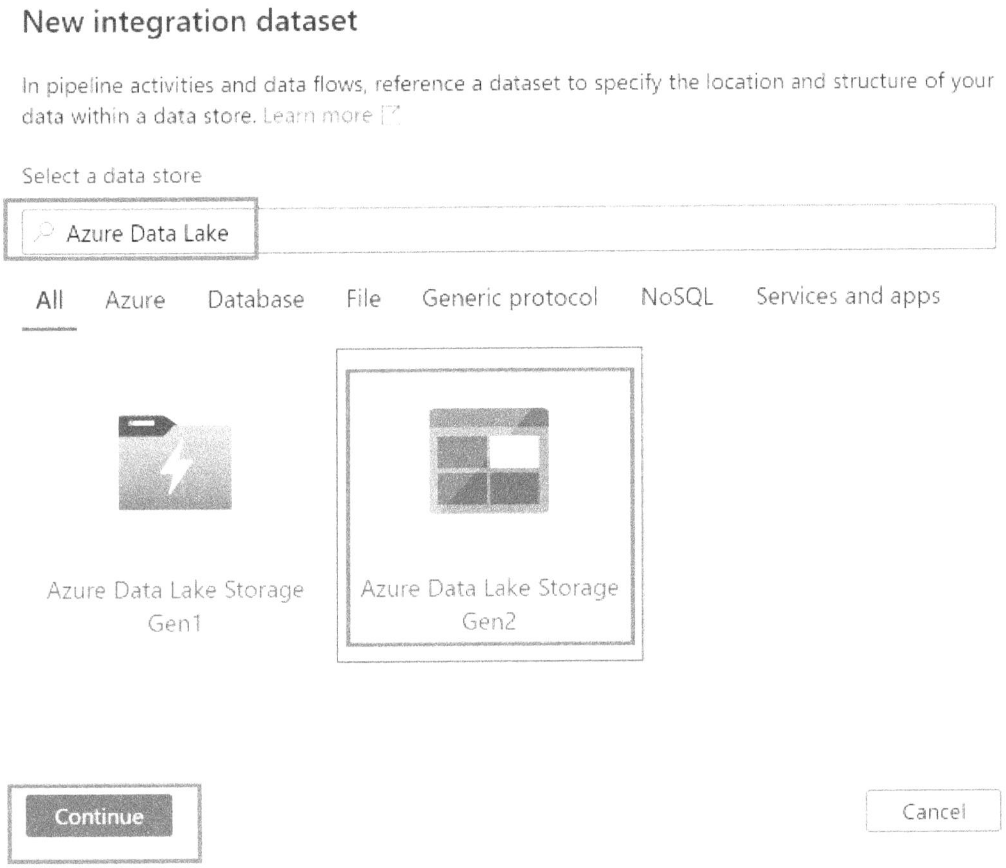

New integration dataset

In pipeline activities and data flows, reference a dataset to specify the location and structure of your data within a data store. Learn more [?

Select a data store

```
🔍 Azure Data Lake
```

All   Azure   Database   File   Generic protocol   NoSQL   Services and apps

Azure Data Lake Storage
Gen1

Azure Data Lake Storage
Gen2

Continue   Cancel

Figure 9.13 – Connecting to Azure Data Lake Gen 2

14. Set the format to **Parquet**:

Select format

Choose the format type of your data

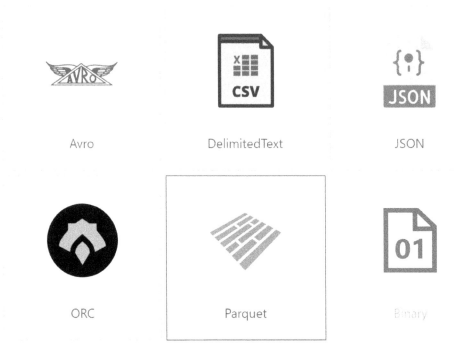

Avro

DelimitedText

JSON

ORC

Parquet

Binary

Figure 9.14 – Connecting to Azure Data Lake Gen 2

15. Provide the dataset name, `transactiontablet1parquet`. Set **Linked service** to **packtadesynapse-WorkspaceDefaultStorage**. Set **File path** to `synapse/transaction_table-t1-parquet/`, the location where we will store the Parquet file. Press **OK**.

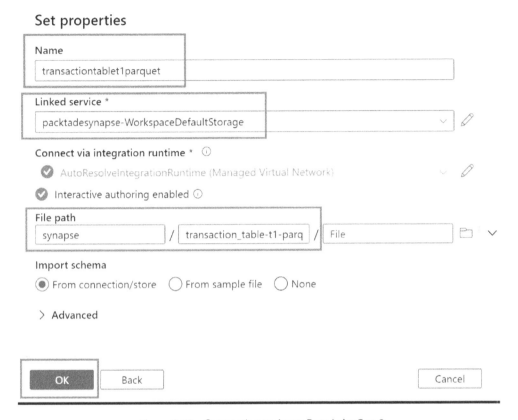

Figure 9.15 – Connecting to Azure Data Lake Gen 2

16. Hit the **Publish all** button. The data flow, datasets, and the pipeline will be published. Once they're published, go to the **Copy_Data_Flow** pipeline:

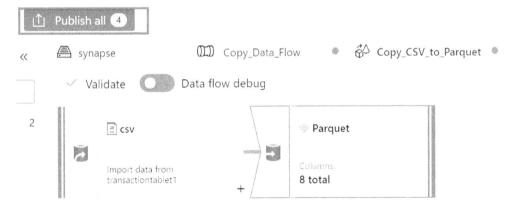

Figure 9.16 – Publishing the pipeline

17. Click on the **Add trigger** button and click **Trigger now** to trigger the execution of the pipeline:

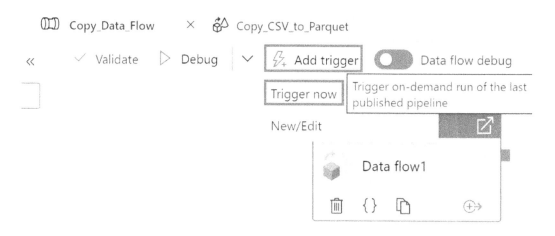

Figure 9.17 – The pipeline execution

18. Once the pipeline successfully runs, go to the storage section by clicking on the data icon (the blue cylinder) on the left-hand side and navigating to the **synapse (Primary)** folder. Right-click on the `transaction_table-t1-parquet` folder and click **Select TOP 100 rows**:

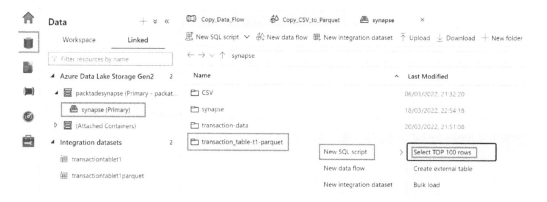

Figure 9.18 – Reading the transferred data

19. Set **File type** to **Parquet format** for querying:

Select TOP 100 rows

📁 transaction_table-t1-parquet

Source folder format settings

Specify the format and layout of your data.

Folder path

https://packatadesynapse.dfs.core.windows.net/synapse/transaction_table-t1-parquet/

File type
Parquet format

Apply   Cancel

Figure 9.19 – Reading Parquet format

20. Hit the **Run** button in the query window. Data is successfully read via serverless SQL pool, confirming that the data transfer successfully completed:

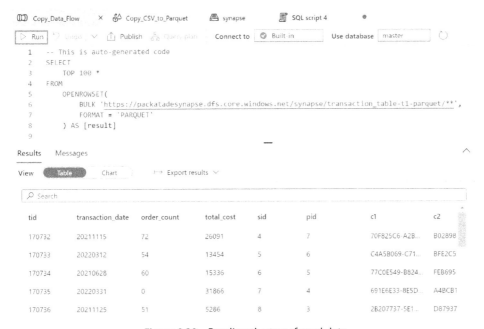

Figure 9.20 – Reading the transferred data

How it works...

The data flow does a simple conversion of the CSV file to a Parquet file. The data flow is linked to a Synapse integration pipeline called **Copy_Data_Flow**. The **Copy_Data_Flow** pipeline uses a data flow activity that calls the **Copy_CSV_to_Parquet** data flow, which does the data transfer. While this task can be done using a simple **Copy activity** as well (instead of data flow), the key advantage of a data flow is being able to perform transformations while moving the data from the source to the sink, which will be covered in more detail in the subsequent recipes in this chapter.

Performing data transformation using activities such as join, sort, and filter

A common scenario in data engineering pipelines is combining two or more files based on a column, filtering by column, sorting the results, and storing them for querying. We will perform the following actions to achieve this:

1. Read two CSV files

2. Use a **join transformation** to combine the two files based on a column

3. Use a **filter transformation** to filter the rows based on a condition

4. **Sort** the filtered data based on a column value and store the result in Parquet format

Getting ready

Create a Synapse Analytics workspace, as explained in the *Provisioning an Azure Synapse Analytics workspace* recipe of *Chapter 8, Processing Data Using Azure Synapse Analytic:*.

1. Download the files – transaction_table-t1.zip and transaction_table-t2.zip – from https://github.com/PacktPublishing/Azure-Data-Engineering-Cookbook-2nd-edition/upload/main/chapter9.

2. Unzip them.

3. In the Synapse Analytics workspace, create a folder named csv (if it doesn't already exist) in the data lake account attached to the Synapse Analytics workspace. Upload the files, transaction_table-t1.csv and transaction_table-t2.csv, to the csv folder.

To see detailed example screenshots for a similar task, follow *steps 1 to 4* in the *How to do it…* section from the *Analyzing data using Serverless SQL pool* recipe of *Chapter 8, Processing Data Using Azure Synapse Analytics*.

How to do it...

In this recipe, we will perform the following:

- Create Azure Data Lake datasets connected to the `transaction_table-t1.csv` and `transaction_table-t2.csv` files

- Create a data flow activity in a pipeline

- Add **two source transformations** for reading CSV files

- Add a join transformation to combine the files based on a column named `tid`

- Add a filter transformation on the `transaction_date` column to filter the data by a particular date

- Add a sort transformation to sort the filtered data by the `total_cost` column

- Create a dataset with Parquet format as destination

- Add a sink and save the result in a dataset created using the Parquet format

The detailed steps to carry this out are as follows:

1. Log in to `portal.azure.com`, go to **All resources,** and search for **packtadesynapse**, the Synapse Analytics workspace created in the *Provisioning an Azure Synapse Analytics workspace* recipe of *Chapter 8, Processing Data Using Azure Synapse Analytics*. Click on the workspace. Click **Open Synapse Studio**:

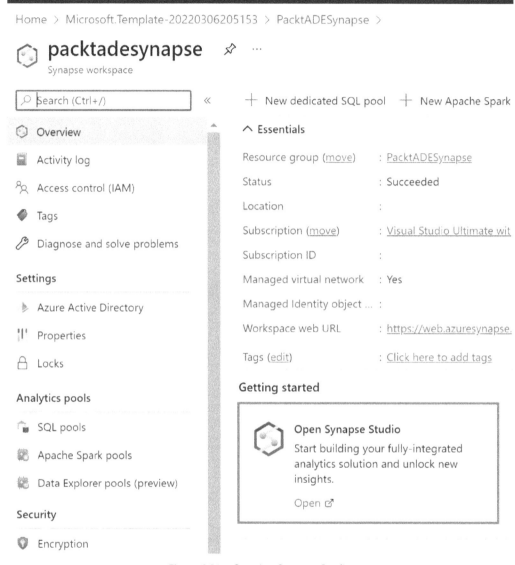

Figure 9.21 – Opening Synapse Studio

2. Click on the blue cylinder (the data symbol) on the left-hand side, which will take you to **Data** section. Click on the **Linked** tab. Expand the data lake account of the Synapse Analytics workspace (**packtadesynapse** for this example). Click on the **synapse (Primary)** container. Click on the **+ New folder** button and create a folder called `transaction_table-transformation-parquet`. We will use this folder as the destination to store the transformed data:

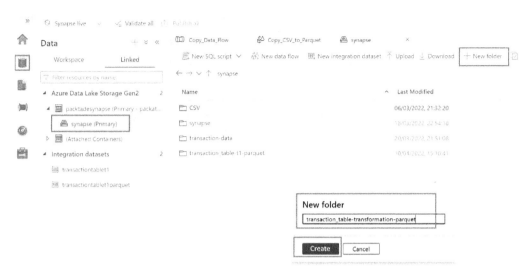

Figure 9.22 – Creating the Synapse Studio folder

3. Create a pipeline named `DataFlow-Transformation`. Add a data flow activity in the pipeline. Create a new data flow named `SortFilterDataFlow` by clicking on the **+ New** button under the data flow activity settings. To see detailed example screenshots for a similar task, follow *steps 3* to *5* in the *How to do it…* section from the *Copying data using a Synapse data flow* recipe:

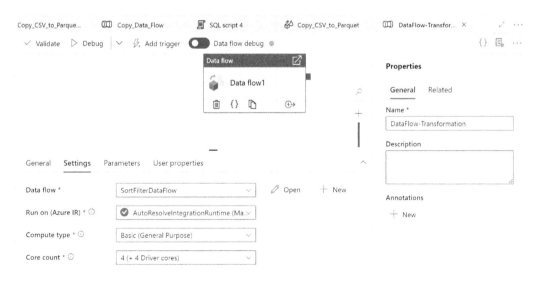

Figure 9.23 – Creating the Dataflow-Transformation pipeline

4. Add a source transformation named `transactiontablet1`, click on the + **New** button, and create a dataset named `transactiontablet1`, as explained in *steps 6* to *10* in the *How to do it…* section from the *Copying data using a Synapse data flow* recipe:

Figure 9.24 – Creating the Dataflow-Transformation pipeline

5. Click on the **Add Source** button.

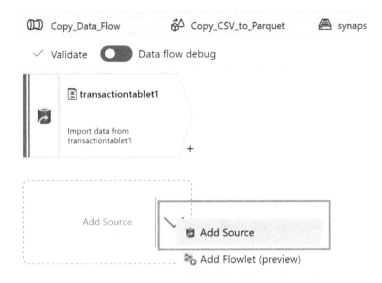

Figure 9.25 – Adding a source

6. Name the source transformation `transactiontablet2`. Click on the + **New** button to create a new dataset to link to `synapse/csv/ transaction_table-t2.csv`:

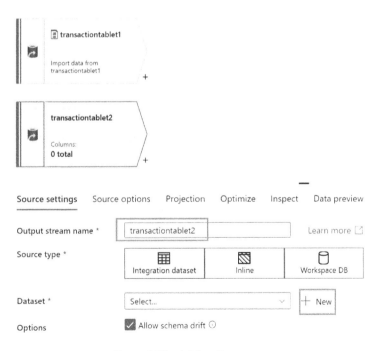

Figure 9.26 – Adding the source

7. Create a dataset named `transactiontablet2` linking to Azure Data Lake Storage Gen2 account attached to the Synapse Analytics workspace, with the location set to `synapse/csv/transaction_table-t2.csv` and the format set to `csv`. Refer to *steps 7* to *10* in the *How to do it...* section from the *Copying data using a Synapse data flow* recipe:

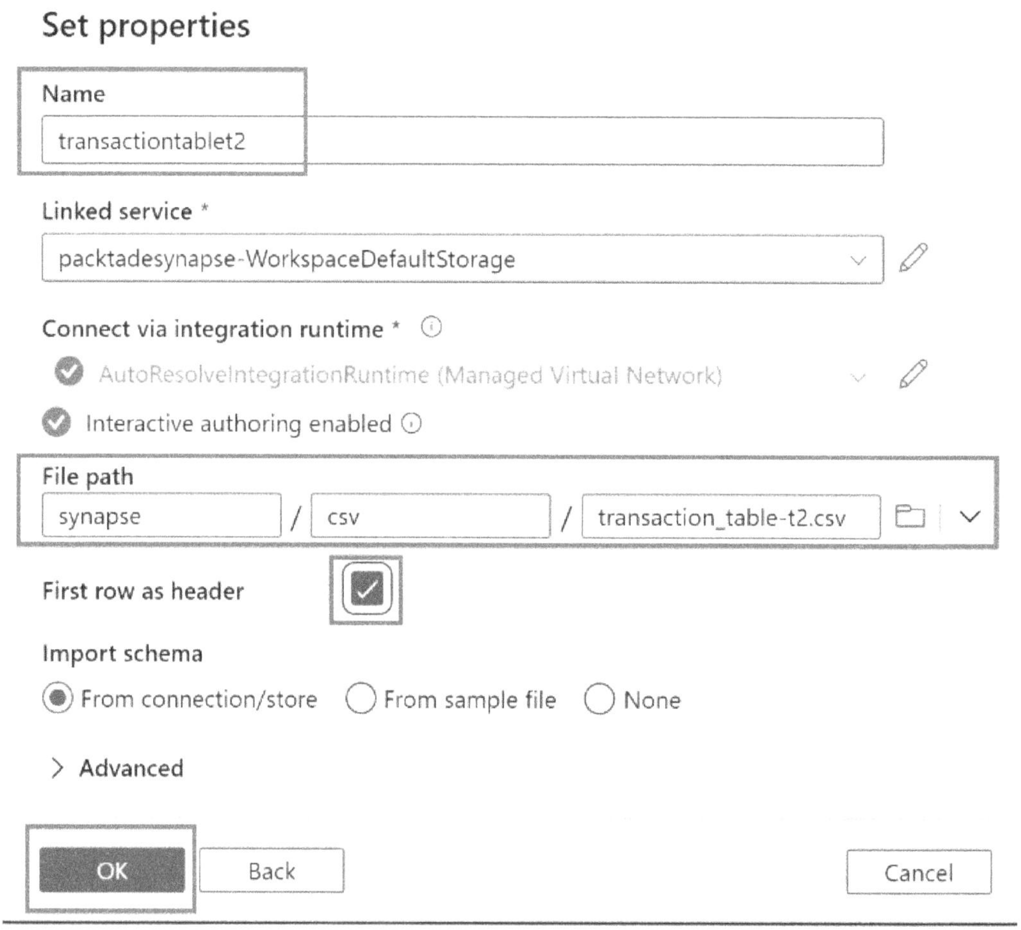

Set properties

Name

transactiontablet2

Linked service *

packtadesynapse-WorkspaceDefaultStorage

Connect via integration runtime * ⓘ

✔ AutoResolveIntegrationRuntime (Managed Virtual Network)

✔ Interactive authoring enabled ⓘ

File path

synapse / csv / transaction_table-t2.csv

First row as header ☑

Import schema

◉ From connection/store   ◯ From sample file   ◯ None

> Advanced

OK   Back   Cancel

Figure 9.27 – Configuring the data source

8. Click on the + button on the **transactiontablet1** source and select the **Join** transformation:

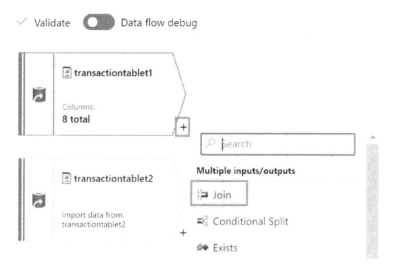

Figure 9.28 – Adding join

9. Set the **Left stream** dataset to **transactiontablet1** and the **Right stream** dataset to **transactiontablet2**. Select **Inner** as the join type and join using the `tid` column, which exists on both the datasets.

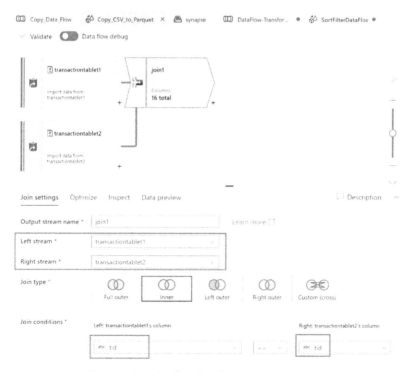

Figure 9.29 – Configuring the join transformation

10. Click on the + button on the right-hand side of the **join1** transformation and select the **Filter** transformation:

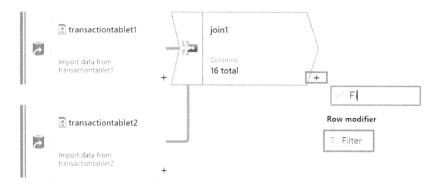

Figure 9.30 – Adding the filter transformation

11. Type `transactiontablet1@transaction_date=="20210915"` into the **Filter on** text box.

Figure 9.31 – Configuring the filter transformation

12. Click on the + button on the right-hand side of the **filter1** transformation and select the **Sort** transformation:

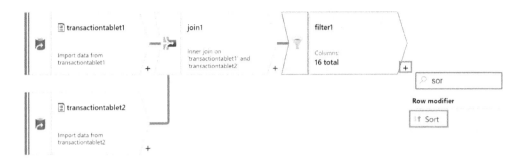

Figure 9.32 – Add sort transformation

13. Under **Sort conditions**, select **transactiontablet1@total_cost** for the **filter1's column** dropdown and set **Order** to **Descending**:

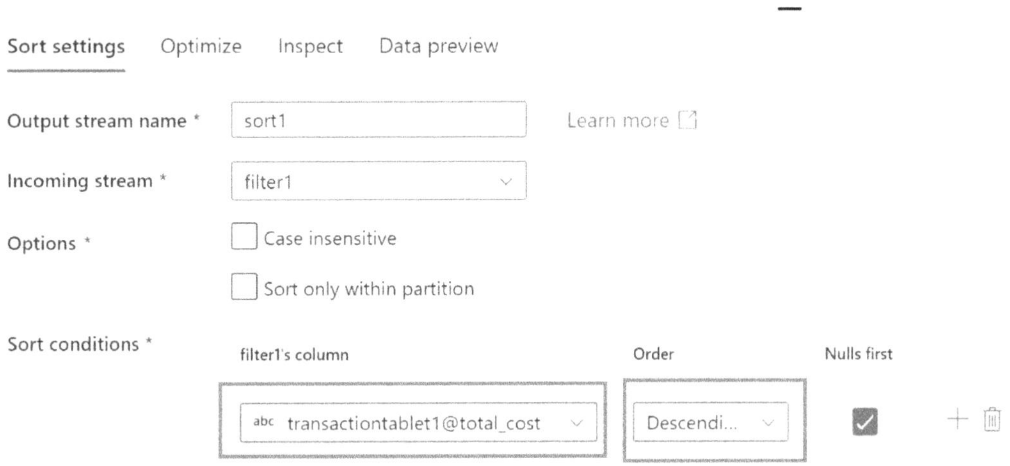

Sort settings   Optimize   Inspect   Data preview

Output stream name *	sort1

Learn more 🗗

Incoming stream *	filter1

Options *	☐ Case insensitive
	☐ Sort only within partition

Sort conditions *

filter1's column	Order	Nulls first
abc transactiontablet1@total_cost	Descendi...	☑ + 🗑

Figure 9.33 – Adding the filter conditions

14. Hit the + button on the right-hand side of the **sort1** transformation and add a sink. Under the **Sink** properties, click on the + **New** button and add a new dataset named `transformationparquet` linking to the Azure Data Lake Gen2 account attached to the Synapse Analytics workspace with the location set to `synapse/transaction_table-transformation-parquet` folder and the format set to Parquet. Refer to *steps 11* to *15* in the *How to do it...* section from *Copying data using a Synapse data flow* recipe to see detailed example screenshots for a similar task. Once complete, the sink should appear as shown in the following screenshot:

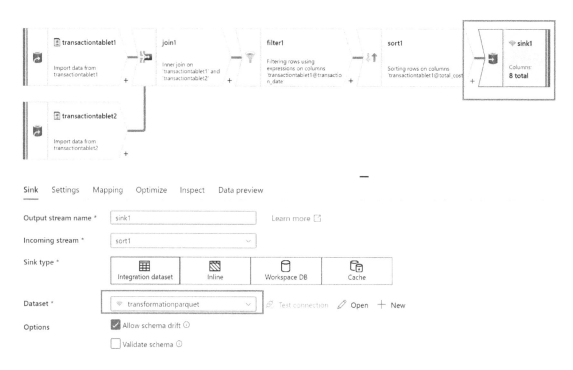

Figure 9.34 – Adding the sink

15. Switch to the **DataFlow-Transformation** pipeline and hit the **Publish all** button. Once published, hit the **Trigger now** button under **Add trigger**. After successful execution, go to the synapse/ transaction_table-transformation-parquet folder in the Synapse Analytics workspace, right-click on it, and select the top 100 rows. Refer to *steps 16* to *20* in the *How to do it…* section from *Copying data using a Synapse data flow* recipe to see detailed example screenshots for a similar task. The result is shown in the following screenshot:

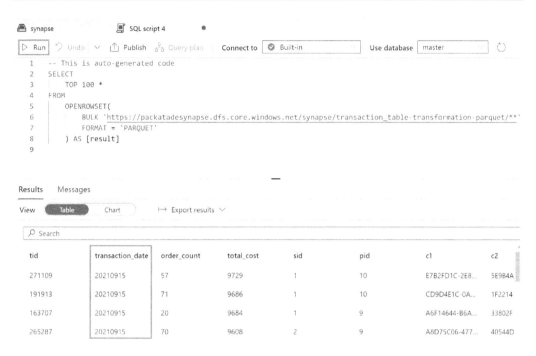

```
                                                                                              •
  synapse                     SQL script 4
  ▷ Run    Undo  ∨    ↑ Publish   Query plan     Connect to  ⊘  Built-in              Use database  master              ↻
  1     -- This is auto-generated code
  2     SELECT
  3         TOP 100 *
  4     FROM
  5         OPENROWSET(
  6             BULK 'https://packatadesynapse.dfs.core.windows.net/synapse/transaction_table-transformation-parquet/**'
  7             FORMAT = 'PARQUET'
  8         ) AS [result]
  9
```

Results   Messages

View   Table   Chart   ↦ Export results ∨

🔍 Search

tid	transaction_date	order_count	total_cost	sid	pid	c1	c2
271109	20210915	57	9729	1	10	E7B2FD1C-2E8...	5E9B4A
191913	20210915	71	9686	1	10	CD9D4E1C-0A...	1F2214
163707	20210915	20	9684	1	9	A6F14644-B6A...	33802F
265287	20210915	70	9608	2	9	A8D75C06-477...	40544D

Figure 9.35 – The result

How it works...

A join transformation helped us to combine the `transaction_table-t1.csv` and `transaction_table-t2.csv` files based on the `tid` column. The combined data from both files was filtered using the filter transformation and the `transaction_table-t1.csv` file's `transaction_date` column was filtered for the date *20210915*. The sort transformation sorted the filtered rows based on the `transaction_table-t1.csv` file's `total_cost` column. A sink transformation was linked to a dataset in Parquet format, which meant that the sorted and filtered data from the CSV file was seamlessly converted and stored in Parquet format too.

Data flows have plenty of transformation options available. You can explore more examples at `https://docs.microsoft.com/en-us/azure/data-factory/data-flow-transformation-overview`.

Monitoring data flows and pipelines

Azure Synapse Analytics provides a user-friendly interface out of the box for monitoring the pipeline and data flow runs. In this recipe, we will track a data flow execution and understand the transformation execution timings.

Getting ready

Create a Synapse Analytics workspace, as explained in the *Provisioning an Azure Synapse Analytics workspace* recipe of *Chapter 8, Processing Data Using Azure Synapse Analytics*.

Complete the *Performing data transformation using activities such as join, sort, and filter* recipe in this chapter to create some data flow runs.

How to do it...

Perform the following steps to monitor data flows:

1. Log in to `portal.azure.com`, go to **All resources**, and search for **packtadesynapse**. Click on the workspace. Click on **Open Synapse Studio**. Click on the monitor button (a speedometer-like circular button) on the left-hand side. By default, it will give you the information about pipeline runs from the last 24 hours:

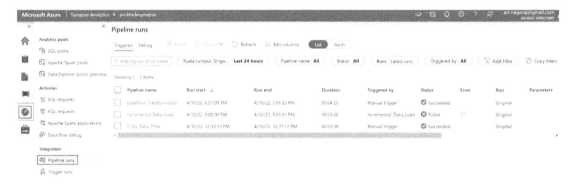

Figure 9.36 – The pipeline runs

2. Click on the **DataFlow-Transformation** pipeline with the most recent run date (for me, it's **4/10/22**).

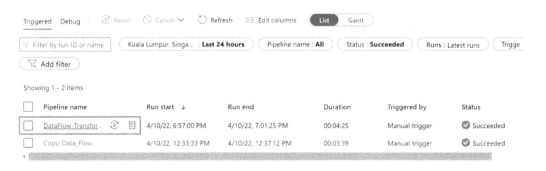

Figure 9.37 – Opening the pipeline execution data

3. Under **Activity runs**, hover your mouse over the **Data flow1** activity and click on the glasses icon:

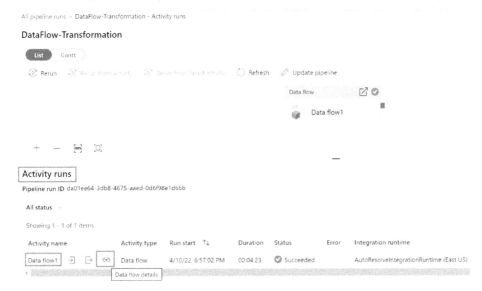

Figure 9.38 – Opening the data flow execution details

4. All the transformations under our activity are displayed. Click on the **Stages** icon to identify the transformation that took the most time:

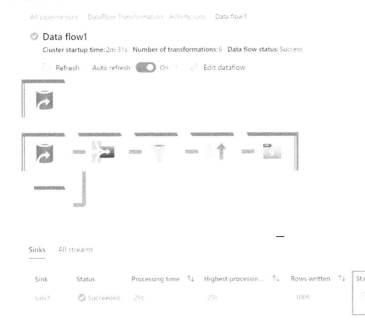

Figure 9.39 – The data flow execution details

5. The total execution time was 25 seconds, of which sink and sort operations took 20 seconds. We also noticed that there were **1009** rows sorted and written to sink. Click **Close**:

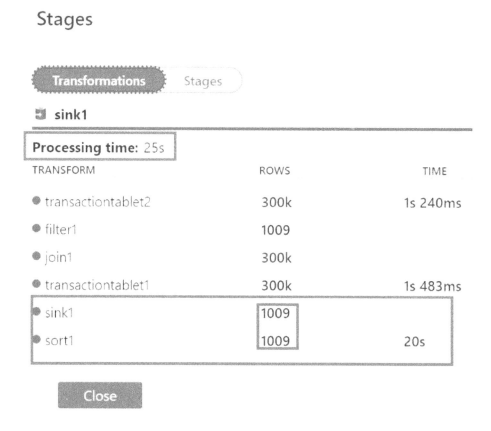

Figure 9.40 – The transformation-level execution details

How it works...

The Azure Synapse Analytics out-of-the-box monitoring solution records details about all pipeline execution runs and details about the activities inside the pipeline too. In this recipe, we saw that we could quickly identify even a slow transformation inside a data flow activity in a matter of a few clicks. The monitoring data is by default stored for 45 days and can be retained for a longer duration by integrating Synapse Analytics diagnostics data with Azure Log Analytics or an Azure Data Lake Storage account.

Configuring partitions to optimize data flows

Data flows, by default, create partitions behind the scenes to make transformation activities such as join, filter, and sort run faster. Partitions split the large data into multiple smaller pieces so that the backend processes in the data flows can divide and conquer their tasks and finish the execution quickly.

In this recipe, we will take a slow-running data flow and adjust the partitioning to reduce the execution time.

Getting ready

Create a Synapse Analytics workspace as explained in the *Provisioning an Azure Synapse Analytics workspace* recipe.

Complete the *Performing data transformation using activities such as join, sort, and filter* and *Monitoring data flows and pipelines* recipes in this chapter.

How to do it...

Perform the following steps to optimize data flows:

1. Log in to `portal.azure.com`, go to **All resources**, and search for **packtadesynapse**. Click on the workspace. Click **Open Synapse Studio**. Click on the monitor button (the speedometer-like circular button) on the left-hand side. Refer to *steps 2* to *5* of the *How to do it...* section from the *Monitoring data flows and pipelines* recipe and perform the following actions:

    I.   Click on the latest run of the **DataFlow-Transformation** pipeline.

    II.   Click on the **Data flow1** activity details.

    III. Click on the **Stages** icon to get the transformation level breakdown of the execution times. Notice that the sort and sink operations consume 80% of the execution time, taking 20 seconds.

2. Click on the **Sort1** transformation in the monitoring window. Notice the following:

    - The total number of rows processed by the transformation is only 1,009

    - There are about 200 partitions

    - Each partition has around 4 to 6 rows

Let's make some changes to partitioning in the sort transformation. Click on the **Edit transformation** button at the top of the page:

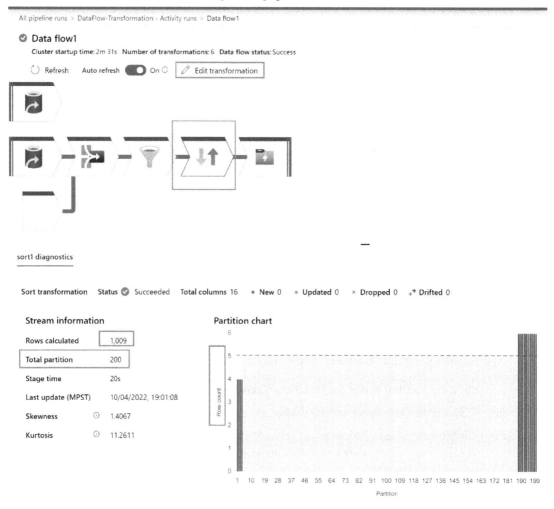

Figure 9.41 – The partition statistics

3. Click on the **sort1** transformation. Go to the **Optimize** section. Select **Single partition** (instead of the existing setting, **Use current partitioning**). Hit the **Publish all** button:

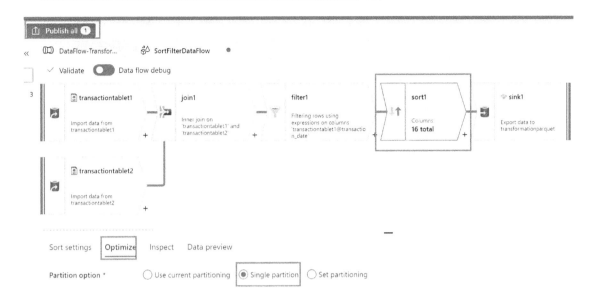

Figure 9.42 – Setting a single partition

4. Go to **DataFlow-Transformation**, click on **Add trigger**, and click **Trigger now**:

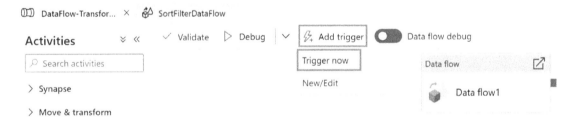

Figure 9.43 – The Trigger now function

5. Check the notification section (the bell icon) on the right-hand corner of the screen. The run completion will be indicated via a notification message. Click **View pipeline run**:

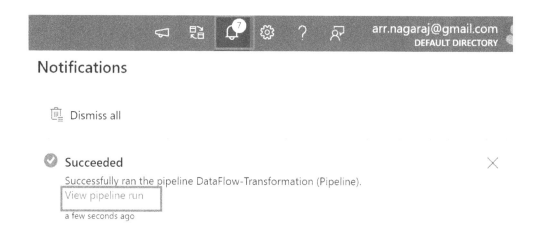

Notifications

Dismiss all

✅ **Succeeded**   ✕

Successfully ran the pipeline DataFlow-Transformation (Pipeline).

View pipeline run

a few seconds ago

Figure 9.44 – Viewing the pipeline run

6. Click on the **Dataflow1** activity details. Click on the **Stages** icon to get the transformation level breakdown of the execution times, as we did previously:

    - Notice that the total execution time has been reduced to 8 seconds

    - The sort and sink executions were reduced to 2.8 seconds, compared to the earlier duration of 20 seconds

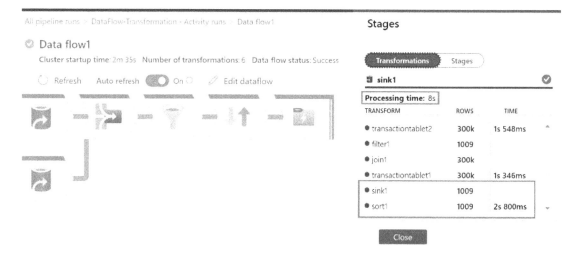

Figure 9.45 – The execution timing of the pipeline run

How it works...

Behind the scenes, Synapse Analytics data flows use tasks in Spark clusters to perform data flow transformations. When observing the partition statistics on the **sink1** transformation in *step 2* of the *How to do it…* section, we noticed that 1,009 rows were split across 200 partitions. Typically, we would like to have at least a few thousand rows per partition (or 10 MB to 100 MB in size). Having 4 to 6 rows per partition makes any transformation slow and hence, the sort operation was slow as well. Having too many partitions implies that the backend jobs spend a lot of time creating many partitions. This becomes overkill when there are just a few rows per partition and most of the time is spent on creating partitions rather than processing the data inside them.

Switching the partition settings to **Single partition** via the sink transformation's **Optimize** setting creates a single partition for all 1,009 rows, performs the sort activity in a single partition, and returns the results quickly. Had there been, say, a few hundred thousand or a few million rows for the sort activity, having multiple partitions would have been a better bet.

Data flows for each transformation, by default, use the setting called **current partitioning**, which implies that the data flow will estimate the optimal number of partitions for that transformation based on the size of the data. Setting a single partition as done in this recipe is *not* recommended for all scenarios – however, it is equally important to track the partition count for transformations and make adjustments if required.

Parameterizing Synapse data flows

Adding parameters to data flows provides flexibility and agility for data flows while performing their transformations. In this recipe, we will create a data flow that accepts a parameter, filters the data based on the value passed in the parameter, and copies the data to Parquet files.

Getting ready

Create a Synapse Analytics workspace as explained in the *Provisioning an Azure Synapse Analytics workspace* recipe in *Chapter 8*, *Processing Data Using Azure Synapse Analytics*.

Complete the *Copying data using a Synapse data flow* recipe in this chapter to create the **Copy_CSV_ to_Parquet** data flow..

How to do it...

In this recipe, we will add a parameter to the **Copy_CSV_to_Parquet** dataflow (which was created in the *Copying data using a Synapse data flow* recipe), and make the **Copy_CSV_to_Parquet** data flow copy selective rows based on the value passed to the parameter by the **Copy_Data_Flow** pipeline (created in the *Copying data using a Synapse data flow* recipe). The **Copy_CSV_to_Parquet** data flow currently copies all 300,000 rows present in the transaction_table-t1.csv file and adds the filter based on parameter that will result in lesser number of rows being copied:

1. Log in to portal.azure.com, go to **All resources**, and search for **packtadesynapse**. Click on the workspace. Click **Open Synapse Studio**. Click on the **Develop** icon (the notebook-like symbol) on the left-hand side of the screen. Expand **Data flows** and click **Copy_CSV_to_ Parquet**. Under **Parameters**, click + **New** and add a new parameter with the name sid. Set **Default value** to 0:

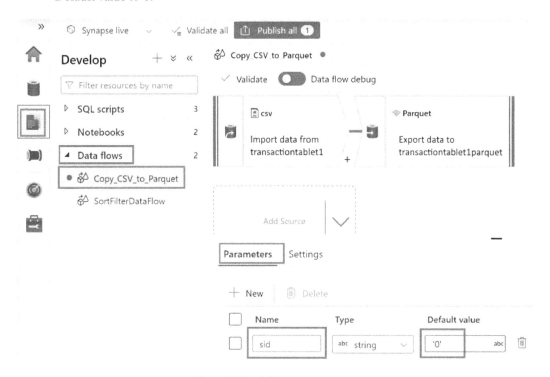

Figure 9.46 – Adding a parameter

2. Click on the + button on the right-hand side of the source transformation named **csv**, search for a filter transformation, and add a filter transformation:

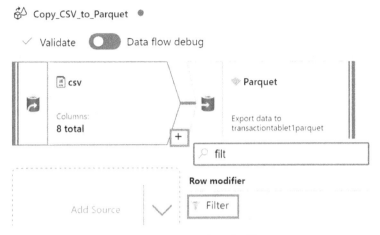

Figure 9.47 – Adding the filter

3. Click on the **Filter on** textbox under **Filter settings**. Click on the **Open expression builder** link found below the textbox:

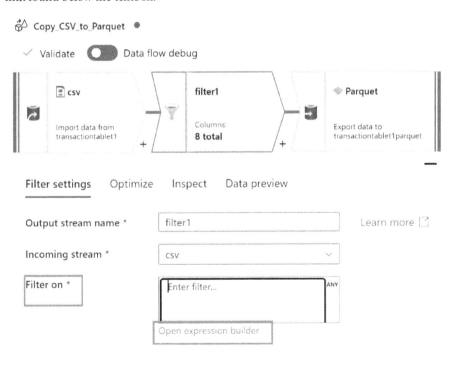

Figure 9.48 – Adding dynamic content

4. In the expression text box, type the column name that we will be filtering the parameter against; in our case, it is sid. Click on == (double equal to operator). Click **Parameters** on the bottom-left-hand side and select the sid parameter. The expression textbox should read as shown in the following screenshot. Click **Save and finish**:

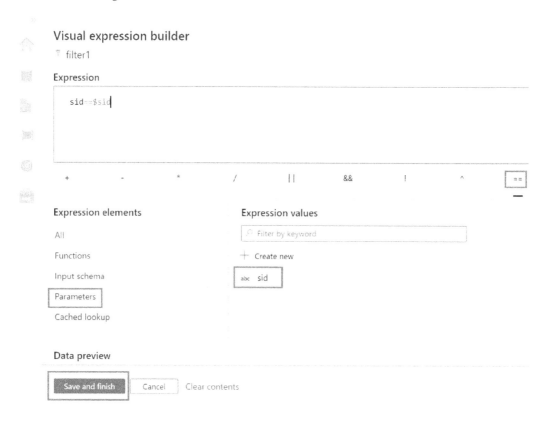

Figure 9.49 – Adding dynamic content

5. Click on the **Integrate** icon (the pipe-like symbol) on the left-hand side of Synapse Studio. Click on the **Copy_Data_Flow** pipeline. Click on the **Data flow1** task and go to the **Parameters** section of the task. The filename parameter we added to the data flow in *step 1* appears here. Click on the **Value** box, select **Pipeline expression**, and check the **Expression** checkbox:

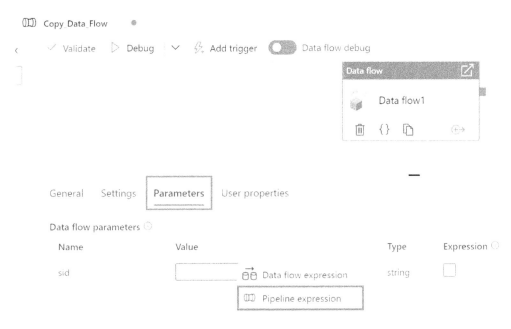

Figure 9.50 – Adding the pipeline expression

6. Type in `'10'` under dynamic content and click **OK**:

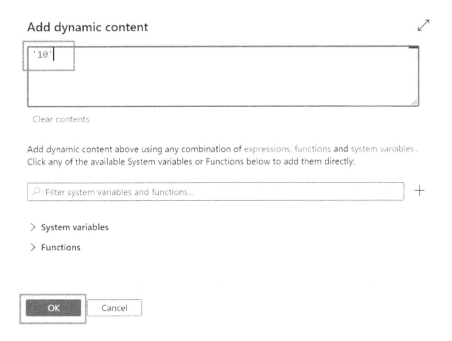

Figure 9.51 – Passing a value to the parameter

7. Publish the pipeline, click on **Add trigger**, and select **Trigger now** to execute the pipeline, as done in *step 17* of the *How to do it…* section in the *Copying data using a Synapse data flow* recipe.

8. Once the run completes, click on the monitoring icon on Synapse Studio. Select the latest run of the **Copy_Data_Flow** pipeline. Click on the **Dataflow1** activity details. Refer to *steps 1* to *4* of the *How to do it…* section in the *Monitoring data flows and pipelines* recipe for detailed example screenshots for a similar task. Notice that the pipeline only processed **28957** rows, compared to 300,000 rows initially present in the `transaction_table-t1.csv` file, as the rows were filtered by parameter:

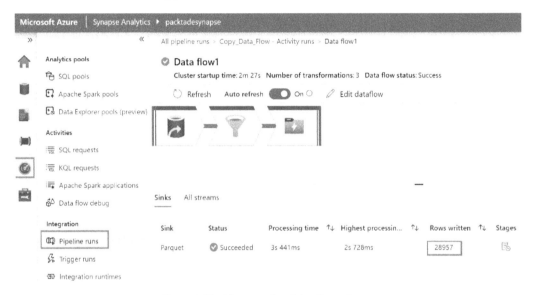

Figure 9.52 – The transfer of the filtered rows

How it works...

We added a parameter named `sid` to the **Copy_CSV_to_Parquet** data flow. We wrote an expression in the filter transformation in the data flow that compared the `sid` parameter with the `sid` column in the data flow. The comparison expression in the filter transformation filtered the rows based on the condition that we defined. We passed the value to the `sid` data flow parameter from the **Copy_Data_Flow** pipeline's data flow activity. The data flow activity exposed the parameters that were present inside the data flow to the pipeline. We passed the value `'10'` from the **Copy_Data_Flow** pipeline to the data flow using the pipeline's data flow activity and all the rows that had a value equal to 10 on the `sid` column were selected and copied in Parquet format to the sink destination.

Handling schema changes dynamically in data flows using schema drift

A common challenge in **extraction, transformation**, **and load** (**ETL**) projects is when the schema changes at the source and the pipelines that are supposed to read the data from the source, transform it, and ingest it to the destination, start to fail. Schema drift, a feature in data flows, addresses this problem by allowing us to dynamically define the column mapping in transformations. In this recipe, we will make some changes to the schema of a data source, use schema drift to detect the changes, and handle changes without any manual intervention gracefully.

Getting ready

Create a Synapse Analytics workspace as explained in the *Provisioning an Azure Synapse Analytics workspace* recipe in *Chapter 8, Processing Data Using Azure Synapse Analytics*.

Complete the *Copying data using a Synapse data flow* recipe in this chapter.

How to do it...

In this recipe, we will be using the **Copy_CSV_to_Parquet** data flow completed in *Copying data using a Synapse data flow*. We will perform the following tasks:

- Add columns to a CSV file used as the source in the **Copy_CSV_to_Parquet** data flow from the *Copying data using a Synapse data flow* recipe

- Observe the new columns that are added being detected by the data flow using the schema drift feature

- Split the originally expected columns and newly detected columns into two different streams

- Use rule-based column mapping to dynamically map the new detected columns

Perform the following steps to achieve this:

1. Download the `transaction_table-t1.zip` file from `https://github.com/PacktPublishing/Azure-Data-Engineering-Cookbook-2nd-edition/upload/main/chapter9` and unzip it. Open the `transaction_table-t1.csv` file on your machine using Excel.

2. Insert a column named `d1` after the `pid` column but before the `c1` column. Fill some random values in for the first 10 rows. Insert a column named `d2` after the last column, `c2`. Fill some random values in for the first 10 rows. Save the file:

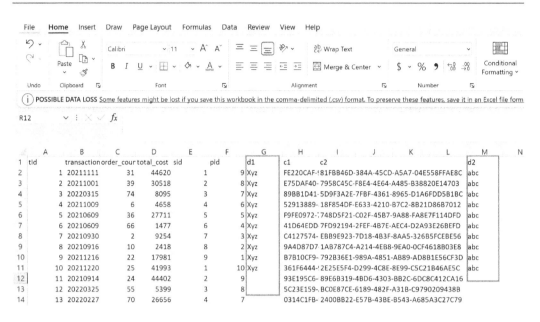

Figure 9.53 – Adding new columns to the data source

3. Upload the `transaction_table-t1.csv` file to the `synapse/csv` folder in the storage account linked with the Synapse workspace. Ensure that you overwrite the existing file. For detailed example screenshots for a similar task, follow *steps 1* to *4* in the *How to do it…* section from the *Analyzing data using serverless SQL pool* recipe of *Chapter 8, Processing Data Using Azure Synapse Analytics*.

4. Within **packtadesynapse** of Synapse Studio, click on the **Develop** icon (the notebook-like symbol) on the left-hand side of the screen. Expand **Data flows** and click **Copy_CSV_to_Parquet**. Turn on the debug mode by clicking on the toggle button at the top:

Figure 9.54 – Turning on the debug mode

5. Click **OK** to enable the debug mode for the data flow. We will use the debug mode to see how data flows through the transformations in this recipe:

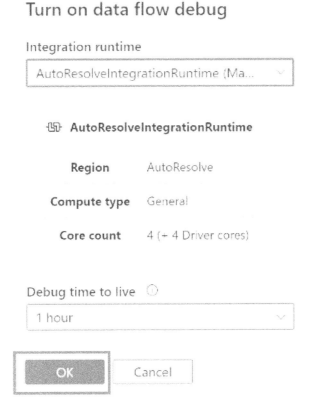

Turn on data flow debug

Integration runtime

AutoResolveIntegrationRuntime (Ma...

🔄 **AutoResolveIntegrationRuntime**

Region   AutoResolve

Compute type   General

Core count   4 (+ 4 Driver cores)

Debug time to live ⓘ

1 hour

OK   Cancel

Figure 9.55 – Turning on the debug mode

6. Click on source transformation (**csv**). Observe that **Allow schema drift** is turned on by default in **Source settings**:

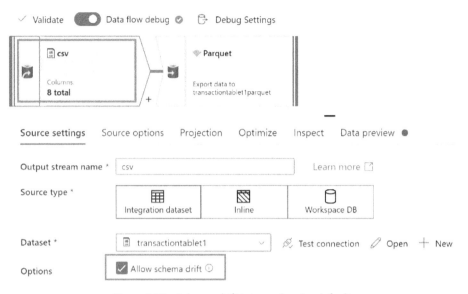

Figure 9.56 – Schema drift is turned on by default

7. Once the debug mode is ready (it typically takes 5 minutes to prepare after turning on), go to **Data preview**. Hit the **Refresh** button. Scroll to the right. Notice that the new columns we added in the Excel sheet, namely d1 and d2, have been automatically added. Ignore the _c9, _c10, and _c11 columns that Excel has added by itself. Also, notice that even though we added the d1 column in the middle of the existing columns, the data flow has positioned it to the end, as it is not an expected column and has been newly added. Click on the **Map drifted** button:

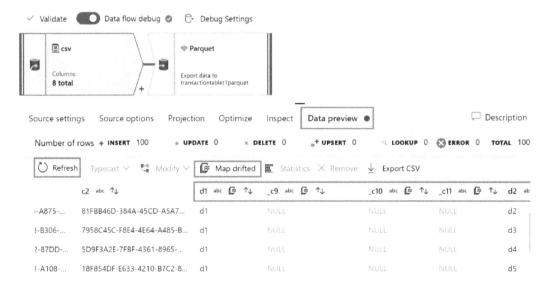

Figure 9.57 – Detecting the new columns

8. Clicking the **Map drifted** button creates a derived column transformation. Click on it and notice that the data flow has automatically created the mappings for the new columns. To make the column mapping dynamic, let's delete the existing column mappings first. Select all the column mappings by checking the checkboxes in the **Columns** list and hitting **Delete**:

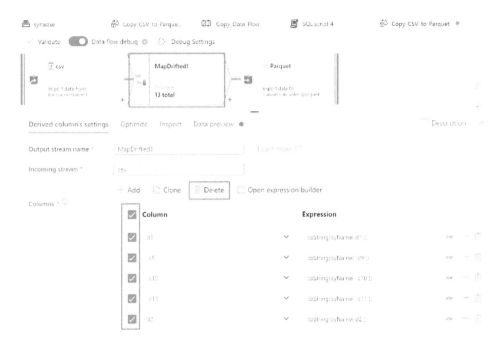

Figure 9.58 – Deleting the column mappings

9. To make column mapping dynamic, let's use the column pattern option. Click **Add** and select **Add column pattern**:

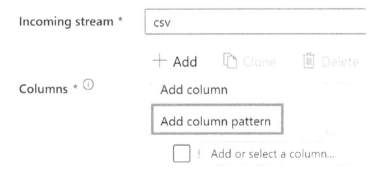

Figure 9.59 – Adding a column pattern

10. Click on the **column1** mapping and hit the delete button. Check the checkbox next to **Each column that matches** and type in `position > 8`. Type `$$` in both textboxes, on the left and right. By specifying `position > 8`, for each column after position number 8 in the list of columns read by the **Mapdrifted1** transformation, the data flow will dynamically create a new column. As all our columns after position 8 (counting the columns from left to right) are derived columns (new columns), we have configured the data mapping to automatically create columns for them. $$ represents the value of the column and by specifying $$ in the left- and right-hand text boxes, we are accepting the values in the columns as they are:

Figure 9.60 – Adding a column pattern

11. A key objective is to separate the derived columns and originally mapped columns into two streams so that the original data transformation from source to sink doesn't break and the expected columns are transferred as planned from the CSV file to the Parquet destination. To achieve this, we can configure a fixed rule mapping for the **Parquet sink transformation**. Click on the **Parquet** sink transformation and go to the **Mapping** section. The original columns are automatically listed. Check the checkbox next to **Input columns** to map all eight original columns that are expected to be copied to the Parquet destination folder:

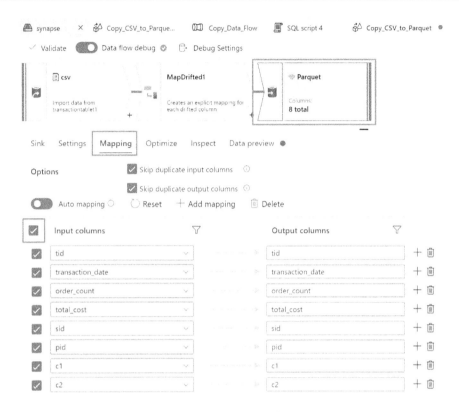

Figure 9.61 – The Parquet sink destination

12. Now, let's get the primary key and the drifted columns into a separate stream. Click on the + symbol on the right-hand side of the **MapDrifted1** transformation again and add **Sink**. This sink will serve as the secondary path for tracking whether there are any drifted columns from the source:

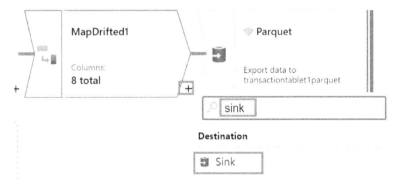

Figure 9.62 – Adding a second sink

13. For the purpose of testing, we set **Sink type** to **Cache**. You can set the sink for the drifted columns to a Parquet dataset (or any integration dataset) if you wish to store the values of the drifted column:

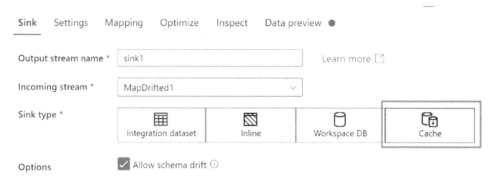

Figure 9.63 – The second sink type

14. Click on the newly added **sink1** transformation (the cache sink) and go to the **Mapping** section. Disable **Auto mapping**. Delete all mapping except the mapping for the **tid** column, as shown in the following screenshot. Once done, click **Add mapping** and select **Rule-based mapping**. Rule-based mapping helps us to dynamically create and define columns based on the column patterns:

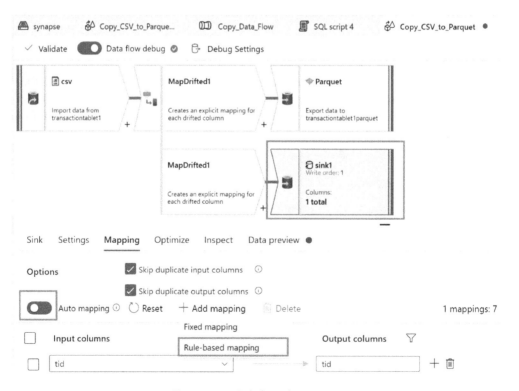

Figure 9.64 – Rule-based mapping

15. For **Rule-based mapping**, specify the column pattern as `position > 8` and the value as `$$`, which implies that columns from any position after 8 will be used as they are, as in *step 10*. Check the **tid** mapping and the **Rule-based mapping** checkboxes:

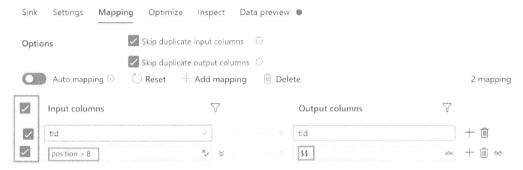

Figure 9.65 – Rule-based mapping for the sink

16. Click **Data preview** and hit the **Refresh** button to see the **tid** column and the derived columns only:

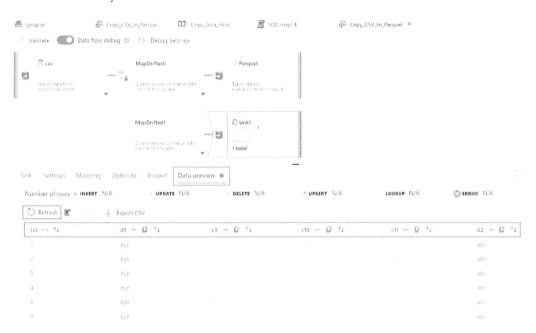

Figure 9.66 – The result at the sink

How it works...

Schema drifting and rule-based mapping offer resilience and flexibility for the pipelines to accommodate almost any kind of change made at the data source. In this recipe, we used the **Map drifted** columns option and a derived columns transformation in *step 7* and *step 8* to identify the unplanned or the new columns for the source. We split the new columns using rule-based mapping to flow to a separate sink (a cache sink for testing) and ensured that the originally expected columns flowed to the Parquet destination using fixed rule mapping. Using the preceding recipe, data engineers can ensure that their pipelines transfer data as expected, but they can also track any new columns that are added to the source and can make adjustments to their pipelines if required.

Building the Serving Layer in Azure Synapse SQL Pool

As introduced in *Chapter 8*, *Processing Data Using Azure Synapse Analytics*, Azure Synapse Analytics comprises three key components – the Synapse integration pipeline to ingest and transform the data, the SQL pool and the Spark pool to process and serve the data, and Power BI integration to visualize the data. The SQL pool, the relational data warehouse engine, is usually the layer that maintains the processed data and will be used as the data serving layer consumed by reporting solutions. So, this chapter will focus on how to load processed data into a dedicated SQL pool and maintain dedicated SQL pools.

In this chapter, we will cover the following main topics:

- Loading data into dedicated SQL pools using PolyBase and T-SQL

- Loading data into dedicated SQL pools using COPY INTO

- Creating distributed tables and modifying table distribution

- Creating statistics and automating the update of statistics

- Creating partitions and archiving data using partitioned tables

- Implementing workload management in an Azure Synapse dedicated SQL pool

- Creating workload groups for advanced workload management

By the end of the chapter, you will have learned how to load data into a SQL pool using the COPY INTO and PolyBase methods, distribute the table using hash/replicate/round-robin algorithms, partition and archive data, maintain table statistics, and perform efficient resource allocation and management using workload management.

Technical requirements

For this chapter, you will need the following:

- A Microsoft Azure subscription

Loading data into dedicated SQL pools using PolyBase and T-SQL

In this recipe, we will load a CSV file into a dedicated SQL pool using PolyBase. PolyBase technology involves creating an external table that links the SQL pool with the file(s) stored in data lake storage. Once the PolyBase table has been created, you will be able to query the external table like any other table and data will be returned from data lake storage seamlessly. This recipe will involve the following tasks:

- Creating a dedicated Synapse SQL pool

- Creating an external table to read CSV files

- Loading the data read from an external table into a regular table stored in a Synapse dedicated SQL pool

Getting ready

To get started, perform the following steps:

1. Log in to `https://portal.azure.com` using your Azure credentials.

2. Create a Synapse Analytics workspace as explained in the *Provisioning an Azure Synapse Analytics workspace* recipe of *Chapter 8*, *Processing Data Using Azure Synapse Analytics*.

3. Download the `transaction-tbl.csv` and `transaction-tbl.parquet` files from `https://github.com/PacktPublishing/Azure-Data-Engineering-Cookbook-2nd-edition/tree/main/chapter10`.

4. In the Synapse Analytics workspace, create a folder named `files` in the data lake account attached to the Synapse Analytics workspace. Upload the `transaction-tbl.csv` and `transaction-tbl.parquet` files to the `files` folder.

For detailed screenshots from a similar task, follow *steps 1* to *4* in the *How to do it…* section from the *Analyzing Data Using Serverless SQL Pool* recipe of *Chapter 8*, *Processing Data Using Azure Synapse Analytics*.

How to do it...

Perform the following steps to load the data:

1. First, let's create a Synapse dedicated SQL pool. Log in to `portal.azure.com`, go to **All resources**, and search for `packtadesynapse`, which is the Synapse Analytics workspace that was created in the *Provisioning an Azure Synapse Analytics workspace* recipe of *Chapter 8, Processing Data Using Azure Synapse Analytics*. Click on the workspace. Click on **Open Synapse Studio**:

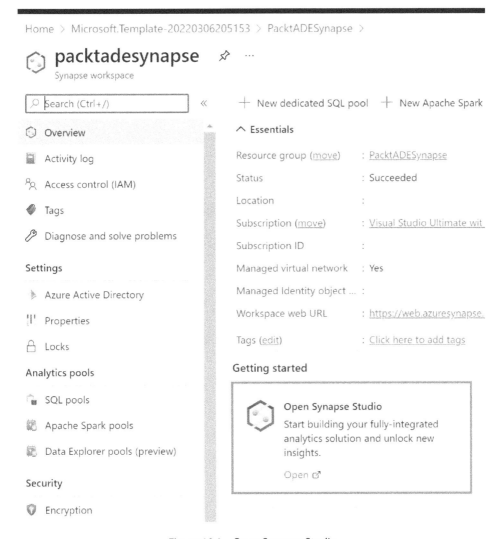

Figure 10.1 – Open Synapse Studio

2. Click on the manage button (the briefcase symbol) on the left-hand side, which will take you to the **Manage** section. Under **SQL pools**, click on **New**:

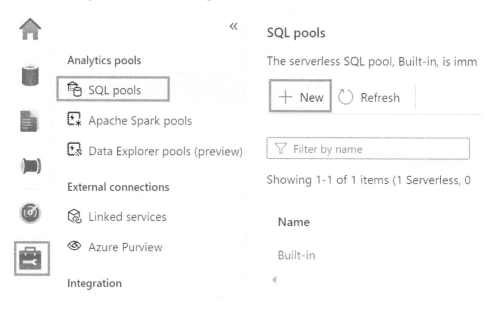

Figure 10.2 – Creating a new SQL pool

3. Create a dedicated SQL pool called `packtadesqlpool`. Set **Performance level** to **DW100c**. Click on **Review + create** and create the SQL pool:

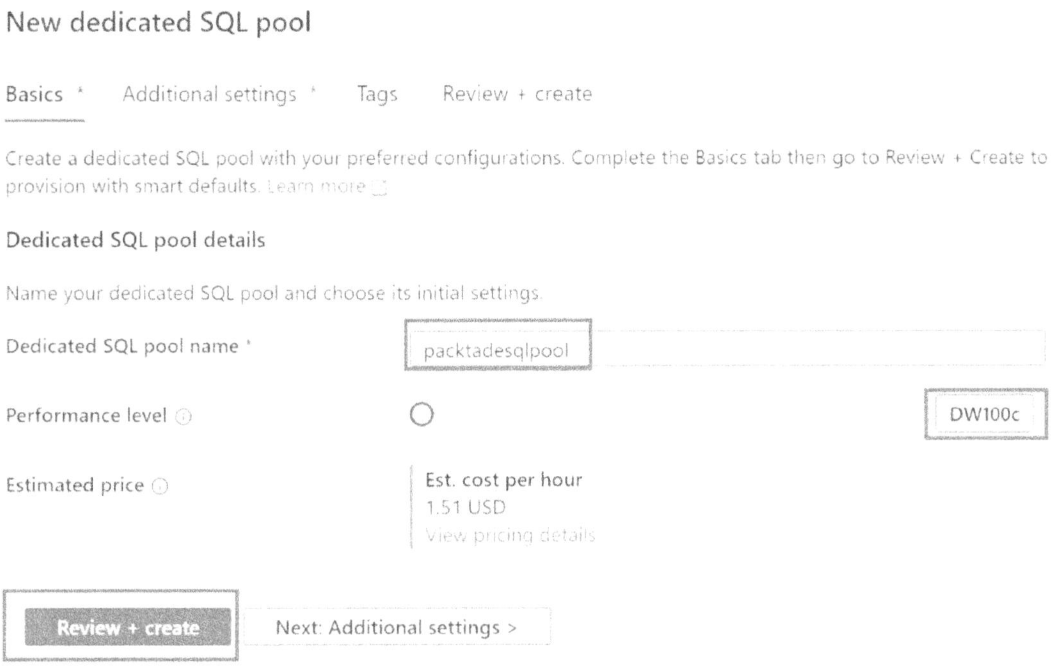

New dedicated SQL pool

Basics *   Additional settings *   Tags   Review + create

Create a dedicated SQL pool with your preferred configurations. Complete the Basics tab then go to Review + Create to provision with smart defaults. Learn more ⌐

Dedicated SQL pool details

Name your dedicated SQL pool and choose its initial settings.

Dedicated SQL pool name *	packtadesqlpool
Performance level ⓘ	◯ DW100c
Estimated price ⓘ	Est. cost per hour 1.51 USD View pricing details

Review + create   Next: Additional settings >

Figure 10.3 – Provisioning the SQL pool

4. Click on the blue cylinder (the data symbol) on the left-hand side, which will take you to the **Data** section. Click on the **Linked** tab. Expand the data lake account of the Synapse workspace (in this example, `packtadesynapse`). Click on the **synapse (Primary)** container. Open the **Files** folder. Right-click on the `transaction-tbl.csv` file. Select **New SQL script** and then select **Create external table**:

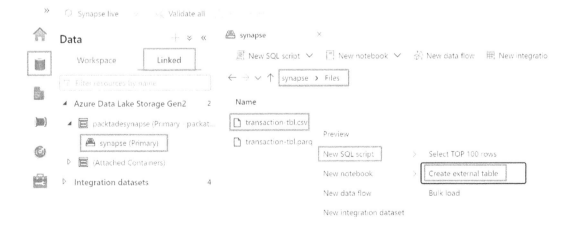

Figure 10.4 – Creating an external table

5.   Select the **Infer column names** option. Click on **Continue**:

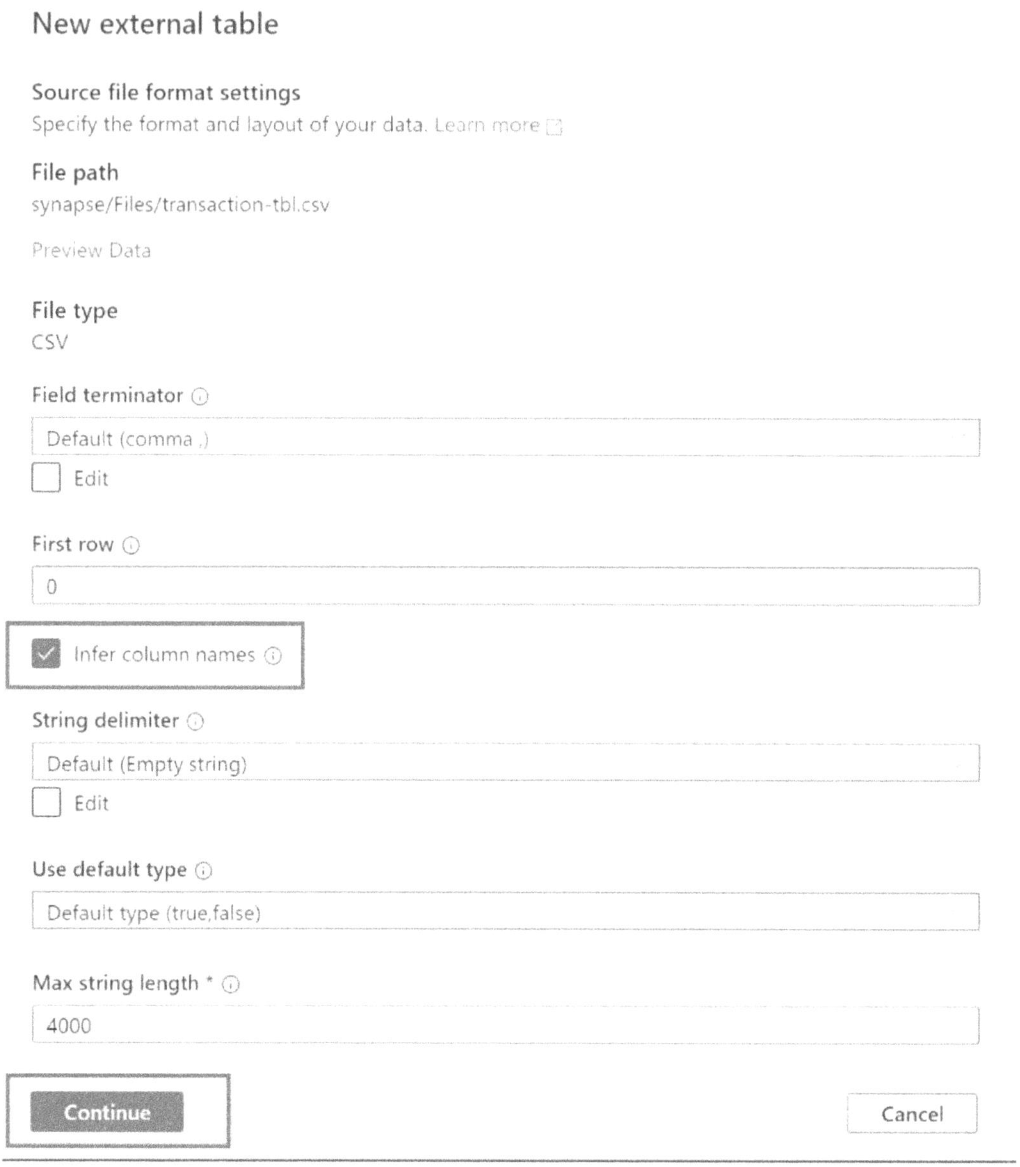

New external table

Source file format settings
Specify the format and layout of your data. Learn more ⬚

File path
synapse/Files/transaction-tbl.csv

Preview Data

File type
CSV

Field terminator ⓘ

Default (comma ,)

☐ Edit

First row ⓘ

0

☑ Infer column names ⓘ

String delimiter ⓘ

Default (Empty string)

☐ Edit

Use default type ⓘ

Default type (true,false)

Max string length * ⓘ

4000

Continue		Cancel

Figure 10.5 – Configuring the external table

6.   Select **packtadesqlpool** as the SQL pool where the external table will be created. Provide a table name of ext_transaction_tbl. Click on **Open script**, and the script for the external table will be generated:

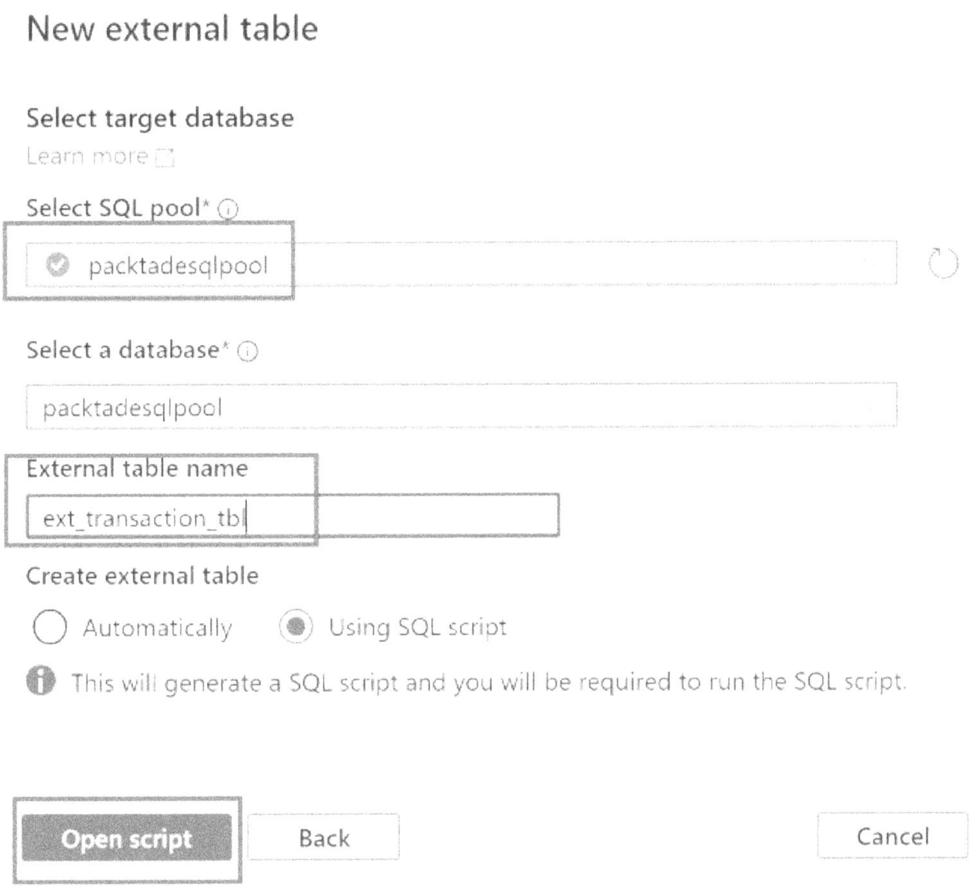

New external table

Select target database
Learn more ☐

Select SQL pool* ⓘ

✅ packtadesqlpool   ↻

Select a database* ⓘ

packtadesqlpool

External table name

ext_transaction_tbl

Create external table

○ Automatically   ● Using SQL script

ⓘ This will generate a SQL script and you will be required to run the SQL script.

Open script   Back   Cancel

Figure 10.6 – Generating the external table script

7. In the CREATE EXTERNAL FILE section, after the USE_TYPE_DEFAULT = FALSE
    line, hit *Enter* and add , FIRST_ROW = 2. This is to ignore the first row since it contains
    the column names:

```
1   IF NOT EXISTS (SELECT * FROM sys.external_file_formats WHERE name = 'SynapseDelimitedTextFormat')
2       CREATE EXTERNAL FILE FORMAT [SynapseDelimitedTextFormat]
3       WITH ( FORMAT_TYPE = DELIMITEDTEXT ,
4             FORMAT_OPTIONS (
5               FIELD_TERMINATOR = ',',
6               USE_TYPE_DEFAULT = FALSE
7               ,FIRST_ROW = 2
8             ))
9   GO
```

Figure 10.7 – Adjustment for the first row

8. In the CREATE EXTERNAL TABLE section, change the data type for columns c1 and c2 to varchar(200) as the uniqueidentifier data type is not yet supported on SQL pools:

```
CREATE EXTERNAL TABLE ext_transaction_tbl (
    [tid] bigint,
    [transaction_date] bigint,
    [order_count] bigint,
    [total_cost] bigint,
    [sid] bigint,
    [pid] bigint,
    [c1] varchar(200),
    [c2] varchar(200)
)
WITH (
LOCATION = 'Files/transaction-tbl.csv',
```

Figure 10.8 – Data type adjustment

9. Hit the **Run** button and verify whether the data has been read successfully from the external table:

Figure 10.9 – Connecting to Azure Data Lake Gen 2

10. Scroll to the bottom of the SQL script window and add the following lines to create a regular table in the dedicated SQL pool and to load the data into a regular table from the external table. The `CREATE TABLE` statement creates a regular table named `dbo.transaction_tbl` stored in a SQL pool. The `CREATE TABLE` statement creates a table with the structure and result set as returned by the `SELECT` query, which follows the `AS` clause. Hit the **Run** button:

```
CREATE TABLE dbo.transaction_
tbl WITH (DISTRIBUTION = ROUND_ROBIN)
AS
Select * from dbo.ext_transaction_tbl;
GO
Select TOP 100 *  from dbo.transaction_tbl
GO
```

The script's result is shown in the following screenshot:

Figure 10.10 – The data loaded and result set returned

You will see that the data has been successfully loaded onto `dbo.transaction_tbl` in the SQL pool, and it is read from the regular SQL pool table.

How it works...

Right-clicking on the CSV file and using the `CREATE EXTERNAL TABLE` option generates most of the SQL script required for creating an external table. For creating external tables against files in a data lake, we need three objects to be created, which are listed here:

- **External File Format** – This configures the file format to be read, which is CSV in our case.
- **External data source** – This specifies which file to be read and where it is located.
- **External table** – This is an actual external table listing the column names and their data type.

Fortunately, all of the preceding objects are automatically handled gracefully by Synapse Studio. Once created, we use the **Create Table As (CTAS)** method to ingest data into a regular table in a SQL pool. `CREATE TABLE <tablename> WITH ( DISTRIBUTION = <> ) AS <SELECT Query>` creates a regular table stored in the dedicated SQL data pool, on the fly, following the structure of the result set of the `SELECT` query. In other words, if the select query contains five columns, a table with five columns matching the data type of the columns returned by the select query will be created. Additionally, the result set data is inserted into the SQL pool table.

While the external table acts as a link to any file(s) stored in the data lake storage account, it is not suitable for running complex queries supporting analytic workloads. Regular tables stored inside a dedicated SQL pool are highly optimized for analytics, and using external tables to ingest the data from a data lake to a dedicated SQL pool is one of the two (the other being using `COPY INTO`) fastest methods available to move data from data lakes to Synapse dedicated SQL pools.

Loading data into a dedicated SQL pool using COPY INTO

Loading data into a dedicated SQL pool using the `COPY INTO` T-SQL statement is a very fast and efficient method. In this recipe, we will load one CSV file and one Parquet file into two dedicated SQL pool tables.

Getting ready

To get started, perform the following steps:

1. Log in to `https://portal.azure.com` using your Azure credentials.

2. Create a Synapse Analytics workspace, as explained in the *Provisioning an Azure Synapse Analytics workspace* recipe of *Chapter 8, Processing Data Using Azure Synapse Analytics*.

3. Download the `transaction-tbl.csv` and `transaction-tbl.parquet` files from `https://github.com/PacktPublishing/Azure-Data-Engineering-Cookbook-2nd-edition/tree/main/chapter10`.

4. In the Synapse Analytics workspace, create a folder named `files` in the data lake account attached to the Synapse Analytics workspace. Upload the `transaction-tbl.csv` and `transaction-tbl.parquet` files to the `files` folder.

    For detailed screenshots from a similar task, follow *steps 1 to 4* in the *How to do it...* section from the *Analyzing data using serverless SQL pool* recipe of *Chapter 8, Processing Data Using Azure Synapse Analytics*.

5. Create a Synapse dedicated SQL pool named `packtadesqlpool`, as described in *steps 1 to 3* in the *How to do it...* section of the *Loading data into dedicated SQL pools using PolyBase using T-SQL* recipe.

How to do it...

1. Let's load a CSV file into a dedicated SQL pool using the `COPY INTO` statement. Log in to `portal.azure.com`, go to **All resources**, and search for `packtadesynapse`, which is the Synapse Analytics workspace that we created in the *Provisioning an Azure Synapse Analytics workspace* recipe of *Chapter 8, Processing Data Using Azure Synapse Analytics*. Click on the workspace. Click on **Open Synapse Studio**:

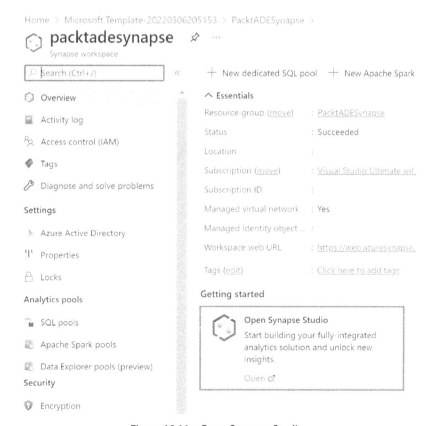

Figure 10.11 – Open Synapse Studio

2. Click on the develop button (the notebook-like symbol) on the left-hand side, which will take you to the **Develop** section. Click on the + button and select **SQL script**:

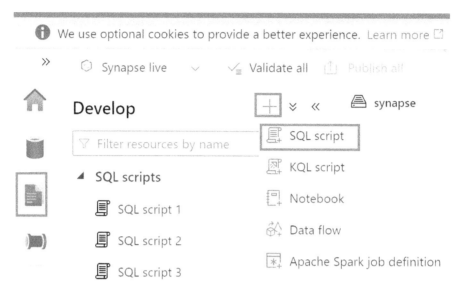

Figure 10.12 – SQL script

3. Paste the following script into the new SQL script window. Select **packtadesqlpool** in the **Connect to** option, and connect to **packtadesqlpool**:

```
CREATE TABLE dbo.transaction_tbl_copy([tid] bigint,
    [transaction_date] date,
    [order_count] bigint,
    [total_cost] bigint,
    [sid] bigint,
    [pid] bigint,
    [c1] nvarchar(200),
    [c2] nvarchar(200))
  WITH (  DISTRIBUTION  = ROUND_ROBIN);
    GO
COPY INTO dbo.transaction_tbl_copy
FROM 'https://packatadesynapse.dfs.core.windows.net/
synapse/Files/transaction-tbl.csv'
WITH
(
FILE_TYPE = 'CSV',
```

```
    MAXERRORS = 10,
    FIRSTROW = 2)
    Select top 100 * from dbo.transaction_tbl_copy
```

The preceding script creates a regular table, transaction_tbl_copy, in a dedicated SQL pool, packtadesqlpool. The COPY INTO statement loads the data from the transaction-tbl.csv file to the transaction_tbl_copy table. The select statement returns the rows by reading from the transaction_tbl_copy table. Update the path of transaction-tbl.csv as per your storage account's name and location, if required:

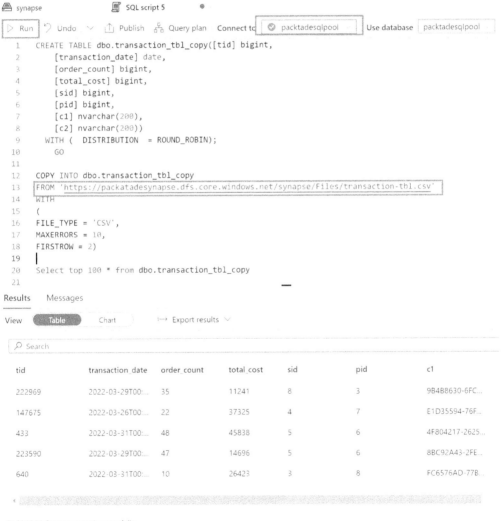

Figure 10.13 – Loading data

4.   Now, let's load the Parquet file. Copy the following script to the bottom of the query window to create a table in the dedicated SQL pool and load the data from the Parquet file. Select the following script and run it. Ensure to update the storage account name and location as per your environment:

```
COPY INTO dbo.transaction_tbl_Parquet
FROM 'https://packatadesynapse.dfs.core.windows.net/
synapse/Files/transaction-tbl.parquet'
 WITH (
    FILE_TYPE = 'Parquet',

    AUTO_CREATE_TABLE = 'ON'
)
Select top 100 * from dbo.transaction_tbl_Parquet
```

The result of the data loading script is shown in the following screenshot:

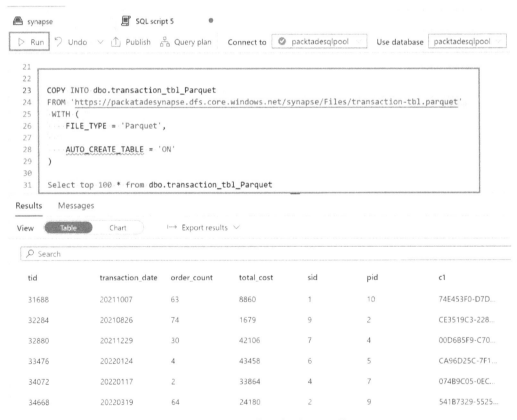

Figure 10.14 – Loading the Parquet file

A new table called `transaction_tbl_parquet` has been successfully created in the dedicated SQL pool and data was read successfully.

How it works...

`COPY INTO` is a simple, easy, and efficient way to load data into dedicated SQL pools. The `COPY INTO` method doesn't involve the creation of any additional objects such as external tables to load data into dedicated SQL pools.

While loading the data from the CSV file, we had to create the table and schema upfront and then issue the `COPY INTO` command to load the data. The schema defined in the script should exactly match the columns in the CSV file for the load to be successful. However, while loading the data to Parquet files, we didn't create a table up front, and we only issued one command, `COPY INTO`, to create the table in the dedicated SQL pool and load the data. We used the `AUTO_CREATE_TABLE = ON` option in the `COPY INTO` command while loading the Parquet files, as the `AUTO_CREATE_TABLE = ON` option is supported for Parquet files and it automatically creates the schema for the dedicated SQL pool table based on the Parquet file structure.

Creating distributed tables and modifying table distribution

A Synapse dedicated SQL pool is a distributed database allowing us to distribute data across multiple nodes and process it in parallel at multiple nodes to make data processing more efficient. To leverage the powers of the distributed database effectively, we need to store the data across the nodes with the correct strategy. A Synapse dedicated SQL pool offers three options while distributing the data, and they are listed as follows:

- **Round-Robin distribution** – This distributes the data equally across the nodes of the SQL pool.
- **Hash distribution** – This distributes the data based on the values of a column in the table.
- **Replicated** – This copies the table to all nodes of the dedicated SQL pool.

The table distribution option is selected at the time of table creation and can't be modified afterward. In this recipe, we will create a table specifying a distribution option, and we will modify the distribution option using a workaround.

Getting ready

To get started, perform the following steps:

1. Log in to `https://portal.azure.com` using your Azure credentials.
2. Create a Synapse Analytics workspace, as explained in the *Provisioning an Azure Synapse Analytics workspace* recipe of *Chapter 8, Processing Data Using Azure Synapse Analytics.*

3. Complete the *Loading data into dedicated SQL pools using PolyBase using T-SQL* recipe to create a dedicated SQL pool named `packtadesqlpool`, an external table named `dbo.ext_transaction_tbl`, and a dedicated SQL pool table named `dbo.transaction_tbl`.

Alternatively, you can use the `Create_External_Table.SQL` script from `https://github.com/PacktPublishing/Azure-Data-Engineering-Cookbook-2nd-edition/tree/main/chapter10` to create `dbo.ext_transaction_tbl` and `dbo.transaction_tbl`.

How to do it...

Perform the following steps to create and modify distributed tables:

1. Log in to `portal.azure.com`, go to **All resources**, and search for `packtadesynapse`. Click on the workspace. Click on **Open Synapse Studio**. Click on the develop button (the notebook-like button) on the left-hand side. Click on the + symbol and select **New SQL script**. Select `packtadesqlpool` in the **Connect to** option. Refer to *steps 1* and *2* in the *How to do it...* section of the *Loading data into dedicated SQL pools using COPY INTO* recipe from this chapter. In the new query window, copy and paste the following script:

```
CREATE TABLE dbo.transaction_tbl_
hash WITH (DISTRIBUTION = HASH(SID))
AS
Select * from dbo.ext_transaction_tbl;
GO
Select top 10 * from dbo.transaction_tbl_hash
```

The preceding script creates a table named `dbo.transaction_tbl_hash`, which is distributed using hash distribution. The `DISTRIBUTION = HASH(SID)` command in the `WITH` clause will create the distributed table based on the values within the `SID` column. The data from the external table, `ext_transaction_tbl`, is loaded into the `dbo.transaction_tbl_hash` table:

Figure 10.15 – Creating the hash table

2. Let's convert `dbo.transaction_tbl_hash` into replicate distribution. The steps involved in doing so are as follows:

    - Create a new table using the CTAS method and load the data from an existing table.

    - Drop the old table.

    - Rename the new table to the old table name.

    Create a new query window, connect to `packtadesqlpool`, and paste the following script:

```
CREATE TABLE dbo.transaction_tbl_
temp WITH ( DISTRIBUTION = REPLICATE)
AS
SELECT * FROM dbo.transaction_tbl_hash;
DROP TABLE dbo.transaction_tbl_hash;
RENAME OBJECT dbo.transaction_tbl_temp to transaction_
tbl_hash
GO

SELECT TOP 10 * FROM dbo.transaction_tbl_hash
GO
```

The preceding script does the following:

- The CREATE TABLE command's WITH clause specifies DISTRIBUTION=REPLICATE as the option and is created using the CTAS syntax using a select query against the dbo.transaction_tbl_hash table. This ensures that the new table created (dbo.transaction_tbl_temp) is based on replicate distribution and it contains the same structure and data as dbo.transaction_tbl_hash.

- Once dbo.transaction_tbl_temp has been created with the same data as dbo.transaction_tbl_hash, the dbo.transaction_tbl_hash table can be dropped.

- Now dbo.transaction_tbl_temp is renamed to dbo.transaction_tbl_hash (the original table name) using the RENAME OBJECT command. The result is verified using the SELECT TOP 10 * command:

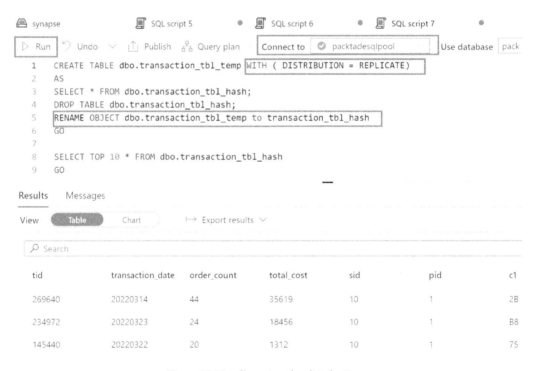

Figure 10.16 – Changing the distribution

Using the preceding method, we have successfully changed the distribution of the table from hash distribution to replicate distribution.

How it works...

Setting the table distribution is a critical task and needs to be done at the time of table creation. To change the table distribution, we need to recreate the table with a new distribution option. In the preceding example, we created a new table based on old the table with a different distribution method, dropped the old table, and renamed the new table to the old name. While the preceding method works, in very large tables, this could be a resource-intensive task as we need to recreate the table again. So, getting the distribution strategy correct at the time of table creation can save a lot of time and effort. Secondly, as the method involves recreating the table and dropping the old table, any additional indexes, statistics, or permissions created in the old table need to be created again in the new table.

Creating statistics and automating the update of statistics

Statistics play a key role in maintaining the optimal performance of the queries in a Synapse dedicated SQL pool. For each query submitted, the query optimizer prepares a query plan that comprises several operations (such as filtering, joining, sorting, and more). Statistics inform the optimizer how many rows are expected to be returned for each of those operations, and based on the statistics input, the optimizer prepares query plans. So, for the query plans to be effective and performance to be optimal, we need good statistics.

In this recipe, we will learn how to create statistics and update statistics.

Getting ready

Create a Synapse Analytics workspace as explained in the *Provisioning an Azure Synapse Analytics workspace* recipe of *Chapter 8, Processing Data Using Azure Synapse Analytics.*

Complete the *Loading data into dedicated SQL pools using PolyBase using T-SQL* recipe to create a dedicated SQL pool named `packtadesqlpool`, an external table named `dbo.ext_transaction_tbl`, and a dedicated SQL pool table named `dbo.transaction_tbl`. Alternatively, you can use the `Create_External_Table.SQL` script from `https://github.com/PacktPublishing/Azure-Data-Engineering-Cookbook-2nd-edition/tree/main/chapter10` to create the `dbo.ext_transaction_tbl` and `dbo.transaction_tbl` tables.

How to do it...

In a Synapse dedicated SQL pool, by default, statistics are created automatically for tables stored in the SQL pool. For the external tables, the statistics are to be created manually. As the data in the table changes, the statistics are to be updated so that they can offer accurate estimates for the query optimizer to prepare efficient query plans. Statistics are not updated automatically for both external tables stored in a data lake and regular tables stored in Synapse dedicated SQL pools. So, in this recipe, we will perform the following tasks:

- Creating statistics for external tables
- Updating the statistics for external tables
- Updating the statistics for regular tables

The following steps can be used to perform these tasks:

1. First, let's create statistics against the external table, dbo.ext_transaction_tbl. Log in to portal.azure.com, go to **All resources**, and search for packtadesynapse. Click on the workspace. Click on **Open Synapse Studio**. Click on the develop button (the notebook-like button) on the left-hand side. Click on the + symbol and select **New SQL script**. Select packtadesqlpool in the **Connect to** option. Refer to *steps 1* and *2* in the *How to do it...* section of the *Loading data into dedicated SQL pools using COPY INTO* recipe for detailed instructions on how to open a new query window. In the new query window, copy and paste the script provided in the following snippet and hit the **Run** button:

   ```
   CREATE STATISTICS ext_transaction_tbl_pid on ext_
   transaction_tbl(pid)
   GO
   DBCC SHOW_STATISTICS(ext_transaction_tbl,ext_transaction_
   tbl_pid)
   GO
   ```

   The CREATE STATISTICS command creates a statistic for the pid column in the ext_transaction external table. The DBCC SHOW_STATISTICS command reads the statistic and displays the expected number of rows for each value in the pid column and confirms that the statistic was successfully created. You can learn more about interpreting the results of SHOW_STATISTICS at https://docs.microsoft.com/en-us/sql/t-sql/database-console-commands/dbcc-show-statistics-transact-sql?view=azure-sqldw-latest:

Figure 10.17 – Creating statistics in an external table

2. Let's try to update the statistic created for an external table. As of writing (May 2022), updating statistics is not supported for external tables. So, as a workaround, we will have to drop and create the statistic on an external table. In a new query window connected to the `packtadesqlpool` database, copy the following script and hit the **Run** button:

```
DROP STATISTICS ext_transaction_tbl.ext_transaction_tbl_
pid
GO
CREATE STATISTICS ext_transaction_tbl_pid on ext_
transaction_tbl(pid)
GO
DBCC SHOW_STATISTICS(ext_transaction_tbl,ext_transaction_
tbl_pid) WITH STAT_HEADER
GO
```

The `DROP STATISTICS` command specifies the table name and the statistic name that needs to be dropped. The `CREATE STATISTICS` command creates the statistic in the column again. The `DBCC SHOW_STATISTICS` command has a `WITH` clause specifying the `STAT_HEADER` option, which provides the date and time at which the statistic was updated or created:

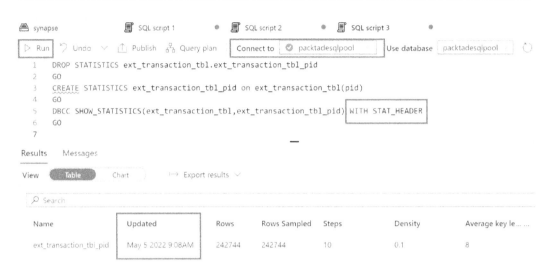

Figure 10.18 – Updating the statistics in the external table

3. As statistics are automatically created in regular tables in a dedicated SQL pool, let's focus on keeping the statistic updated. Download the `update_statistics_regular_table.SQL` file from `https://github.com/PacktPublishing/Azure-Data-Engineering-Cookbook-2nd-edition/blob/main/chapter10/update_statistics_regular_table.sql`. Click on the + button (the same button used to create a new query window) on the left-hand side and select **Import**:

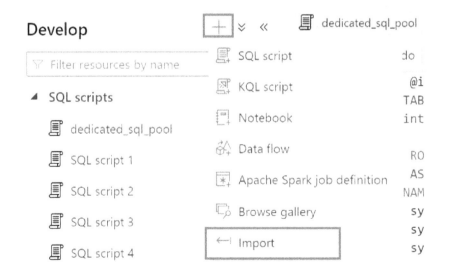

Figure 10.19 – Importing the SQL script

4.   Select the downloaded `update_statistics_regular_table.SQL` script, connect to `packtadesqlpool`, and click on the **Run** button:

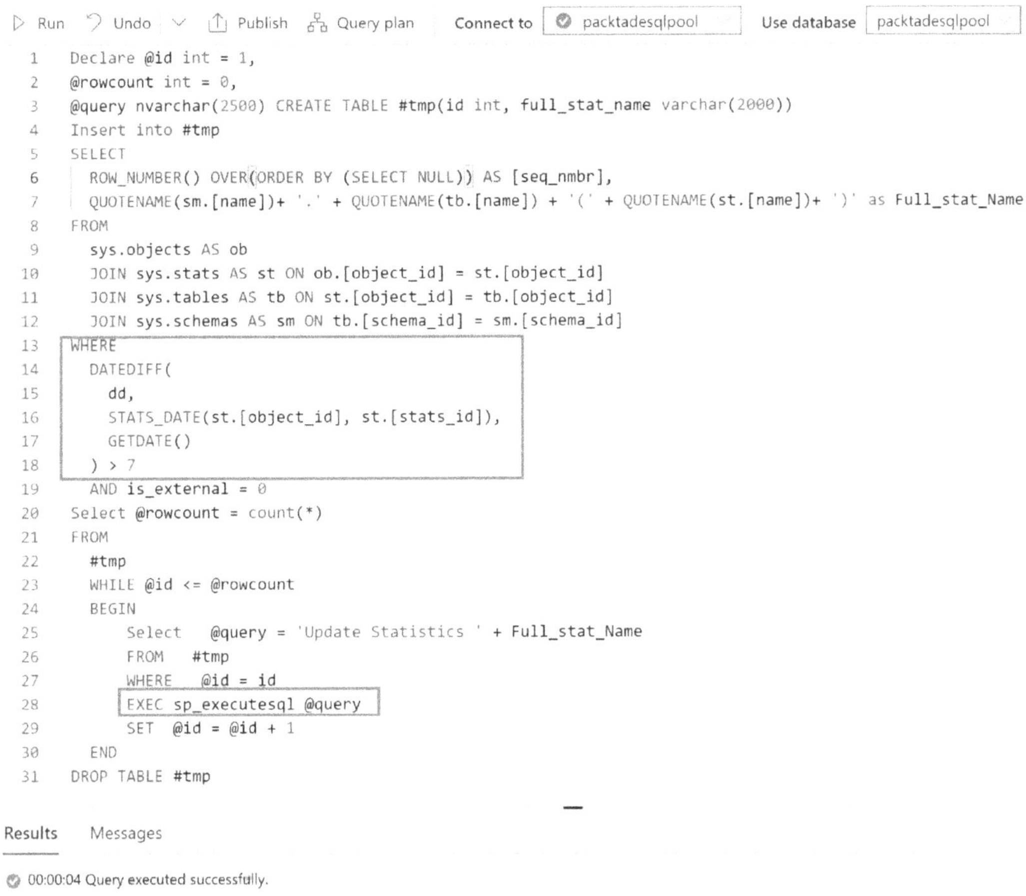

```
▷ Run    ↺ Undo  ∨   ⬆ Publish  ⬚ Query plan    Connect to  ✅ packtadesqlpool         Use database  packtadesqlpool

 1   Declare @id int = 1,
 2   @rowcount int = 0,
 3   @query nvarchar(2500) CREATE TABLE #tmp(id int, full_stat_name varchar(2000))
 4   Insert into #tmp
 5   SELECT
 6     ROW_NUMBER() OVER(ORDER BY (SELECT NULL)) AS [seq_nmbr],
 7     QUOTENAME(sm.[name])+ '.' + QUOTENAME(tb.[name]) + '(' + QUOTENAME(st.[name])+ ')' as Full_stat_Name
 8   FROM
 9     sys.objects AS ob
10     JOIN sys.stats AS st ON ob.[object_id] = st.[object_id]
11     JOIN sys.tables AS tb ON st.[object_id] = tb.[object_id]
12     JOIN sys.schemas AS sm ON tb.[schema_id] = sm.[schema_id]
13   WHERE
14     DATEDIFF(
15       dd,
16       STATS_DATE(st.[object_id], st.[stats_id]),
17       GETDATE()
18     ) > 7
19     AND is_external = 0
20   Select @rowcount = count(*)
21   FROM
22     #tmp
23     WHILE @id <= @rowcount
24     BEGIN
25         Select    @query = 'Update Statistics ' + Full_stat_Name
26         FROM    #tmp
27         WHERE    @id = id
28         EXEC sp_executesql @query
29         SET  @id = @id + 1
30     END
31   DROP TABLE #tmp
```

Results   Messages

✅ 00:00:04 Query executed successfully.

Figure 10.20 – Updating the statistics for regular tables

The script in `update_statistics_regular_table.SQL` identifies all the statistics in regular tables that have not been updated for the last 7 days. The script identifies them using system views, such as `sys.tables`, `sys.stats`, and a function called `STATS_DATE`, which returns the last date the statistic was updated.

The syntax for updating the statistic is `Update Statistic <Schema Name>.<Table Name>.<Statistic Name>`. The script checks whether the data returned by the `STATS_DATE` function is older than 7 days, prepares the `Update Statistic` command for each old statistic, and stores the command in a temporary table, `#tmp`. The script loops through each row in temporary table `#tmp` using a `WHILE` loop, extracts the `Update Statistic` command, and dynamically executes the `Update Statistic` command for each old statistic using the `sp_executesql` command.

How it works...

Creating statistics for external tables is straightforward. Updating statistics against external tables is not supported, and you might need to drop and recreate statistics to keep the statistics against the external table up to date. Additionally, we need to remember that for external tables, as statistics are not created automatically, we need to create them for the key columns (columns in the WHERE clause, JOIN conditions, sort operations, and GROUP BY clause) and maintain them diligently to keep the performance of the queries against external tables optimal.

For regular tables stored in a dedicated SQL pool, statistic creation is automatic by default. To keep the statistic updated, we used the script in the update_statistics_regular_table.SQL file. This script automatically identifies all the statistics in your database that haven't been updated for the last 7 days and updates each of the old statistics. The script could be created as a stored procedure and executed from the SQL pool stored procedure task in the Synapse integration pipeline and scheduled on a weekly basis to fix all outdated statistics.

Creating partitions and archiving data using partitioned tables

Partitioned tables split the table into smaller segments based on a particular column's values. Splitting the table into multiple partitions makes data loading and data archival faster and can also improve query performance. In this recipe, we will perform the following tasks:

- Creating a partitioned table and loading the data
- Archiving data older than 6 months using partition commands

Getting ready

Create a Synapse Analytics workspace as explained in the *Provisioning an Azure Synapse Analytics workspace* recipe of *Chapter 8, Processing Data Using Azure Synapse Analytics*.

Create a Synapse dedicated SQL pool named packtadesqlpool and an external table named ext_transaction_tbl, as described in *steps 1* to *9* in the *How to do it...* section of the *Loading data into dedicated SQL pools using PolyBase using T-SQL* recipe. Alternatively, you can use the Create_External_Table.SQL script from https://github.com/PacktPublishing/Azure-Data-Engineering-Cookbook-2nd-edition/tree/main/chapter10 to create the dbo.ext_transaction_tbl and dbo.transaction_tbl tables.

How to do it...

In this recipe, we will create a partitioned table called `Transaction_Partitioned`. We will create a partition based on the `transaction_date` column. We will partition the table into seven partitions in the following fashion:

- The first partition will contain transaction data from before October 2021.

- The second to sixth partitions will contain data from October 1, 2021, to March 1, 2022. Each partition will contain one month of data.

- The seventh partition will contain data after March 1, 2022.

Once the table has been partitioned as described in the preceding list, we will archive data (copy the data to another table and delete it from the original table) older than October 1, 2021, using a method called partition switching. The steps are provided as follows:

1. Download the `table_partition_boundary.sql` file from `https://github.com/PacktPublishing/Azure-Data-Engineering-Cookbook-2nd-edition/blob/main/chapter10/table_partition_boundary.sql`.

2. Log in to `portal.azure.com`, go to **All resources**, and search for `packtadesynapse`. Click on the workspace. Click on **Open Synapse Studio**. Click on the develop button (the notebook-like button) on the left-hand side. Click on the + symbol and select **Import**:

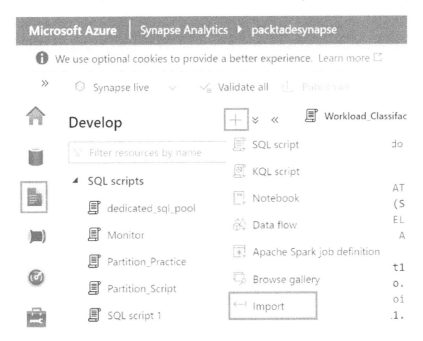

Figure 10.21 – Updating the statistics for the regular tables

3. Select the `table_partition_boundary.sql` file, which was downloaded in *step 1*. Connect to the `packtadesqlpool` database and run the script. The script creates a view called `table_partition_boundary`, which will be used in this recipe to get the table partition details:

Figure 10.22 – Creating the table partition boundary view

4. Let's create a partitioned table, `Transaction_Partitioned`, as described in the third step. Click on the + button and select **SQL Script** to open a new query window. Connect to **packtadesqlpool**. Copy and paste the following script into the query window and run the script:

```
CREATE TABLE [dbo].[Transaction_Partitioned]
WITH
(
    DISTRIBUTION = ROUND_ROBIN,
    CLUSTERED COLUMNSTORE INDEX,
    PARTITION    (   [transaction_
date] RANGE RIGHT FOR VALUES
                    (20211001,20211101,20211201,20220101,
20220201,20220301
                    )
            )
)
AS
```

```
Select * from dbo.ext_transaction_tbl
GO
Select min(transaction_date) as min_trans_
dt,max(transaction_date) as max_trans_dt
from Transaction_Partitioned;
GO
```

This script creates a partitioned table named `Transaction_Partitioned`. In the `Create table` command of the `Partition` clause, we have specified the `transaction_date` column, which implies that the table will be partitioned based on the `transaction_date` column. In the `RANGE RIGHT FOR VALUES` clause, six dates have been specified, which implies that the table will have seven partitions. The values specified in the `RANGE RIGHT FOR VALUES` clause are called **boundary values**. The first boundary value in the list is `20211001` (October 1, 2021), which would mean that all rows with `transaction_date` less than `20211001` (before October 1, 2021) will be stored in partition 1. The rows between `20211101` (November 1, 2021) and `20211201` (December 1, 2021) will be stored in partition 2 and similarly up till partition 6. All the rows after `20220301` (March 1, which is the last boundary value specified on the list) will be stored in partition 7. The script result is provided in the following screenshot:

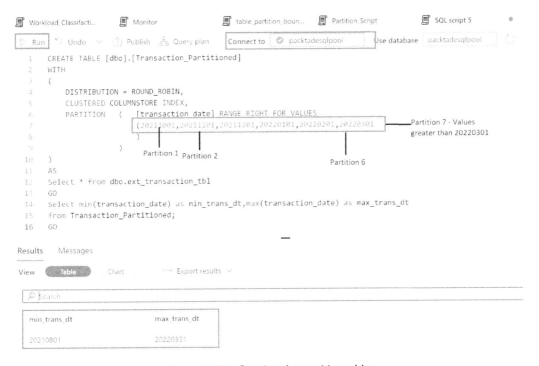

Figure 10.23 – Creating the partition table

5. Let's check the `table_partition_boundary` view to see the partition details. Copy and paste the following query into the same query window and run the query to verify the partition:

```
Select * from dbo.table_partition_boundary
GO
```

```
14    Select min(transaction_date) as min_trans_dt,max(transaction_date) as max_trans_dt
15    from Transaction_Partitioned;
16    GO
17    Select * from dbo.table_partition_boundary
18    GO
```

sults   Messages

iew   Table   Chart   ⟼ Export results ∨

🔎 Search

name	boundary_id	value	rows
Transaction_Partitioned	1	20211001	34677
Transaction_Partitioned	2	20211101	34677
Transaction_Partitioned	3	20211201	34677
Transaction_Partitioned	4	20220101	34677
Transaction_Partitioned	5	20220201	34677
Transaction_Partitioned	6	20220301	34677
Transaction_Partitioned	(NULL)	(NULL)	34682

Figure 10.24 – Verifying the partition details

We can see seven rows representing seven partitions. The `value` column represents the boundary value and the `boundary_id` column is like the partition number. The seventh partition has NULL as its boundary value, as it contains all values greater than `20220301` and, hence, won't have a boundary. The `rows` column represents the approximate row count in each partition.

6. Let's move all the rows from partition 1 to another table as it contains rows from before October 1, 2021. To achieve that, we need to create an empty table with the exact same structure as our original table (`Transaction_Partitioned`). The empty table should be partitioned with a single boundary value that needs to be moved. Copy the following script to the same query window and run the query:

```
CREATE TABLE [dbo].[Transaction_Partitioned_before_oct]
WITH
```

```
(
    DISTRIBUTION = ROUND_ROBIN,
    CLUSTERED COLUMNSTORE INDEX,
    PARTITION    (    [transaction_
date] RANGE RIGHT FOR VALUES
                    (20211001
                    )
                )
)
AS
Select * from dbo.Transaction_Partitioned where 1 = 2;
```

Here is the script in the query window:

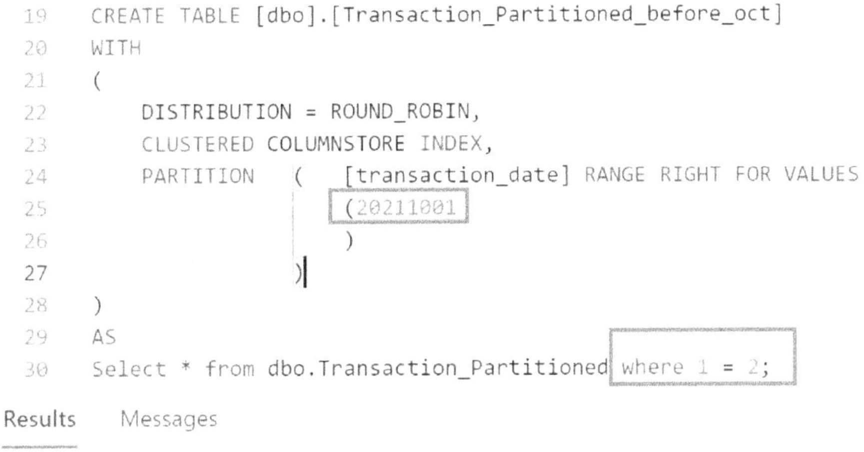

```
19    CREATE TABLE [dbo].[Transaction_Partitioned_before_oct]
20    WITH
21    (
22        DISTRIBUTION = ROUND_ROBIN,
23        CLUSTERED COLUMNSTORE INDEX,
24        PARTITION    (    [transaction_date] RANGE RIGHT FOR VALUES
25                        (20211001
26                        )
27                    )|
28    )
29    AS
30    Select * from dbo.Transaction_Partitioned where 1 = 2;
```

Results   Messages

Figure 10.25 – Creating an archival table

The CREATE Table command uses a select query from the Transaction_Partitioned table but with a WHERE clause, 1 = 2, to ensure only the structure of the table is copied into the new Transaction_Partitioned_before_oct table. Only the boundary value that needs to be archived is specified in the partition boundary value list of the new table.

7. Now, let's move partition 1 from the Transaction_Partitioned table to the Transaction_Partitioned_before_oct table. This can be done using the SWITCH command in the partition tables. Copy and paste the following commands into the same query window and run them:

```
ALTER TABLE Transaction_
Partitioned SWITCH PARTITION 1 to Transaction_
Partitioned_before_oct PARTITION 1
```

```
GO
Select name,boundary_id,value,rows
from dbo.table_partition_boundary
where value = 20211001
GO
```

The ALTER TABLE command moves partition 1 from Transaction_Partitioned to the Transaction_Partitioned_before_oct table's partition 1. For the move to be successful, three key conditions need to be met:

- The destination table and the source table should have the same structure.

- The boundary values of the partition being moved and the destination partition should match.

- The destination partition should be empty.

As we had met all the mentioned conditions, the move was successful. The move can be verified using a Select query against the table_partition_boundary view:

Figure 10.26 – Moving data to the archival table

We can see that all the rows from the first partition of Transaction_Partitioned have been now moved to the Transaction_Partitioned_before_oct table as the rows column indicates a zero for the first partition of Transaction_Partitioned.

8. As we had moved partition 1 from Transaction_Partitioned and it is empty, it makes sense to remove the partition. This can be done using the MERGE command in the partitioned tables. The MERGE command works if the partition is empty. Copy and paste the command into the same query window and run the query:

```
ALTER TABLE Transaction_
Partitioned MERGE RANGE('20211001')
```

```
GO
Select name,boundary_id,value
from dbo.table_partition_boundary
where name = 'Transaction_Partitioned'
```

In the ALTER TABLE command, we used the MERGE command and specified the empty boundary value of 20211001 (October 1) as the input for the RANGE clause. This caused the partition with a boundary value of 20211001 (October 1) to be merged with partition 2 (the next partition in line with a boundary value of 20211101, November 1). To verify, we queried the table_partition_boundary view and six partitions. Partition 1's boundary value has changed to 20211101 (November 1), and the partition with a boundary value of 20211001 (October 1) was removed:

```
48    ALTER TABLE Transaction_Partitioned MERGE RANGE('20211001')
49    GO
50    Select name,boundary_id,value
51    from dbo.table_partition_boundary
52    where name = 'Transaction_Partitioned'
53
54
```

Figure 10.27 – Removing the empty partition

With the preceding steps, we have moved a partition to an archive table and removed the empty partition from the original table.

How it works...

Performing archival operations is extremely challenging in tables involving millions or billions of rows. Archiving the data using traditional `delete` and `insert` statements can be extremely time-consuming, running into hours or even days. It can also block other queries for long periods. However, if you use partition commands such as `SWITCH/MERGE`, as we have done here, the movement of data works seamlessly, that is, in a matter of seconds. This is made possible because behind the scenes, there are no physical data movements for the preceding steps, and the engine remaps the partition to the archival table provided the tables involved adhere to the requirements of the `SWITCH` partition command.

Implementing workload management in an Azure Synapse dedicated SQL pool

How often have we seen a scenario where we have queries coming from a less important system hogging most of the resources and causing fewer resources to be available for business-critical queries, eventually resulting in dissatisfied customers/stakeholders? Wouldn't it be nice to classify the queries based on their business needs and reserve and allocate resources accordingly?

Workload management in a Synapse dedicated SQL pool helps to classify the queries based on the user account used to run the query, the user role, and the application used and map them to resource classes. Resource classes are predefined resource pools with resource limits set as a percentage of the total number of resources. By defining rules that classify the queries and mapping them to resource classes, we could reserve resources based on the business needs and ensure that critical queries get the right amount of resources most of the time.

In this recipe, we will classify the query in a resource class and verify the resource allocated to the query.

Getting ready

Create a Synapse Analytics workspace as explained in the *Provisioning an Azure Synapse Analytics workspace* recipe of *Chapter 8, Processing Data Using Azure Synapse Analytics*.

Complete the *Loading data into dedicated SQL pools using PolyBase using T-SQL* recipe to create a dedicated SQL pool named `packtadesqlpool`, an external table named `dbo.ext_transaction_tbl`, and a dedicated SQL pool table named `dbo.transaction_tbl`. Alternatively, you can use the `Create_External_Table.SQL` script from `https://github.com/PacktPublishing/Azure-Data-Engineering-Cookbook-2nd-edition/tree/main/chapter10` to create the `dbo.ext_transaction_tbl` and `dbo.transaction_tbl` tables.

How to do it...

In this recipe, we will be performing the following tasks:

- Scaling up the data warehouse to **DWU 300** as it's required for this recipe
- Creating a SQL account, running a query, and measuring the resource allocated

- Creating a workload classifier function and mapping the classifier function to the workload resource class.

- Rerunning the query executed in the previous step and verifying the resource allocation

Follow these steps:

1. Log in to `portal.azure.com`, go to **All resources**, and search for `packtadesynapse`. Click on the workspace. Click on **Open Synapse Studio**. Click on the develop button (the notebook-like button) on the left-hand side. Click on the + symbol and select **New SQL script**. Select `packtadesqlpool` in the **Connect to** option. Set the database as `Master`. Refer to *steps 1* and *2* in the *How to do it...* section of the *Loading data into dedicated SQL pools using COPY INTO* recipe for detailed screenshots on how to open a new query window. In the new query window, copy and paste the script provided in the following script to scale up the Synapse instance:

    ```
    ALTER DATABASE packtadesqlpool
    MODIFY (SERVICE_OBJECTIVE = 'DW300c');
    ```

    The following is a screenshot of the execution:

Figure 10.28 – Scaling up the database

As the database needs to scale up, it will pause and resume after a few minutes.

2. Let's create a user account named `AppUser` and run a heavy query using that user account. Connect to the `packatadesqlpool` database. Copy and paste the script into the same query window and run the query:

    ```
    CREATE USER AppUser without login
    GO
    GRANT SELECT ON dbo.transaction_tbl to AppUser
    GO
    EXECUTE AS USER = 'AppUser'
    Select t1.pid,t1.c1,t2.c2,sum(t2.order_count)
    ```

```
FROM dbo.transaction_tbl t1
inner join dbo.transaction_tbl t2 on t1.transaction_
date = t2.transaction_date
WHERE t1.tid < 1000
Group by t1.pid,t1.c1,t2.c2
order by sum(t2.order_count)
REVERT;
GO
```

The `EXECUTE AS USER` command switches the context to the `AppUser` account. The select query takes a few minutes to run. The `REVERT` command puts the context back into the original account:

Figure 10.29 – Running a query without workload management

3. While the query is running, click on the + button and open a new query window. Execute the script as follows:

```
SELECT req.request_id, classifier_name, group_name,
command,resource_allocation_percentage
  FROM sys.dm_pdw_exec_requests req
  Inner join sys.dm_pdw_exec_sessions ses on req.session_
id = ses.session_id   and req.request_id = ses.request_id
    ORDER BY submit_time DESC
```

We can notice that the script was, by default, classified into a resource class called `smallrc` and was granted 8.25 % of the resources:

Figure 10.30 – Checking the number of resources consumed

4. Let's create a classifier function and rerun the same expensive query. Copy and paste the following script into the same query window and run it:

```
CREATE WORKLOAD CLASSIFIER WC_AppUser WITH
( WORKLOAD_GROUP = 'mediumrc'
 ,MEMBERNAME = 'AppUser'
)
GO
EXECUTE AS USER = 'AppUser'
```

```
Select t1.pid,t1.c1,t2.c2,sum(t2.order_count)
FROM dbo.transaction_tbl t1
inner join dbo.transaction_tbl t2 on t1.transaction_
date = t2.transaction_date
WHERE t1.tid < 1000
Group by t1.pid,t1.c1,t2.c2
order by sum(t2.order_count)
REVERT;
GO
```

We have created a workload classifier function named WC_AppUser. In the CREATE WORKLOAD CLASSIFIER function, we specify the workload group as mediumrc. Mediumrc is a built-in resource class that provides 10% of resource allocation for queries in databases at the DWU 300 service level:

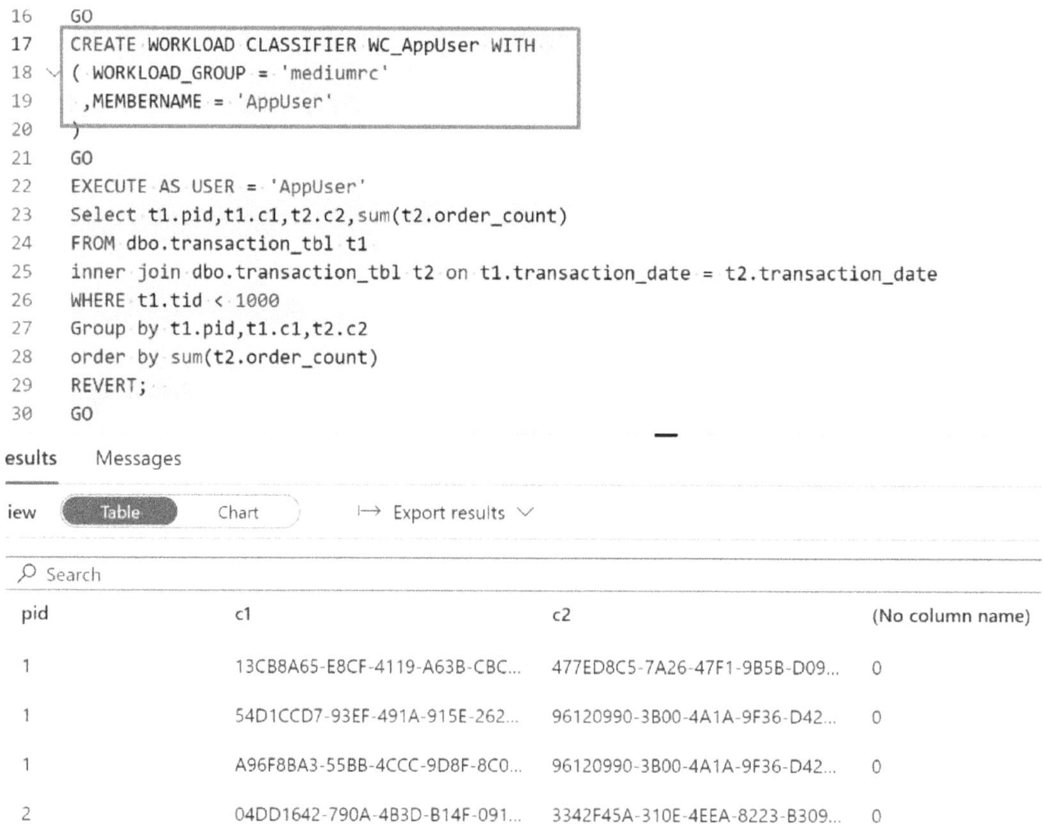

```
16    GO
17    CREATE WORKLOAD CLASSIFIER WC_AppUser WITH
18  ∨ ( WORKLOAD_GROUP = 'mediumrc'
19      ,MEMBERNAME = 'AppUser'
20    )
21    GO
22    EXECUTE AS USER = 'AppUser'
23    Select t1.pid,t1.c1,t2.c2,sum(t2.order_count)
24    FROM dbo.transaction_tbl t1
25    inner join dbo.transaction_tbl t2 on t1.transaction_date = t2.transaction_date
26    WHERE t1.tid < 1000
27    Group by t1.pid,t1.c1,t2.c2
28    order by sum(t2.order_count)
29    REVERT;
30    GO
```

esults   Messages

iew   Table   Chart   ⊢→ Export results ∨

🔍 Search

pid	c1	c2	(No column name)
1	13CB8A65-E8CF-4119-A63B-CBC...	477ED8C5-7A26-47F1-9B5B-D09...	0
1	54D1CCD7-93EF-491A-915E-262...	96120990-3B00-4A1A-9F36-D42...	0
1	A96F8BA3-55BB-4CCC-9D8F-8C0...	96120990-3B00-4A1A-9F36-D42...	0
2	04DD1642-790A-4B3D-B14F-091...	3342F45A-310E-4EEA-8223-B309...	0

Figure 10.31 – Running a query with the classifier function

5. Click on the + button, select **SQL Script**, and open a new query window. Then, execute the same script used in *step 3* to monitor the resource allocation:

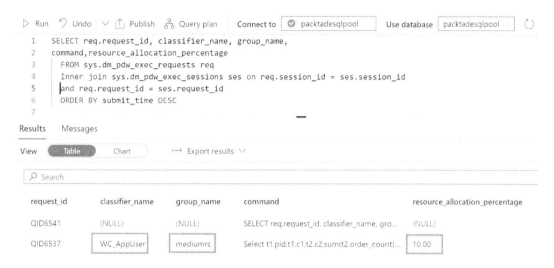

Figure 10.32 – Checking the resource allocation

We can see that the query has been classified using the WC_AppUser classifier function and has been allocated 10% of resources using the mediumrc resource class.

How it works...

A Synapse dedicated SQL pool offers four dynamic resource classes, namely smallrc, mediumrc, largerc, and xlargerc. The smallrc, mediumrc, largerc, and xlargerc resource classes offer 8%, 10%, 22%, and 70% of resource usage, respectively, at DWU 300C. The allocation percentages for these resource classes vary marginally depending on the DWU. Detailed information on the percentage allocation of resources is provided at https://docs.microsoft.com/en-us/azure/synapse-analytics/sql-data-warehouse/resource-classes-for-workload-management.

Mapping the query to a resource class by classifying it using a user account helps us to set resource limits on workloads from different user accounts. You could map user accounts to higher resource classes such as largerc and xlargerc for more critical and resource-intensive queries.

Creating workload groups for advanced workload management

Workload groups in Synapse dedicated pools allow you to define resource pools with custom resource allocation percentages compared to the predefined percentages offered by resource classes. Additionally, workload groups offer the flexibility to set minimum and maximum resource percentages for the entire pool and also for each request. Setting a minimum resource percentage for a workload group ensures there will always be a percentage of resource reserved for the queries mapped to the resource group. In this recipe, we will define a custom resource group with minimum and maximum resource percentages for the pool and for each request, and we will also classify it in such a way that the resource allocation changes depending on the time of query execution.

Getting ready

Create a Synapse Analytics workspace as explained in the *Provisioning an Azure Synapse Analytics workspace* recipe of *Chapter 8, Processing Data Using Azure Synapse Analytics*.

Complete the *Loading data into dedicated SQL pools using PolyBase using T-SQL* recipe to create a dedicated SQL pool named `packtadesqlpool`, an external table named `dbo.ext_transaction_tbl`, and a dedicated SQL pool table named `dbo.transaction_tbl`. Alternatively, you can use the `Create_External_Table.SQL` script from `https://github.com/PacktPublishing/Azure-Data-Engineering-Cookbook-2nd-edition/tree/main/chapter10` to create the `dbo.ext_transaction_tbl` and `dbo.transaction_tbl` tables.

Complete the *Implementing workload management in an Azure Synapse dedicated SQL pool* recipe to create a workload classifier function and map it to the workload resource class.

How to do it...

In this recipe, we will be performing the following tasks:

- Creating a workload group with minimum and maximum resource percentages for the pool and per request.

- Creating a classifier function based on user account and login time.

- Testing queries at different times to see the resource allocation.

Detailed steps are provided as follows:

1. Log in to portal.azure.com, go to **All resources**, and search for packtadesynapse. Click on the workspace. Click on **Open Synapse Studio**. Click on the develop button (the notebook-like button) on the left-hand side. Click on the + symbol and select **New SQL script**. Select packtadesqlpool in the **Connect to** option. Copy and paste the following script and then hit the **Run** button:

```
CREATE WORKLOAD GROUP WG_AppUser_offpeak WITH
( REQUEST_MIN_RESOURCE_GRANT_PERCENT = 12
,REQUEST_MAX_RESOURCE_GRANT_PERCENT = 20
 ,MIN_PERCENTAGE_RESOURCE = 24
 ,CAP_PERCENTAGE_RESOURCE = 40
)
GO
CREATE WORKLOAD CLASSIFIER WC_AppUser_offpeak WITH
( WORKLOAD_GROUP = 'WG_AppUser_offpeak'
 ,MEMBERNAME = 'AppUser'
 ,START_TIME = '13:30'
 ,END_TIME = '23:00'
)
```

The preceding script creates a workload group called WG_AppUser_offpeak. The WG_AppUser_offpeak configuration is explained as follows:

*   REQUEST_MIN_RESOURCE_GRANT_PERCENT of 12% implies any query classified to map to WG_AppUser_offpeak will get a minimum of 12% of total resources in the database.

*   REQUEST_MAX_RESOURCE_GRANT_PERCENT of 20% implies a query classified to map to WG_AppUser_offpeak can get a maximum of 20% of total resources in the database.

*   MIN_PERCENTAGE_RESOURCE of 24% implies 24% of total resources in the database will be reserved for the WG_AppUser_offpeak pool.

*   CAP_PERCENTAGE_RESOURCE of 40% indicates the total resource usage of all the queries in WG_AppUser_offpeak can reach up to 40%.

Additionally, the script creates a classifier function named WC_AppUser_offpeak, which maps the AppUser account with the WG_AppUser_offpeak workload group. The classifier needs to map the queries from the AppUser account to WC_AppUser_offpeak only during off-peak hours between 9:30 p.m. and 7 a.m. Singapore time. The START_TIME and END_TIME values are set in UTC time (8 hours behind Singapore time) and, hence, are specified as 13:30 and 23:00, respectively:

Figure 10.33 – Configuring the workload management

2. Let's run the expensive query using `AppUser` before the off-peak time, which starts at 13:30 hours UTC. Copy the following script and run it in a new query window before the start time of the off-peak period:

```
EXECUTE AS USER = 'AppUser'
Select getdate(), t1.pid,t1.c1,t2.c2,sum(t2.order_count)
FROM dbo.transaction_tbl t1
inner join dbo.transaction_tbl t2 on t1.transaction_
date = t2.transaction_date
WHERE t1.tid < 1000
Group by t1.pid,t1.c1,t2.c2
order by sum(t2.order_count)
REVERT;
GO
```

The output of the script is shown in the following screenshot:

Figure 10.34 – Running a query before the off-peak period

The first column returns the current time in UTC, which is 13:29, just before the start of the off-peak period.

3. While the query is running, click on the + button, select **SQL Script**, and open a new query window. Run the following script to check the resource class and the resources allocated for the query:

```
SELECT req.request_id, classifier_name, resource_class,
command,resource_allocation_percentage,submit_time
  FROM sys.dm_pdw_exec_requests req
  Inner join sys.dm_pdw_exec_sessions ses on req.session_
id = ses.session_id
  and req.request_id = ses.request_id
  ORDER BY submit_time DESC
```

The output of the script is shown in the following screenshot:

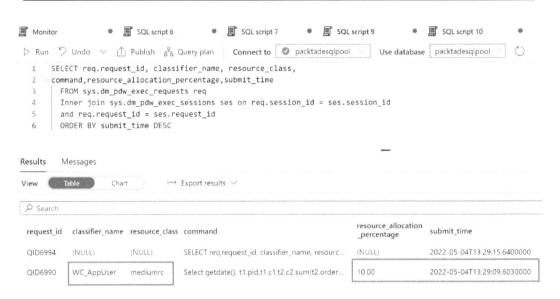

Figure 10.35 – Monitoring the query before the off-peak period

The preceding script execution shows that the script has been classified using the WC_AppUser classifier function (which was created in the *Implementing workload management in an Azure Synapse dedicated SQL pool* recipe) and has been allocated 10% of resources via the mediumrc resource class. The submit_time column shows the time as 13:29, which is just before the off-peak period that starts at 13:30 hours.

4. Let's run the select query again using the AppUser account but this time inside the off-peak period. Copy the script from *step 2* and rerun it. The Getdate() function returns 13:44, which is in the off-peak period defined:

Figure 10.36 – Running a query during the off-peak period

5. Click on the + button, select **SQL Script**, and open a query window. Run the monitoring script used in *step 3* to check the resource allocation:

Figure 10.37 – Running a query during the off-peak period

We noticed that as the `submit_time` query was in the off-peak period, it was mapped to the `WG_AppUser_offpeak` workload group. The query is allocated 20% of resources, as the maximum resource per query of the `WG_AppUser_offpeak` workload group is 20%.

How it works...

Using workload groups, we have more flexibility in allocating resources per query level and for the entire resource pool, too. In this example, there were two classifier functions, namely `WG_AppUser_offpeak` and `WG_AppUser`, that were mapped to the `AppUser` account. `WG_AppUser_offpeak` had an additional mapping based on query submission time; therefore, when the classifier function matched both the user account and query submission time, the `WG_AppUser_offpeak` classifier was selected and appropriate resources were allocated.

It is important to note that allocating more resources per query is good for getting heavy tasks done but would limit the number of users accessing the resource pool concurrently. It is essential to strike a balance between concurrency and resources, and workload groups help us to strike that balance. In the preceding example, the application needs to support more users during peak hour operations; therefore, we allocated 10% of resources for the query via the `WG_AppUser` workload group. However, maintenance jobs during off-peak hours require additional resources, so 20% of resources were allocated via `WG_AppUser_offpeak`. We had set a cap on total resources in the `WG_AppUser_offpeak` workload group to 40%, which implies it had the capacity to support just another query with 20% resource allocation. However, that still works, as during off-peak hours, the resource pool is not expected to support many concurrent users.

11

Monitoring Synapse SQL and Spark Pools

As introduced in *Chapter 8, Processing Data Using Azure Synapse Analytics*, Azure Synapse Analytics is comprised of three key components – a Synapse integration pipeline to ingest and transform the data, SQL pool and Spark pool to process and serve the data, and Power BI integration to visualize the data. Monitoring the health of Synapse SQL and Spark pools is an integral part of the work of a data engineer managing large data engineering projects.

In this chapter, you will learn how to integrate Synapse SQL and Spark pools with Azure Log Analytics, identify long-running Spark jobs and SQL queries using a Log Analytics workspace, monitor Synapse health from Azure Monitor, and check the table and index health of SQL dedicated pool tables using **Dynamic Management Views (DMVs)**.

In this chapter, we will cover the following recipes:

- Configuring a Log Analytics workspace for Synapse SQL pools
- Configuring a Log Analytics workspace for Synapse Spark pools
- Using Kusto queries to monitor SQL and Spark pools
- Creating workbooks in a Log Analytics workspace to visualize monitoring data
- Monitoring table distribution, data skew, and index health using Synapse DMVs
- Building monitoring dashboards for Synapse with Azure Monitor

Technical requirements

For this chapter, you will need a Microsoft Azure subscription.

Configuring a Log Analytics workspace for Synapse SQL pools

A Log Analytics workspace is an Azure service that's used to store the diagnostic logs of several Azure services in a single place. With a Log Analytics workspace, you can store the performance metric data of a Synapse SQL pool, such as how much DWU was used, I/O usage, queued queries, query-level consumption details, and more. In this recipe, we will configure a Log Analytics workspace for a Synapse SQL pool.

Getting ready

To get started, log in to `https://portal.azure.com` using your Azure credentials:

- Create a Synapse Analytics workspace, as explained in the *Provisioning an Azure Synapse Analytics workspace* recipe of *Chapter 8, Processing Data Using Azure Synapse Analytics*.

- Create a dedicated Synapse SQL pool database, as explained in *steps 1 to 3* in the *How to do it...* section of the *Loading data into a dedicated SQL pool using PolyBase and T-SQL* recipe of *Chapter 10, Building the Serving Layer in an Azure Synapse SQL Pool*.

How to do it...

First, let's create a Log Analytics workspace:

1. Go to the `portal.azure.com` home page and click on **Create a resource**. Search for `Log Analytics` and select **Log Analytics Workspace**. Click on the **Create** button:

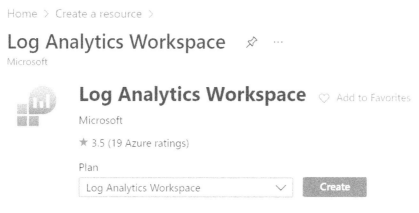

Figure 11.1 – Creating a Log Analytics workspace

2. Set the resource group name as `PacktADESynapse`. Set the instance name as `PacktADELogAnalytics`. Pick the same location as that of your Synapse workspace. Then, click on **Review + Create**:

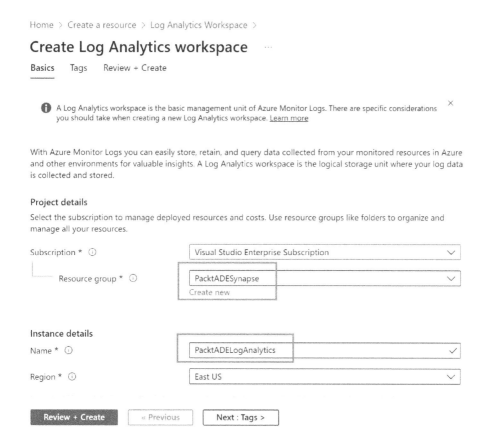

Figure 11.2 – Providing a Log Analytics workspace name

3. Log in to portal.azure.com, go to **All resources**, and search for packtadesqlpool, which is the dedicated Synapse SQL pool you created in the *Loading data into a dedicated SQL pool using PolyBase and T-SQL* recipe of *Chapter 10, Building the Serving Layer in Azure Synapse SQL Pool*. Click on **SQL pool** and search for **Diagnostics Settings**:

Figure 11.3 – Open dedicated SQL pool diagnostics

4. Click on + **Add diagnostic setting**:

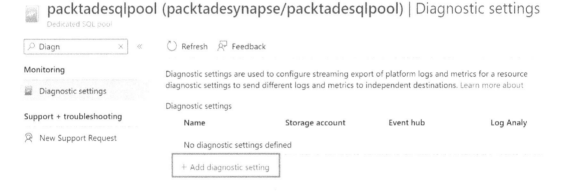

Figure 11.4 – Adding diagnostics

5. Set **Diagnostic setting name** to Synapsededicatedpooldiag. Then, select all the options under **Categories**. Click on the **Send to Log Analytics Workspace** checkbox and select **PacktADELogAnalytics** for **Log Analytics workspace**. Then, click **Save**:

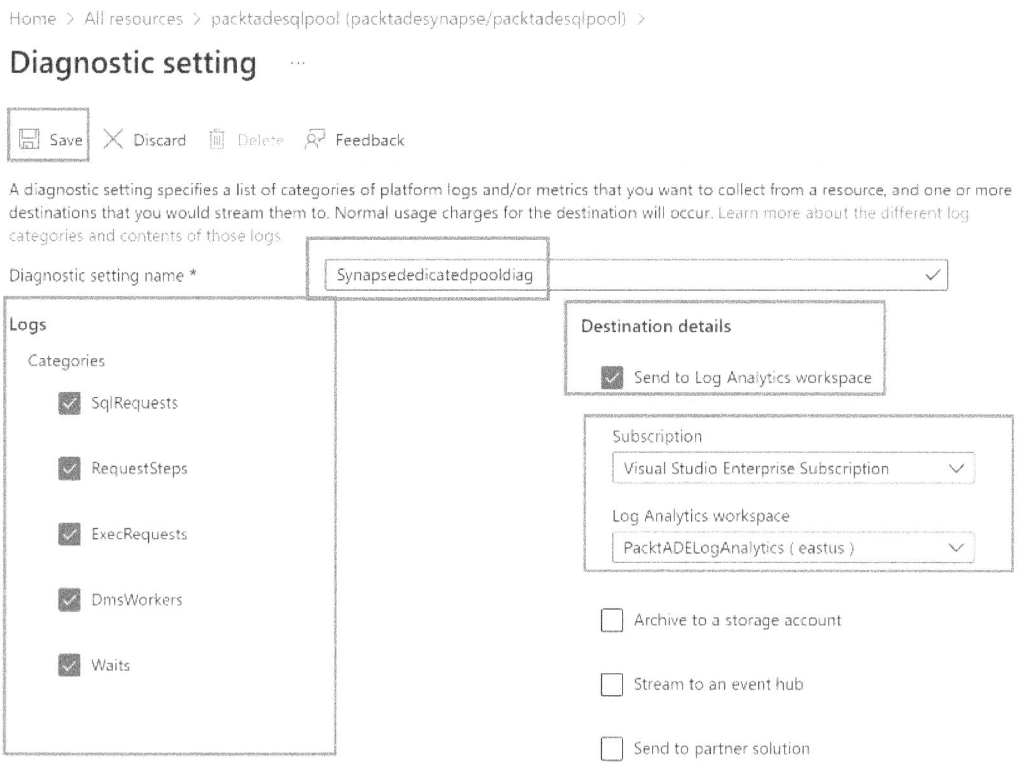

Home > All resources > packtadesqlpool (packtadesynapse/packtadesqlpool) >

Diagnostic setting ...

🖫 Save   ✕ Discard   🗑 Delete   𝒬 Feedback

A diagnostic setting specifies a list of categories of platform logs and/or metrics that you want to collect from a resource, and one or more destinations that you would stream them to. Normal usage charges for the destination will occur. Learn more about the different log categories and contents of those logs

Diagnostic setting name *   Synapsededicatedpooldiag   ✓

Logs

Categories

☑ SqlRequests

☑ RequestSteps

☑ ExecRequests

☑ DmsWorkers

☑ Waits

Destination details

☑ Send to Log Analytics workspace

Subscription

Visual Studio Enterprise Subscription   ⌄

Log Analytics workspace

PacktADELogAnalytics (eastus)   ⌄

☐ Archive to a storage account

☐ Stream to an event hub

☐ Send to partner solution

Figure 11.5 – Adding diagnostics

How it works...

As a first step, we created a Log Analytics workspace and then we linked the diagnostic setting in the dedicated SQL pool to it. As the queries are executed in the dedicated SQL pool, performance metrics data about the queries will be recorded in the Log Analytics workspace. The granularity/depth of the performance metric data recorded depends on the log categories selected while configuring the diagnostic settings of the dedicated SQL pool. We selected all five categories of logs available to gather detailed information about the SQL pool's performance. A brief description of the log categories is as follows:

- **SqlRequests**: Provides information about query steps at the node or distribution level for SQL requests

- **RequestSteps**: Provides step-level information for each recorded query

- **ExecRequests**: Provides information about all the queries that are executed in the SQL pool

- **DmsWorkers**: Provides information about the worker process moving data across nodes

- **Waits**: Provides information about various wait categories (**CPU/Memory/Locks/IO/Network**) that queries were spending time in

Accessing the data in a Log Analytics workspace and finding key performance insights will be covered in the subsequent recipes of this chapter.

Configuring a Log Analytics workspace for Synapse Spark pools

In this recipe, we will explore integrating a Log Analytics workspace with Synapse Spark pools.

Getting ready

To get started, log in to `https://portal.azure.com` using your Azure credentials:

- Create a Synapse Analytics workspace, as explained in the *Provisioning an Azure Synapse Analytics workspace* recipe of *Chapter 8, Processing Data Using Azure Synapse Analytics*.

- Create a Spark pool cluster, as explained in the *Provisioning and configuring Spark pools* recipe of *Chapter 8, Processing Data Using Azure Synapse Analytics*.

- Create a Log Analytics workspace (if you haven't already done so), as explained in *steps 1* and *2* in the *How to do it…* section of the *Configuring a Log Analytics Workspace for Synapse dedicated SQL Pools* recipe in this chapter.

How to do it…

Follow these steps to integrate Synapse Spark pools with a Log Analytics workspace:

1. Download the `spark_loganalytics_conf.txt` file from `https://github.com/PacktPublishing/Azure-Data-Engineering-Cookbook-2nd-edition/blob/main/chapter11/`.

2. Log in to `portal.azure.com`, go to **All resources**, and search for **PacktADELogAnalytics**, which is the Log Analytics workspace you created in the *Configuring a Log Analytics workspace for Synapse dedicated SQL pools* recipe of this chapter. Click on **Agents Management** and copy the **Workspace ID** and **Primary key** information on the right:

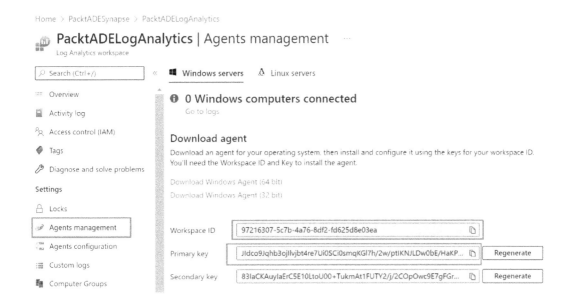

Figure 11.6 – Workspace details

3. Open the spark_loganalytics_conf.txt file you downloaded in *step 1*. Paste the workspace ID after spark.synapse.logAnalytics.workspaceId (the second line) in the file. Paste the primary key after spark.synapse.logAnalytics.secret (the third line) in the file. Then, save the file:

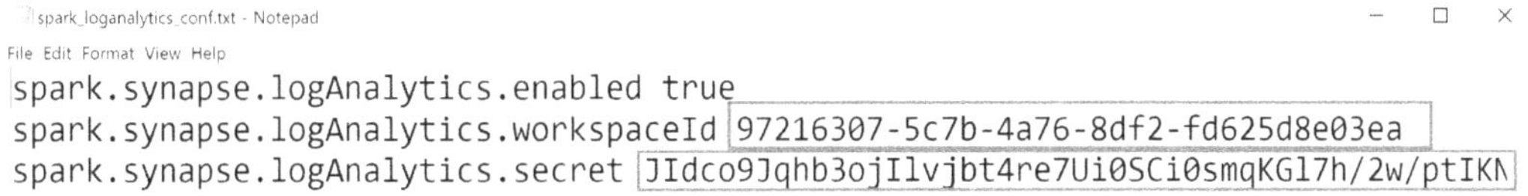

```
spark_loganalytics_conf.txt - Notepad                                    —   □   ×
File  Edit  Format  View  Help
spark.synapse.logAnalytics.enabled true
spark.synapse.logAnalytics.workspaceId 97216307-5c7b-4a76-8df2-fd625d8e03ea
spark.synapse.logAnalytics.secret JIdco9Jqhb3ojIlvjbt4re7Ui0SCi0smqKGl7h/2w/ptIKN
```

Figure 11.7 – Adding workspace details to the file

4. Log in to portal.azure.com, go to **All resources**, and search for **packtsparkpool**, which is the Spark pool you created in the *Provisioning and configuring Spark pools* recipe of *Chapter 8, Processing Data Using Azure Synapse Analytics*. Click on **Spark configuration** under **Settings**. Then, click on **Upload spark config file**:

Home > PacktADESynapse > packtsparkpool (packtadesynapse/packtsparkpool)

packtsparkpool (packtadesynapse/packtsparkpool) | Spark configuration
Apache Spark pool

🔍 conf ✕	«	↑ Upload spark config file ↺ Refresh
🔍 Access control (IAM)		User-provisioned spark config
Settings		Name Size
🔘 Spark configuration		No user-provided config file currently uploaded. You can upload "spark config file".

Figure 11.8 – Configuring a Spark pool

5. Click on the folder-like icon and select the `spark_loganalytics_conf.txt` file you saved in *step 3*. Click on the **Upload** button:

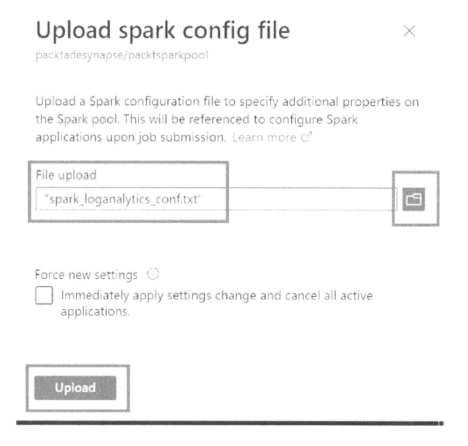

Upload spark config file   ✕

packtadesynapse/packtsparkpool

Upload a Spark configuration file to specify additional properties on the Spark pool. This will be referenced to configure Spark applications upon job submission. Learn more ☐

File upload

"spark_loganalytics_conf.txt"   📁

Force new settings ○

☐ Immediately apply settings change and cancel all active applications.

Upload

Figure 11.9 – The Upload spark config file screen

6. Once uploaded, the configuration file's name will be reflected on the Spark pool configuration:

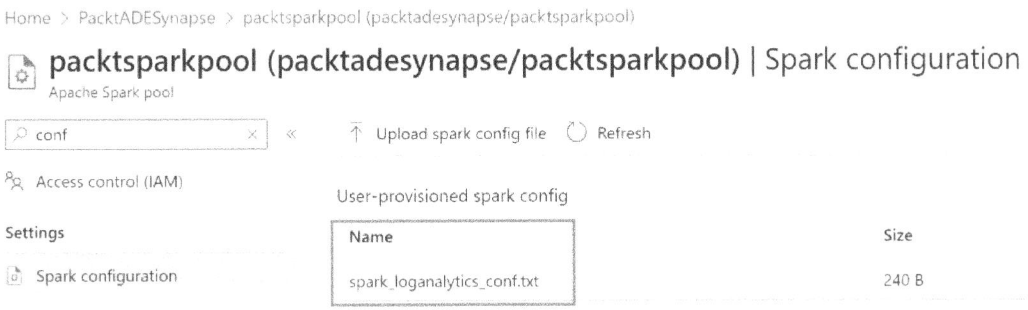

Home > PacktADESynapse > packtsparkpool (packtadesynapse/packtsparkpool)

packtsparkpool (packtadesynapse/packtsparkpool) | Spark configuration
Apache Spark pool

🔍 conf	✕	≪	⬆ Upload spark config file	↻ Refresh

<table>
<tr><td>𝐑 Access control (IAM)</td><td colspan="2">User-provisioned spark config</td></tr>
<tr><td>Settings</td><td>Name</td><td>Size</td></tr>
<tr><td>📄 Spark configuration</td><td>spark_loganalytics_conf.txt</td><td>240 B</td></tr>
</table>

Figure 11.10 – Updated Spark pool

How it works...

Log Analytics for Synapse Spark pools was enabled via the `spark_longanalytics_conf.txt` file's first line – that is, `spark.synapse.logAnalytics.enabled true`. Using the credentials of the Log Analytics workspace specified in `spark_longanalytics_conf.txt`, the Spark pool will log all the performance metrics of all jobs and notebooks executed in the Spark pool in the Log Analytics workspace. Analyzing the performance metric data of a Spark pool and identifying key insights will be covered in the next recipe.

Using Kusto queries to monitor SQL and Spark pools

In the previous two recipes (*Configuring a Log Analytics workspace for Synapse dedicated SQL pools* and *Configuring a Log Analytics workspace for Synapse Spark pools*), we created a Log Analytics workspace and configured the diagnostics logs of Synapse-dedicated SQL and Spark pools to be written to it. In this recipe, we will use Kusto queries to query a Log Analytics workspace to gain insights into the diagnostic logs collected.

Getting ready

To get started, log in to `https://portal.azure.com` using your Azure credentials:

- Complete the *Configuring a Log Analytics workspace for Synapse dedicated SQL Pools* and *Configuring a Log Analytics workspace for Synapse Spark pools* recipes covered earlier in this chapter to configure a Log Analytics workspace for Synapse SQL and Spark pool.

- Complete the *Loading data into a dedicated SQL pool using PolyBase and T-SQL* recipe of *Chapter 10, Building the Serving Layer in Azure Synapse SQL Pool*, to create a dedicated SQL pool named `packtadesqlpool`, an external table named `dbo.ext_transaction_tbl`, and a dedicated SQL pool table named `dbo.transaction_tbl`. Alternatively, to create the `dbo.ext_transaction_tbl` and `dbo.transaction_tbl` tables, you may use the `Create_External_Table.sql` script available in `https://github.com/PacktPublishing/Azure-Data-Engineering-Cookbook-2nd-edition/tree/main/chapter10`.

How to do it...

In this recipe, we will do the following:

- Run Kusto queries in a Log Analytics workspace to find long-running SQL queries in a dedicated SQL pool at a particular time.

- Run Kusto queries in a Log Analytics workspace to find high CPU-consuming Spark jobs.

Perform the following steps to find long-running SQL queries in the dedicated SQL pool:

1. The first step is to create a workload on a Synapse SQL pool:

    I.   Download the `SQLPool_Queries.sql` script from `https://github.com/PacktPublishing/Azure-Data-Engineering-Cookbook-2nd-edition/blob/main/chapter11/SQLPool_Queries.sql`.

    II.   Log in to `portal.azure.com`, go to **All resources**, and search for **packtadesynapse**. Click on the workspace and click on **Open Synapse Studio**. Then, click the develop button (the notebook-like button) on the left. Click on the + symbol and select **Import**:

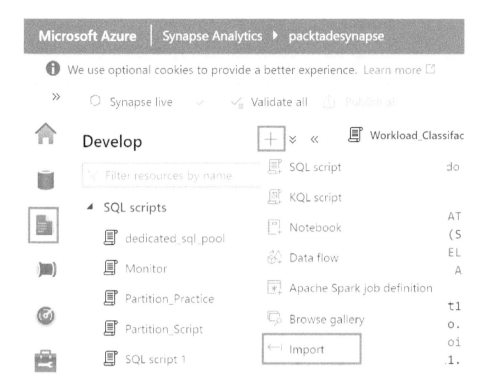

Figure 11.11 – Importing the SQL script

III.   Select the SQLPool_Queries.sql file you downloaded in *step I*. Connect to the
       packtadesqlpool database and run the script. The script will return 100 queries in 1 to
       2 minutes:

Figure 11.12 – Running the SQL workload

2. After the `SQLPool_Queries.sql` file finishes executed, wait 10 to 15 minutes and then go to `portal.azure.com`, go to **All resources**, and search for **PacktADELogAnalytics**, the Log Analytics workspace, and click on it. Click on **Logs**. Close the **Queries** popup using the **close** button at the top right:

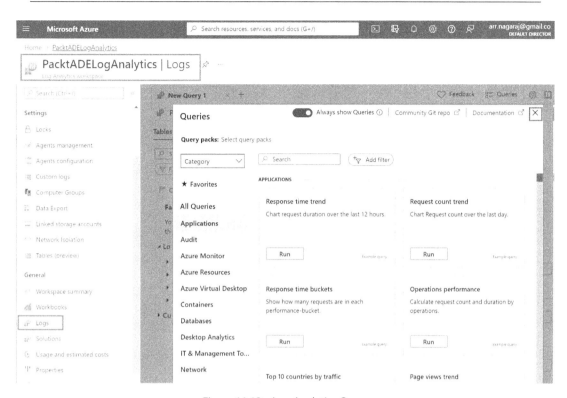

Figure 11.13 – Log Analytics Query

3. Kusto is the language that's used to query Log Analytic Workspace. Copy and paste the Kusto Query provided here. Provide the appropriate values for the Start_Time, End_Time, and DatabaseName variables. Start_Time and End_Time are in UTC:

```
let Start_Time = datetime(2022-05-21 05:00:00);
let End_Time = datetime(2022-05-21 06:00:00);
let DatabaseName = "packtadesqlpool";
SynapseSqlPoolExecRequests
| where TimeGenerated    between (Start_Time ..End_
Time) and StartTime between (datetime(2000-05-20)..
TimeGenerated)
| where Label !="health_checker"
| where Status contains "Running"
| where _ResourceId endswith DatabaseName
| extend duration_sec = datetime_diff("second",
TimeGenerated,StartTime)
| summarize duration_sec = max(duration_sec), Command
```

```
= any(Command),Label = any(Label),ResourceClass =
any(ResourceClass),QueryPlan = any(ExplainOutput),Status
= any(Status),Source = any(SourceSystem) by RequestId
| order by duration_sec
| limit 10
```

This script provides the top 10 longest-running active queries for the time specified and provides the duration the query ran for and the resource class it was assigned:

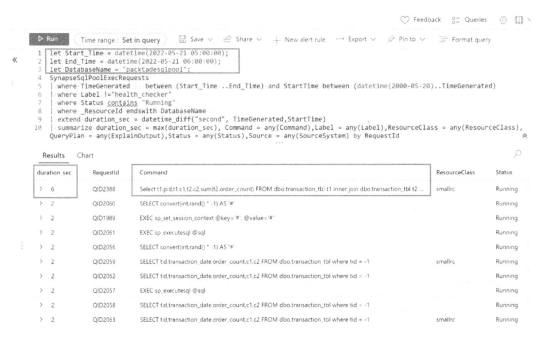

Figure 11.14 – Long-running SQL queries

Perform the following steps to find the highest CPU-consuming Spark jobs:

1. First, we need to run a few spark notebooks so that we have some diagnostic data recorded in our Log Analytics workspace. So, download the sparkpool_notebook1.ipynb and sparkpool_notebook2.ipynb notebooks from https://github.com/PacktPublishing/Azure-Data-Engineering-Cookbook-2nd-edition/blob/main/chapter11.

2. In **Synapse Studio**, click the develop button (the notebook-like button) on the left. Click on the + symbol, select **Import**, and select the sparkpool_notebook1.ipynb file that was downloaded. Similarly, import the sparkpool_notebook2.ipynb file.

3. Go to the **sparkpool_notebook1** notebook. Select **packtsparkpool** from the **Attach to** dropdown. Then, click **Run all**:

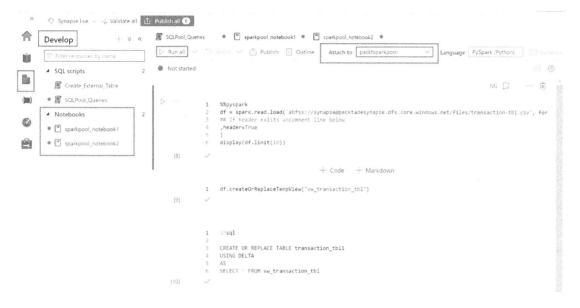

Figure 11.15 – Import notebook

4. Similarly, run the **sparkpool_notebook2** notebook. Both notebooks should finish running in 3 to 5 minutes.

5. Once the run completes, wait 10 to 15 minutes. You will need to wait this long as it can take up to 15 minutes for the notebook's execution data to reflect in the Log Analytics workspace. Go to **PacktADELogAnalytics**, the Log Analytics workspace. Click on **Logs**. Close the **Queries** popup (if it appears) using the close button at the top right. Then, copy and paste the following Kusto script in the new query window:

```
let CpuData =
SparkMetrics_CL
| where workspaceName_s == "packtadesynapse" and
clusterName_s == "packtsparkpool"
| where name_s contains_cs "executor.cpuTime"
| extend cputime = count_d / 1000000
| summarize sum(cputime) by
TimeGenerated,applicationName_s;

CpuData
| summarize cpu_time_ms = max(sum_cputime) by
bin(TimeGenerated,10m),applicationName_s
| sort by cpu_time_ms desc
```

```
| project  applicationName_s,cpu_time_
ms,bin(TimeGenerated,10m)
| limit   10
```

Set **Time range** to **Last hour** and click the **Run** button. The execution results are shown in the following screenshot:

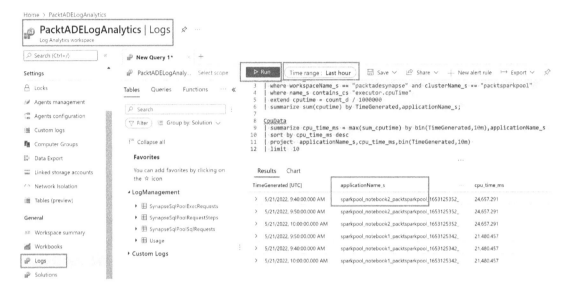

Figure 11.16 – KQL script for the Spark pool's CPU

The script has a `where` clause that has filters against the following three columns:

- `workspaceName_s`: Takes the Synapse workspace's name

- `clusterName_s`: Takes the Spark pool's name

- `name_s`: Filters for CPU time

The `name_s` column contains several key performance metrics related to CPU, memory, data movement, and many more. In this script, we focused on CPU execution time. The script gives the most expensive notebooks for every 10-minute window, measured by CPU time consumed. The `applicationName_s` column reveals the notebook name or the job name that ran on the Spark pool. This script shows that `sparkpool_notebook2` took 3 seconds longer than `sparkpool_notebook1`.

6. I have uploaded another KQL script, `shuffle_operation.txt`, in `https://github.com/PacktPublishing/Azure-Data-Engineering-Cookbook-2nd-edition/blob/main/chapter11` to track Spark jobs or notebooks that moved the most amount of data across the nodes. If you are interested, you can try it out to monitor Spark jobs to maximize data movement via shuffle operations.

How it works...

Configuring a Log Analytics workspace with a Synapse Spark pool and a SQL pool ensures the diagnostic logs are recorded in the Log Analytics workspace. The preceding two scripts that we covered showcase how you could use Kusto scripts in a Log Analytics workspace to find the top SQL queries or Spark jobs. There are additional scripts provided in `https://github.com/microsoft/Azure_Synapse_Toolbox/tree/master/Log_Analytics_queries` that you can use to explore Log Analytics workspace usage further for Synapse-dedicated SQL pool monitoring.

Creating workbooks in a Log Analytics workspace to visualize monitoring data

Workbooks help you visually explore data stored in Log Analytics workspaces. Workbooks make exploring data stored in a Log Analytics workspace easier since you don't need to write Kusto queries to read the data. In this recipe, we will create two workbooks – one each for monitoring Synapse Spark and SQL pools.

Getting ready

To get started, log in to `https://portal.azure.com` using your Azure credentials:

- Complete the *Configuring a Log Analytics workspace for Synapse dedicated SQL pools* and *Configuring a Log Analytics workspace for Synapse Spark pools* recipes covered earlier in this chapter.

- Download the `SQLPool_Queries.sql` script from `https://github.com/PacktPublishing/Azure-Data-Engineering-Cookbook-2nd-edition/blob/main/chapter11/SQLPool_Queries.sql` and run it in **Packtadesqlpool**, as explained in *step 1* of the *How to do it...* section of the *Using Kusto queries to monitor Spark and SQL pools* recipe.

- Download the `sparkpool_notebook1.ipynb` and `sparkpool_notebook2.ipynb` notebooks from `https://github.com/PacktPublishing/Azure-Data-Engineering-Cookbook-2nd-edition/blob/main/chapter11`, import them into the **Packtadesynapse** workspace, and execute that workspace against **packtadesparkpool**, as explained in *step 4* of the *How to do it...* section of the *Using Kusto queries to monitor Spark and SQL pools* recipe.

How to do it...

First, let's configure a workbook for monitoring Synapse-dedicated SQL pools:

1. Download the `DedicatedSQLPool.workbook` file from `https://github.com/PacktPublishing/Azure-Data-Engineering-Cookbook-2nd-edition/blob/main/chapter11`.

2. Log in to `portal.azure.com`, go to **All resources**, and search for **PacktADELogAnalytics**, the Log Analytics workspace. Under the **General** section, click on **Workbooks**. Click on **Empty** under the **Quick start** section to create an empty workbook:

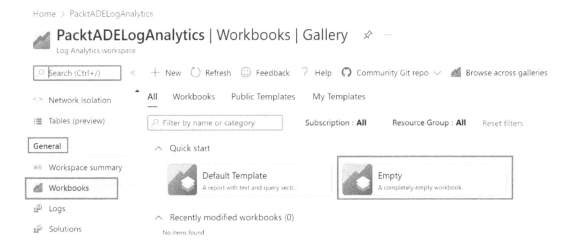

Figure 11.17 – Creating an empty workbook

3. Click on the </> (**Advanced Editor**) icon on the right:

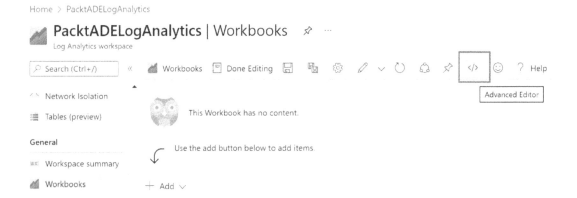

Figure 11.18 – Workbook Advanced Editor

4. Erase whatever content is in the text area and keep it empty:

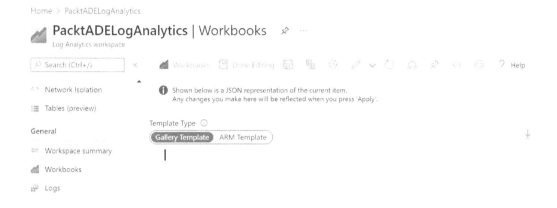

Figure 11.19 – Emptying Advanced Editor

5. Open the `DedicatedSQLPool.workbook` file you download in *step 1* using Notepad. Copy the complete content and paste it into Advanced Editor in the Log Analytics workbook. Click on **Apply**:

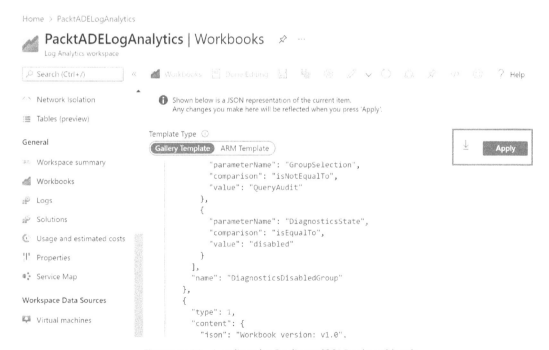

Figure 11.20 – Loading the DedicatedSQLPool workbook

6. Set `LogAnalyticsSubscription` to be your subscription name, `LogAnalyticsWorkspace` as `PacktADELogAnlaytics`, and `DatabaseResourceName` as `packtadesynapse/packtadesqlpool`. Click the **Done Editing** button:

Figure 11.21 – Workbook – Done Editing

7. You should see the Log Analytics data being visualized. Click the **Save** icon at the top. Set **Title** to `DedicatedSQLPool` and click the **Save** button once more:

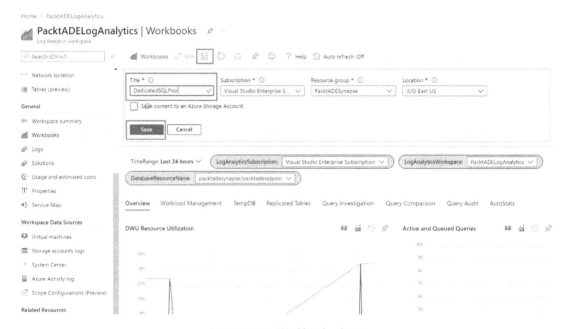

Figure 11.22 – Workbook – Save

8. Let's do a basic performance check using the **DedicatedSQLPool** workbook:

   I. Observe the DWU usage spike at 9:07 A.M. via the **DWU Resource Utilization** tile in the **Overview** tab:

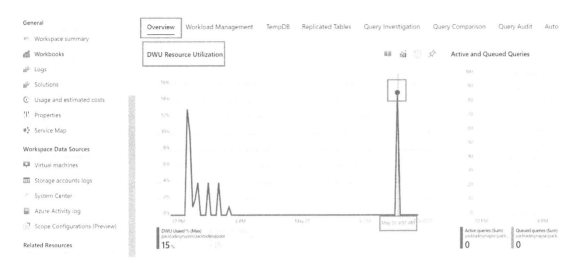

Figure 11.23 – DWU spike

II.   To find the queries that were active at 9:07 A.M., scroll down to the **All Queries** section in the same tab. Notice that **QID3410** was active. Let's get the query plan for query **QID3410**:

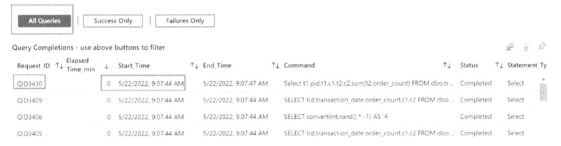

Figure 11.24 – All Queries

III.   Switch over to the **TempDB** tab:

Figure 11.25 – The TempDB tab

IV.   Scroll down to **20 Largest Query Steps by Most Rows Moved**. Let's click on **QID3410**, which we noted down earlier:

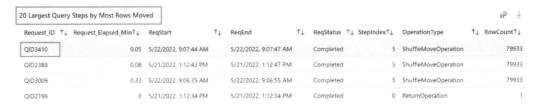

20 Largest Query Steps by Most Rows Moved								
Request_ID ↑↓	Request_Elapsed_Min↑↓	ReqStart ↑↓	ReqEnd ↑↓	ReqStatus ↑↓	StepIndex↑↓	OperationType ↑↓	RowCount↑↓
QID3410	0.05	5/22/2022, 9:07:44 AM	5/22/2022, 9:07:47 AM	Completed	5	ShuffleMoveOperation	79933
QID2388	0.08	5/21/2022, 1:12:43 PM	5/21/2022, 1:12:47 PM	Completed	5	ShuffleMoveOperation	79933
QID3009	0.33	5/22/2022, 9:06:35 AM	5/22/2022, 9:06:55 AM	Completed	5	ShuffleMoveOperation	79933
QID2196	0	5/21/2022, 1:12:34 PM	5/21/2022, 1:12:34 PM	Completed	0	ReturnOperation	1

Figure 11.26 – 20 Largest Query Steps by Most Rows Moved

V.   Scroll down further to the **Query Plan** section. The query plan for **QID3410** will be listed clearly. Here, we can see that **ShuffleMoveOperation** and **ReturnOperation** took the longest:

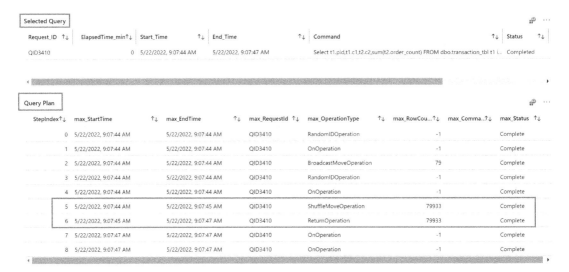

Selected Query					
Request_ID ↑↓	ElapsedTime_min↑↓	Start_Time ↑↓	End_Time ↑↓	Command ↑↓	Status ↑↓
QID3410	0	5/22/2022, 9:07:44 AM	5/22/2022, 9:07:47 AM	Select t1.pid,t1.c1,t2.c2,sum(t2.order_count) FROM dbo.transaction_tbl t1 i...	Completed

Query Plan								
StepIndex↑↓	max_StartTime ↑↓	max_EndTime ↑↓	max_RequestId ↑↓	max_OperationType ↑↓	max_RowCou...↑↓	max_Comma...↑↓	max_Status ↑↓
0	5/22/2022, 9:07:44 AM	5/22/2022, 9:07:44 AM	QID3410	RandomIDOperation	-1		Complete
1	5/22/2022, 9:07:44 AM	5/22/2022, 9:07:44 AM	QID3410	OnOperation	-1		Complete
2	5/22/2022, 9:07:44 AM	5/22/2022, 9:07:44 AM	QID3410	BroadcastMoveOperation	79		Complete
3	5/22/2022, 9:07:44 AM	5/22/2022, 9:07:44 AM	QID3410	RandomIDOperation	-1		Complete
4	5/22/2022, 9:07:44 AM	5/22/2022, 9:07:44 AM	QID3410	OnOperation	-1		Complete
5	5/22/2022, 9:07:44 AM	5/22/2022, 9:07:45 AM	QID3410	ShuffleMoveOperation	79933		Complete
6	5/22/2022, 9:07:45 AM	5/22/2022, 9:07:47 AM	QID3410	ReturnOperation	79933		Complete
7	5/22/2022, 9:07:47 AM	5/22/2022, 9:07:47 AM	QID3410	OnOperation	-1		Complete
8	5/22/2022, 9:07:47 AM	5/22/2022, 9:07:47 AM	QID3410	OnOperation	-1		Complete

Figure 11.27 – Query Plan

Now, let's create a workbook for monitoring Spark pools.

9.   Download the `SparkPool.workbook` file from `https://github.com/PacktPublishing/Azure-Data-Engineering-Cookbook-2nd-edition/blob/main/chapter11`.

10.   Follow *steps 2* to *5* to copy the contents of the `SparkPool.Workbook` file to the Advanced Editor area of the new workbook to create Spark pools. Set the workspace's name as `PacktADESynapse` and the Spark pool as `packtsparkpool`. Set **App Livy Id | Name** for any notebook or job run you would like to check. Then, click the **Done Editing** button:

Figure 11.28 – Configuring a Spark workbook

11. Click the **Save** icon. Then, set **Title** to **Sparkpool**. Click the **Save** button once more to save the workbook:

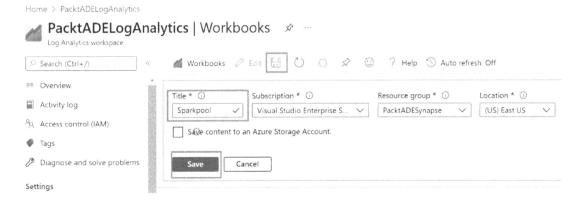

Figure 11.29 – Configuring the Spark workbook

12. Switch to the **Application Summary** tab. The **Application Summary** tab indicates how many executors (or how many worker nodes) were used to run the notebook and how many internal jobs were executed and their statuses:

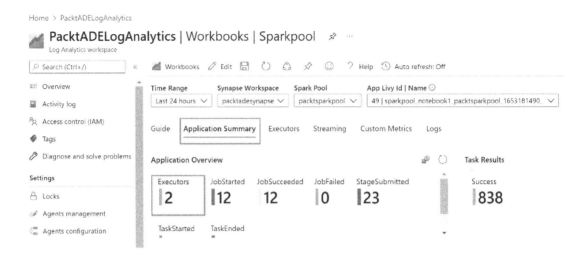

Figure 11.30 – Sparkpool workbook overview

13. Click on the **Executors** tab. The **Executors** tab provides insights into CPU usage, memory usage, and data processed by each node:

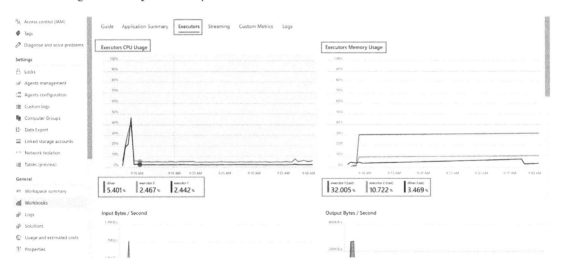

Figure 11.31 – Sparkpool workbook overview

How it works...

Workbooks are like templatized Kusto scripts combined with visuals that cover all the key metrics required for monitoring the health of Azure services. Workbooks use Kusto scripts to read the data in a Log Analytics workspace and present it in simple visuals for easier interpretation. The workbooks used in this recipe can be found at `https://github.com/microsoft/Azure_Synapse_Toolbox/tree/master/Monitor_Workbooks`, which is part of the official GitHub page maintained by Microsoft. This GitHub page contains other workbooks for serverless SQL pools and Synapse pipelines. We recommended that you explore these while following the configuration method explained in this recipe.

Monitoring table distribution, data skew, and index health using Synapse DMVs

In distributed databases such as Synapse dedicated SQL pools, the table is distributed across multiple nodes by design. When the rows in the table are not evenly distributed across the nodes, data distribution is said to be skewed. Data skew scenarios can have an impact on query performance. This recipe will provide a script based on Synapse **Dynamic Management Views** (**DMVs**) that you can use to monitor table skew.

Tables in a dedicated SQL pool have a column store index created by default. Column store indexes store the rows of the table in columnar format, which is optimized for processing analytics workloads. Each column store index in a table is subdivided into segments. A column store segment can be of three states – **Open**, **Closed**, or **Compressed**. For the column store index to be effective, its segments need to meet the following conditions:

- The number of segments in an open or closed state should be minimal.
- Each column store segment should have at least 100,000 rows.
- Segments in a compressed state should have at least a million rows.

In this recipe, we will develop a script to check for these conditions and verify the column store index's health.

Getting ready

To get started, log in to `https://portal.azure.com` using your Azure credentials:

- Create a Synapse Analytics workspace, as explained in the *Provisioning an Azure Synapse Analytics workspace* recipe of *Chapter 8, Processing Data Using Azure Synapse Analytics*.

- Download the `transaction-tbl.csv` file from `https://github.com/PacktPublishing/Azure-Data-Engineering-Cookbook-2nd-edition/tree/main/chapter10`. In the Synapse Analytics workspace, create a folder named `files` in the Data Lake account attached to it. Upload the `transaction-tbl.csv` file to the `files` folder. For detailed screenshots for a similar task, follow *steps 1 to 4* in the *How to do it...* section of the *Analyzing data using a serverless SQL pool* recipe of *Chapter 8, Processing Data Using Azure Synapse Analytics*.

- Complete the *Loading data into a dedicated SQL pool using PolyBase and T-SQL* recipe of *Chapter 10, Building the Serving Layer in Azure Synapse SQL Pool*, to create a dedicated SQL pool named `packtadesqlpool`, an external table named `dbo.ext_transaction_tbl`, and a dedicated SQL pool table named `dbo.transaction_tbl`. Alternatively, you may use the `Create_External_Table.sql` script at `https://github.com/PacktPublishing/Azure-Data-Engineering-Cookbook-2nd-edition/tree/main/chapter10` to create the `dbo.ext_transaction_tbl` and `dbo.transaction_tbl` tables.

How to do it...

In this recipe, we will be performing two major tasks:

- Detecting tables with skewed data distributions using Synapse DMVs
- Detecting tables with poor column store indexes

Perform the following steps to detect tables with skewed data:

1. First, let's create a skewed table. Log in to `portal.azure.com`, go to **All resources**, and search for `packtadesynapse`. Click on the workspace. Then, click on **Open Synapse Studio**. Click the develop button (the notebook-like button) on the left. Click the + button and select **SQL Script**. Finally, connect to the **packtadesqlpool** database and run the following script:

```
CREATE table dbo.transaction_tbl_skew WITH (DISTRIBUTION
= HASH(pid_skew))
AS
SELECT tid,sid, case when pid > 2 then 8 else pid end as
pid_skew,
transaction_date,
order_count,c1,c2
FROM dbo.transaction_tbl
```

This script creates a table that's distributed based on the `pid_skew` column, which will mostly contain a value of 8, causing the table to be distributed in an uneven or skewed fashion. The script's execution is shown in the following screenshot:

Figure 11.32 – Creating a skewed table

Download the Table-Skew.sql file from https://github.com/PacktPublishing/ Azure-Data-Engineering-Cookbook-2nd-edition/blob/main/chapter11/ Table-Skew.sql. Open Synapse Studio and click the develop button (the notebook-like button) on the left. Click on the + symbol and select **Import**:

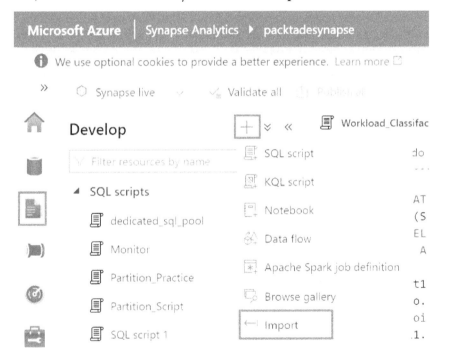

Figure 11.33 – Importing the SQL script

2. Select the Table_skew.sql file you downloaded in *step 2*. Connect to the **packtadesqlpool** database and run the script, as shown in the following screenshot:

Figure 11.34 – Table skew

A table in a Synapse-dedicated SQL pool will be split into a maximum of 60 distributions (parts) by default, irrespective of the number of nodes in the Synapse data warehouse instance (for example, if the Synapse instance has four nodes, each node will have 15 distributions of the table; if the Synapse instance has 10 nodes, each node will have six distributions). The script uses the `sys.dm_pdw_nodes_db_partition_stats` DMV, which contains several rows in each distribution. To measure the skew, the script compares the lowest and highest row count per distribution using the following formula:

*Table Skew Percentage = (Highest row count per distribution - Lowest row count per distribution) * 100 / Highest row count at a distribution*

The higher *Table Skew Percentage* is, the bigger the difference in row count across distributions, and hence the bigger the table skew will be. For tables with *Table Skew Percentage* over 25%, verify the choice of distribution column and recreate the table using a different distribution column, if required.

Now, let's explore a script for checking the quality of the column store index:

1. Download the `Clustered_Index_Health_Check.sql` file from `https://github.com/PacktPublishing/Azure-Data-Engineering-Cookbook-2nd-edition/blob/main/chapter11/Clustered_Index_Health_Check.sql`. Open Synapse Studio and click the develop button (the notebook-like button) on the left. Then, click on the + symbol and select **Import**:

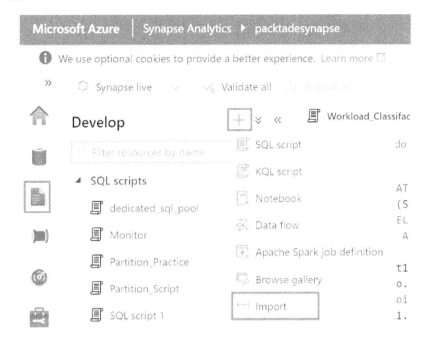

Figure 11.35 – Importing the SQL script

2. Select the `Clustered_Index_Health_Check.sql` file you downloaded in *step 1*. Connect to the **packtadesqlpool** database and run the script, as shown in the following screenshot:

Figure 11.36 – Column store health check script

This script identifies that the tables are too small to be stored using the column store format as they only contain a few thousand rows per segment. The script recommends that the table be converted into a heap table.

The script identifies tables with poor-quality column store indexes and offers recommendations to fix them. The script checks for the following conditions and offers the corresponding recommendations:

Condition Verified	Recommendation	Reason
Checks if the average rows per segment are less than 100,000 rows	Make the table into a heap table, which will store the table without the column store index.	Column store indexes are ideal for large tables with at least a few million rows in total. For small tables, column store indexes would be inefficient.
Checks if there are more than 10 segments that are in an open state	Verify if the table is partitioned, causing too many small segments, and fix the partition.	When the state of the column store segment is open, it remains in an uncompressed format, offering sub-optimal performance for analytic queries. If there are more than 1 million rows per segment, the engine would automatically compress the segment. It is recommended that you verify the reason for small segments – for example, due to the table partitioning strategy or data skewness.
Checks if there are more than 10 segments in a closed state	Run `ALTER TABLE <Table name> Reorg` to fix it.	If a segment is in a closed state, it implies it is ready to be compressed and is waiting for the system's background process to move it to a compressed state. You could force the background process to compress by performing a table reorg operation.
Checks if any compressed segments have less than 1 million rows (non-optimized segments)	Recreate the table using the `Create TABLE AS` command or rebuild the index with a higher resources class/more resources. You could scale up the DWU units of synapse the instance during the rebuild operation and scale them down once you've finished.	A compressed segment with less than 1 million rows offers sub-optimal performance. This is possible if there were not enough resources when the table was created or when the index was rebuilt.

Building monitoring dashboards for Synapse with Azure Monitor

By default, Azure Monitor stores the key performance counters of Synapse SQL pools and Sparks pool. Creating a dashboard using Azure Monitor's performance counters allows you to observe all the key performance counters at a glance and deduct any health-related issues on the Synapse workspace.

Getting ready

The prerequisites for this recipe are as follows:

- Create a Synapse Analytics workspace, as explained in the *Provisioning an Azure Synapse Analytics workspace* recipe of *Chapter 8, Processing Data Using Azure Synapse Analytics*.

- Complete the *Loading data into a dedicated SQL pool using PolyBase and T-SQL* recipe of *Chapter 10, Building the Serving Layer in Azure Synapse SQL Pool* to create a dedicated SQL pool named `packtadesqlpool`, an external table named `dbo.ext_transaction_tbl`, and a dedicated SQL pool table named `dbo.transaction_tbl`.

- Download the `SQLPool_Queries.sql` script from `https://github.com/PacktPublishing/Azure-Data-Engineering-Cookbook-2nd-edition/blob/main/chapter11/SQLPool_Queries.sql`. Run the script, as explained in *step 1* of the *How to do it...* section of the *Using Kusto queries to monitor SQL and Spark pools* recipe.

- Create a Spark pool cluster, as explained in the *Provisioning and configuring Spark pools* recipe of *Chapter 8, Processing Data Using Azure Synapse Analytics*.

- Download the `sparkpool_notebook1.ipynb` and `sparkpool_notebook2.ipynb` notebooks from `https://github.com/PacktPublishing/Azure-Data-Engineering-Cookbook-2nd-edition/blob/main/chapter11`, import them into the **Packtadesynapse** workspace, and execute it against **packtadesparkpool**, as explained in *step 4* of the *How to do it...* section of the *Using Kusto queries to monitor Spark and SQL pools* recipe.

How to do it...

Let's create a dashboard that will monitor the health of Synapse SQL pools and Spark pools by performing the following steps:

1. Log in to `portal.azure.com`, go to **All resources**, and search for `packtadesqlpool`. Search for **Metrics** under **Monitoring** and click on it. In the drop-down for **Metric**, select **DWU used percentage**:

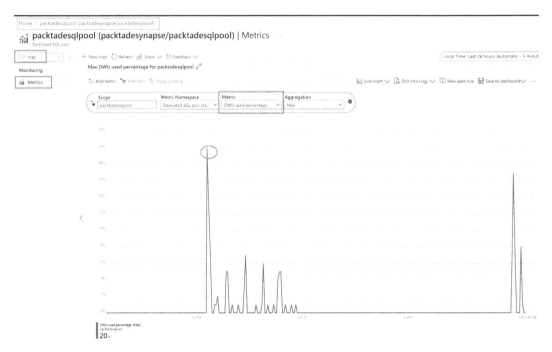

Figure 11.37 – DWU used percentage

The **DWU used percentage** metric indicates the percentage of total **Datawarehouse Units** (**DWUs**) used at a particular time. **DWU used percentage** is the key metric to identify if the data warehouse can manage the workload. A value of 100 for **DWU used percentage** would indicate that the data warehouse is running at its fullest capacity.

2. At the top right, click on **Save to dashboard** and select **Pin to dashboard**:

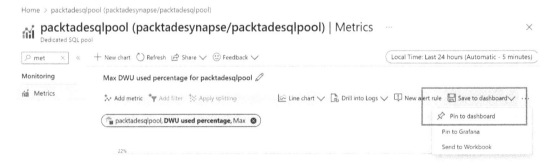

Figure 11.38 – Pin to dashboard

3. Click on the **Create new** tab. Select the type as **Shared**. Then, set **Dashboard name** to SynapseMonitoring. Uncheck the **Publish to 'dashboards' resource group** option and set **Resource group** to **PacktADESynapse**, the same resource group we used to create the Synapse Analytics workspace. Then, click on **Create and pin**:

Figure 11.39 – Create and pin

4. Click on + **New chart**. Set **Metric** to **Connections** and click on **Save to dashboard**:

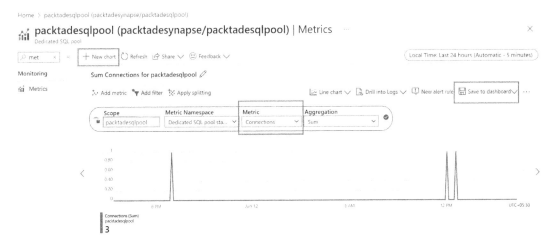

Figure 11.40 – Adding a new chart

5.   Select **Pin to dashboard**:

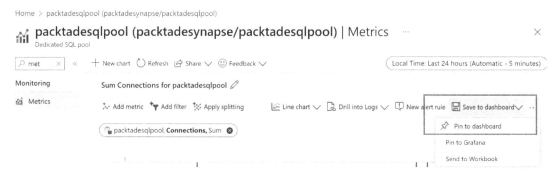

Figure 11.41 – Pin to dashboard

6.   Select the **Existing** tab and set **Type** to **Shared**. Set **Dashboard** to **SynapseMonitoring**, the same dashboard we created in *step 3*. Then, click on **Pin**:

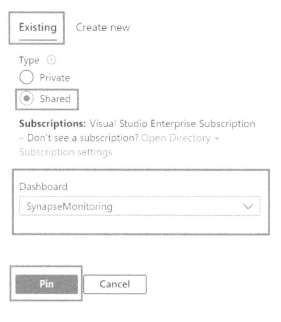

Figure 11.42 – Pin to dashboard

7.   Repeat *steps 4* to *6* to create new charts for the following metrics and add them to the **SynapseMonitoring** dashboard:

- **Queued queries**: To track queries that are queued due to a lack of concurrency slots/resources

- **Active queries**: To track the number of active queries in the dedicated SQL pool

- **Memory used percentage**: To track the total memory used in the Synapse-dedicated SQL pool

- **Local tempdb used percentage**: To track the tempdb usage percentage in the dedicated SQL pool

8. Now, let's add Sparkpool metrics to the dashboard. Go to **All resources** in the Azure portal and search for packtsparkpool. Search for **Metrics** under **Monitoring** and click on it. In the drop-down for **Metric**, select **vCores allocated**. Click on **Save to dashboard** and pin it to the **SynapseMonitoring** dashboard, as explained in *steps 4 to 7*:

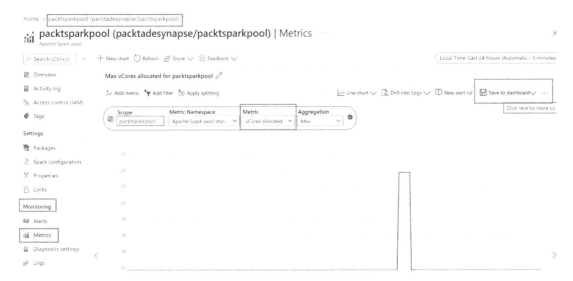

Figure 11.43 – vCores allocated

9. Repeat *steps 4 to 7* to create new charts for the following Sparkpool metrics and add them to the **SynapseMonitoring** dashboard:

- **Active Apache Spark Pool application**: To track the number of active Spark jobs

- **Memory Allocated**: To track the memory used by the Spark pool

10. Go to **All resources** in the Azure portal and search for SynapseMonitoring. A resource of the **Shared Dashboard** type in the **PacktAdeSynapse** resource group should be listed. Click on it:

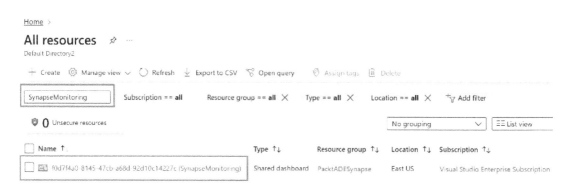

Figure 11.44 – Dashboard

11. Click on **Go to dashboard**:

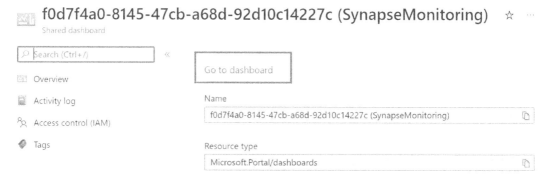

Figure 11.45 – Go to dashboard

12. All the charts we added will appear. Click on **Edit**:

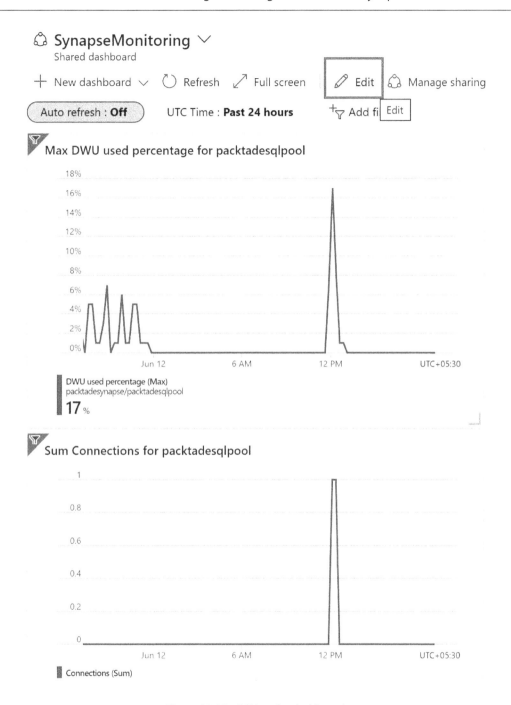

Figure 11.46 – Editing the dashboard

13. Close the **Tile Gallery** area:

Figure 11.47 – Closing the Tile Gallery area

14. Resize, drag, and reposition the charts based on their importance and the relevance of their metrics. Then, click **Save**:

Figure 11.48 – Saving the dashboard

15. Click on **Auto refresh: Off** and set it to **Every 5 minutes**:

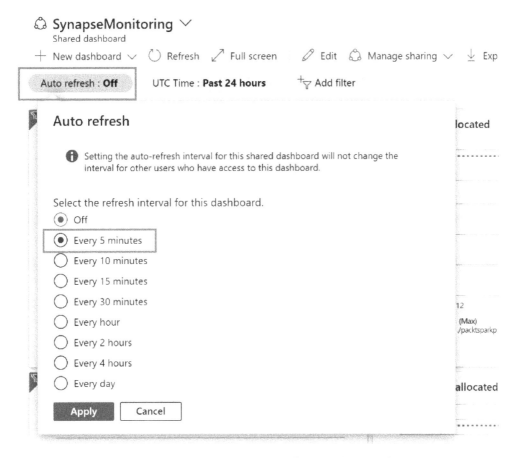

Figure 11.49 – Choosing when to refresh the dashboard

16. The dashboard is ready to be consumed. You may click on the **Manage sharing** button and then select the **Access control** option to grant permission to other users in your organization to view or edit the dashboard:

Figure 11.50 – Access control dashboard

How it works...

By default, Azure Monitor collects the health metrics of all Azure services and stores them for 30 days. Health metrics collected by Azure Monitor can be consolidated and viewed using dashboards. Having the key metrics across services in one place helps you correlate performance problems too. Setting the refresh frequency helps keep the dashboard updated automatically. Dashboards can be easily shared with anyone in your organization.

Optimizing and Maintaining Synapse SQL and Spark Pools

In the previous chapter, we learned how to monitor Synapse dedicated SQL and Spark pools. As a natural follow-up to that, in this chapter, we will cover techniques that can optimize the performance of Synapse SQL and Spark pools. In addition to optimization techniques, we will also share maintenance techniques such as pausing Synapse dedicated SQL pools and configuring a longer backup retention duration for Synapse dedicated SQL pools.

By the end of this chapter, you will have learned how to optimize Synapse dedicated SQL pool by reading a query plan and fixing its distribution, building a replication cache, and configuring result set caching. You will also know how to improve Spark pool performance using Z-ordering and partitioning techniques. In addition to optimization techniques, you will have learned about various maintenance techniques to automatically pause Synapse dedicated SQL pool during quiet periods, configure a backup retention that is longer than the default 7 days available in Synapse dedicated SQL pool, and run maintenance operations such as `OPTIMIZE` and `VACUUM`.

In this chapter, we'll cover the following recipes:

- Analyzing a query plan and fixing table distribution
- Monitoring and rebuilding a replication table cache
- Configuring result set caching in Azure Synapse dedicated SQL pool
- Auto pausing Synapse dedicated SQL pool
- Configuring longer backup retention for a Synapse SQL database
- Optimizing Delta tables in a Synapse Spark pool lake database
- Optimizing query performance in Synapse Spark pools

Technical requirements

For this chapter, you will need the following:

- A Microsoft Azure subscription
- PowerShell

Analyzing a query plan and fixing table distribution

Synapse dedicated SQL Pool is a distributed database engine that is comprised of one control node and one or more compute nodes. A control node is like the brain of the database engine and all the queries that are submitted by client applications are received by the control node. The control node splits the work that needs to be done to process the query and assigns it to compute nodes in Synapse dedicated SQL Pool. The compute nodes process the work assigned to them and return the partial results to the control node. The control node combines the results from compute nodes, performs any additional processing, and returns the final output to the client application. The number of compute nodes depends on the **data warehouse units** (**DWUs**) of the Synapse dedicated SQL pool, with SQL pools with higher DWUs getting more compute nodes. For example, a DWU1500 SQL pool has three compute nodes, while a DWU30000 SQL pool has 60 compute nodes. The architecture of a Synapse dedicated SQL pool with three compute nodes (DWU1500c) can be seen in the following diagram:

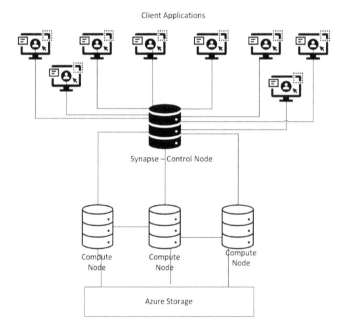

Figure 12.1 – Synapse dedicated SQL pool architecture

As the data is processed in multiple compute nodes, the engine moves the data across nodes to process the queries submitted. Typically, data movement across nodes is the most expensive operation in distributed query processing. Data movement can be reduced by distributing the data across nodes correctly, which can be achieved by selecting the correct distribution strategy for the table.

In this recipe, we will analyze a query plan, identify the data movement, and change the table distribution to improve query performance.

Getting ready

To get started, log into `https://portal.azure.com` using your Azure credentials.

Create a Synapse Analytics workspace, as explained in the *Provisioning an Azure Synapse Analytics workspace* recipe of *Chapter 8, Processing Data Using Azure Synapse Analytics*.

Complete the *Loading data into a dedicated SQL pool using PolyBase and T-SQL* recipe of *Chapter 10, Building the Serving Layer in Azure Synapse SQL Pool*, to create a dedicated SQL pool named **packtadesqlpool**, an external table named `dbo.ext_transaction_tbl`, and a dedicated SQL pool table named `dbo.transaction_tbl`. Alternatively, to create the `dbo.ext_transaction_tbl` and `dbo.transaction_tbl` tables, you may use the `Create_External_Table.SQL` script available at `https://github.com/PacktPublishing/Azure-Data-Engineering-Cookbook-2nd-edition/tree/main/chapter10`.

How to do it...

1. Log into `portal.azure.com`, go to **All resources**, and search for `packtadesynapse`, the Synapse Analytics workspace you created in the *Provisioning an Azure Synapse Analytics workspace* recipe of *Chapter 8, Processing Data Using Azure Synapse Analytics*. Click on the workspace, then **Open Synapse Studio**:

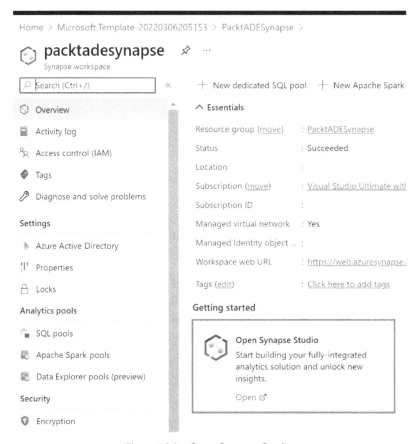

Figure 12.2 – Open Synapse Studio

2. Click on the **Develop** button (the notebook-like symbol) on the left, which will take you to the **Develop** section. Click on the + button and select **SQL script**:

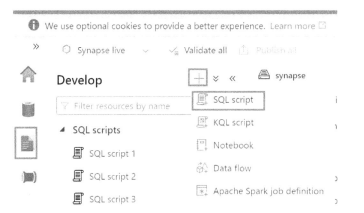

Figure 12.3 – SQL script

3. In this recipe, to showcase the impact of data movement, we need to scale up the Synapse dedicated SQL pool to DWU 500. Paste the following script in the new **SQL script** window:

```
ALTER DATABASE packtadesqlpool
MODIFY (SERVICE_OBJECTIVE = 'DW500c');
```

Select **packtadesqlpool** under **Connect to** and **master** under **Use database**. Then, click on the **Run** button to execute the script:

Figure 12.4 – Scaling up

4. Select **packtadesqlpool** under **Use database** and copy and paste the following script in the same query window. The scripts will create two tables called transaction_tbl_t1 and transaction_tbl_t2. Now, transaction_tbl_t1 is a hash that's been distributed based on the pid column, while transaction_tbl_t2 is a hash that's been distributed based on the tid column:

```
CREATE table dbo.transaction_tbl_
t1 WITH (DISTRIBUTION = HASH(pid))
AS
SELECT *
FROM dbo.transaction_tbl

CREATE table dbo.transaction_tbl_
t2 WITH (DISTRIBUTION = HASH(tid))
AS
SELECT *
FROM dbo.transaction_tbl
```

Click on the **Run** button to execute the script:

```
SQL script 1          Chapter12_Query_plan

Run    Undo      Publish    Query plan    Connect to    packtadesqlpool        Use database    packtadesqlpool

1    ALTER DATABASE packtadesqlpool MODIFY (SERVICE_OBJECTIVE = 'DW500c');
2
3    CREATE table dbo.transaction_tbl_t1 WITH (DISTRIBUTION = HASH(pid))
4    AS
5    SELECT *
6    FROM dbo.transaction_tbl
7
8    CREATE table dbo.transaction_tbl_t2 WITH (DISTRIBUTION = HASH(tid))
9    AS
10   SELECT *
11   FROM dbo.transaction_tbl
12
```

Figure 12.5 – Creating a table

5. Copy and paste the following script in the same query window. There are two SELECT queries here. The first is an expensive SELECT query, while the second query provides the query plan of the first query. Ensure that you select both queries and run them together as a single batch:

```
Select
t1.sid, AVG(t2.total_cost),max(t1.order_count) as max_
ord_cnt, t2.transaction_date
from dbo.transaction_tbl_t1 t1 inner join dbo.
transaction_tbl_t2 t2
on t1.tid = t2.tid
Group by t1.sid,t2.transaction_date;


Select request_id,operation_type,step_index,row_
count,total_elapsed_time,command from sys.dm_pdw_request_
steps
where request_id in ( Select request_id from sys.dm_pdw_
exec_requests where command like '%Select
t1.sid, AVG(t2.total_cost),max(t1.order_count) as max_
ord_cnt%'
and session_id in ( select session_id from sys.dm_pdw_
exec_sessions s  ) )
and start_time between dateadd(ss,-
20,getdate()) and getdate()
order by total_elapsed_time desc
```

Click on the **Run** button to execute the script:

Figure 12.6 – Getting the query plan

6. The preceding output indicates that **operation_type ShuffleMoveOperation** was the most expensive operation in the query as it had the highest elapsed time of 605 milliseconds and was moving ~242,000 rows across all nodes. **ShuffleMoveOperation** is an internal operation of the Synapse engine, where it attempts to move the data across the nodes to process the query. Click on the **Export results** button and select **CSV** to export the result as a comma-separated file:

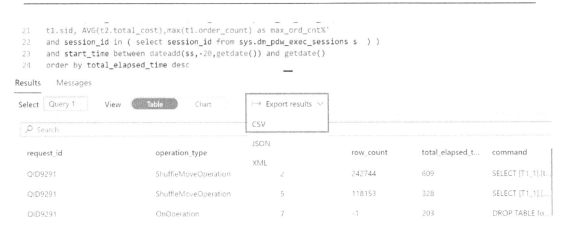

```
21    t1.sid, AVG(t2.total_cost),max(t1.order_count) as max_ord_cnt%'
22    and session_id in ( select session_id from sys.dm_pdw_exec_sessions s  ) )
23    and start_time between dateadd(ss,-20,getdate()) and getdate()
24    order by total_elapsed_time desc
```

Figure 12.7 – Exporting the query as a CSV file

7. Open the downloaded CSV file in Excel and expand the command column. The first row in the command column contains the internal query used by the engine, which caused a large shuffle movement. Expand the command column; notice that the query selecting the columns required the transaction_tbl_t1 table. This implies that the engine was shuffling the data of the transaction_tbl_t1 table across all nodes to process the join condition:

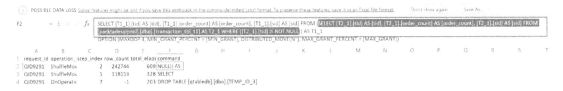

Figure 12.8 – Shuffle operation

8. If we observe the join condition of the expensive SELECT query we submitted in *step 5*, we will see that the transaction_tbl_t1 and transaction_tbl_t2 tables have been joined on the SELECT query using column tid. However, in *step 4*, we created the tables in the following way:

- transaction_tbl_t1 by hash distributing it using the pid column

- transaction_tbl_t2 by hash distributing it using the tid column

A hash distribution stores the data on the nodes of the Synapse dedicated SQL pool based on the value of the column used for hash distribution. In our case, `transaction_tbl_t1` was distributed using the `pid` column, while `transaction_tbl_t2` was distributed based on the `tid` column. To reduce the processing time and the time spent on the shuffle movement operation, the table should be hash distributed using the column that was used for the join operation in the `SELECT` query. As our query has been joined using the `tid` column, let's change the distribution of the `transaction_tbl_t1` table using the `tid` column. Please note that `transaction_tbl_t2` has already been distributed using the `tid` column, so no changes are required. Execute the following script to drop and recreate `transaction_tbl_t1` using the `tid` column:

```
Drop table dbo.transaction_tbl_t1;
CREATE table dbo.transaction_tbl_
t2 WITH (DISTRIBUTION = HASH(tid))
AS
SELECT *
FROM dbo.transaction_tbl
```

Click on the **Run** button to execute the script:

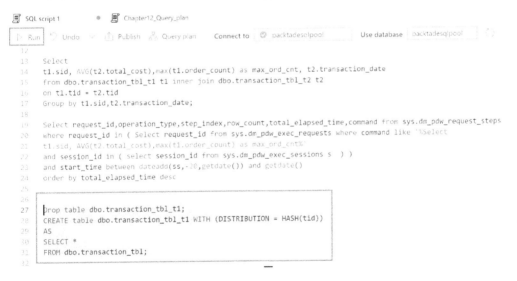

Figure 12.9 – Table recreation

9. Rerun the two SELECT queries (the SELECT query joining transaction_tbl_t1 and transaction_tbl_t2 and the SELECT query fetching the query plan) we ran in *step 5*. You will notice that the major **ShuffleMoveOperation** that moved ~242,000 rows across all nodes no longer exists in the query plan and that the query processing time has also improved significantly:

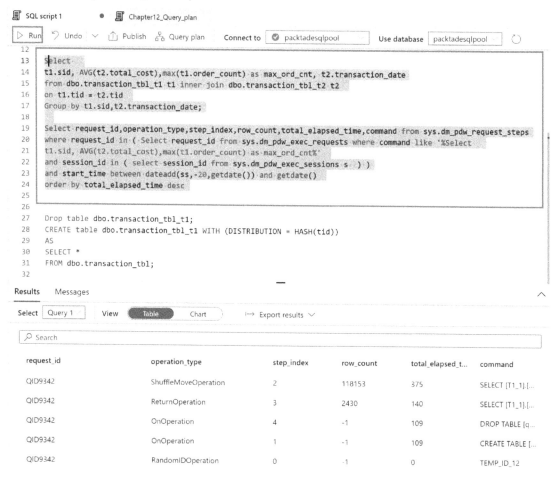

Figure 12.10 – Query plan after redistribution

10. Execute the following script to scale it down to DW100:

```
ALTER DATABASE packtadesqlpool
MODIFY (SERVICE_OBJECTIVE = 'DW100c');
```

How it works...

Aligning the column used for distributing the table with the column used for the join condition ensures that the query optimizer can perform most of the processing at the individual nodes without having to move all the rows of the table across all nodes. Before changing the distribution of the `transaction_tbl_t1` table, we performed a shuffle operation on the SELECT query, which moved 242,722 rows. This is the count of all the rows in `transaction_tbl_t1`. `transaction_tbl_t1` was moved completely across the nodes to make the join work. However, once the `transaction_tbl_t1` table was distributed using the `tid` column, the join condition was fully performed concurrently at each node without us having to move all the rows. Only the subset of rows that were required for processing subsequent operations after the join operation (such as the GROUP BY and ORDER BY clauses) were moved using the shuffle operation, thereby reducing the processing time. It is important to note that while it is nearly impossible to remove all shuffle operations in large queries, eliminating full table movement using the correct choice of distribution column is critical for ensuring the optimal performance of the query.

Monitoring and rebuilding a replication table cache

Replication tables in Synapse dedicated SQL pool are tables that are copied to all the compute nodes and the control node of a Synapse dedicated SQL pool. So, if a Synapse dedicated SQL pool contains three compute nodes, then four copies of the replicated table are present in the database. A replication table prevents the Synapse engine from broadcasting a table across all compute nodes during query processing. However, the replication table would be effective and would prevent broadcast operations when the replication table cache is in the **Ready** state.

This recipe will teach you how to monitor the state of a replication table cache and rebuild the cache for all the replication tables in your Synapse dedicated SQL pool database.

Getting ready

To get started, log into `https://portal.azure.com` using your Azure credentials.

Create a Synapse Analytics workspace, as explained in the *Provisioning an Azure Synapse Analytics workspace* recipe of *Chapter 8, Processing Data Using Azure Synapse Analytics*.

Complete the *Loading data into a dedicated SQL pool using PolyBase and T-SQL* recipe of *Chapter 10, Building the Serving Layer in Azure Synapse SQL Pool*, to create a dedicated SQL pool named **packtadesqlpool**, an external table named **dbo.ext_transaction_tbl**, and a dedicated SQL pool table named **dbo.transaction_tbl**. Alternatively, to create the **dbo.ext_transaction_tbl** and **dbo.transaction_tbl** tables, you may use the `Create_External_Table.SQL` script available at `https://github.com/PacktPublishing/Azure-Data-Engineering-Cookbook-2nd-edition/tree/main/chapter10`.

How to do it...

Follow these steps to create a replication table, monitor the state of the replication table cache, and rebuild the replication table cache to move it into a **Ready** state:

1. Scale up the **packtadesqlpool** instance to DWU500, as described in *steps 1 to 3* of the *How to do it...* section of the *Analyzing a query plan and fixing table distribution* recipe.

2. Select **packtadesqlpool** under **Use database** and copy and paste the following script in the same query window. This script will create a replicated table called dbo.supplier:

    ```
    Create table dbo.supplier WITH ( DISTRIBUTION = REPLICATE)
     AS
     Select distinct sid
    [sid],
    'S' + '-' + convert(varchar(2),sid)  as supplier_name
     FROM [dbo].[transaction_tbl_t1]
    ```

    Click on the **Run** button to create the dbo.supplier table:

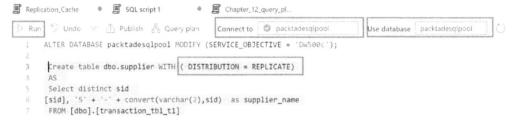

Figure 12.11 – Creating a replicated table

3. Run the following script to check the replication table cache of all replicated tables in any Synapse database. The script queries the **sys.pdw_replicated_table_cache_state Dynamic Management View (DMV)** to find the replication cache's status:

    ```
    SELECT '[' + sch.[name] + '].[' + t.
    [name] + '];' AS table_name, c.[state] , p.[distribution_
    policy_desc]
            FROM sys.tables t
            JOIN sys.pdw_replicated_table_cache_state c
              ON c.object_id = t.object_id
            JOIN sys.pdw_table_distribution_properties p
    ```

```
        ON p.object_id = t.object_id
    JOIN sys.schemas sch
        ON t.schema_id = sch.schema_id
    WHERE p.[distribution_policy_desc] = 'REPLICATE'
    ORDER BY c.[state], table_name
```

Click the **Run** button to execute the script and check the replication cache's status. It will be listed as **NotReady**:

Figure 12.12 – Checking the replication table's status

4. Copy and paste the following two SELECT queries into the query window and execute them together. The first SELECT query joins the dbo.transaction_tbl and dbo.supplier tables and returns some results. The second query prints the execution plan of the first query:

```
    Select s.supplier_name,sum(total_cost)
    FROM [dbo].[transaction_tbl] t INNER JOIN dbo.supplier s
on t.sid = s.sid
```

```
Group by s.supplier_name

Select request_id,operation_type,step_index,row_count,
total_elapsed_time,command from sys.dm_pdw_request_steps
where request_id in ( Select request_id from sys.dm_pdw_
exec_requests where command like '%Select s.supplier_
name,sum(total_cost)%'
and session_id in ( select session_id from sys.dm_pdw_
exec_sessions s  ) )
and start_time between dateadd(ss,-
20,getdate()) and getdate()
order by step_index
```

Select both queries and click on the **Run** button to execute them:

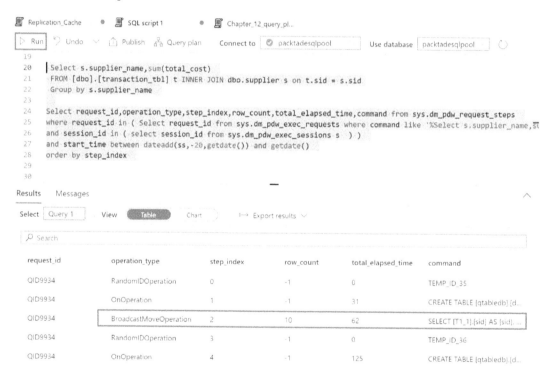

Figure 12.13 – Query plan of the replication table in the NotReady state

You will notice that the query plan has a **BroadcastMoveOperation**, which implies that the `dbo.supplier` table was broadcasted across all nodes, even though it was a replicated table. This is the result of the replicated table being in the **NotReady** state.

5. Perform the following steps to rebuild all the replication tables in the **NotReady** state in the Synapse dedicated SQL Pool database:

   I. Download the `Replication_Cache_Rebuild.SQL` script from `https://github.com/PacktPublishing/Azure-Data-Engineering-Cookbook-2nd-edition/blob/main/chapter12/Replication_Cache_Rebuild.sql`.

   II. Click on the **Develop** button (the notebook-like button) on the left. Click on the + symbol and select **Import**:

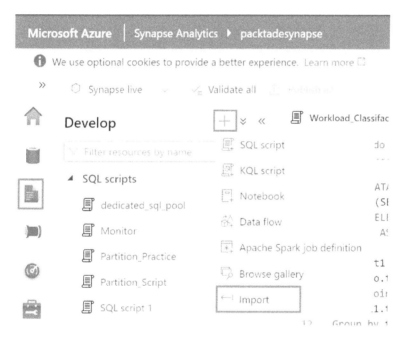

Figure 12.14 – Importing the SQL script

Select the `Replication_Cache_Rebuild.SQL` file you downloaded in *step 1*. Connect to the **packtadesqlpool** database and run the script, as shown in the following screenshot:

```
Replication_Cache    ×    SQL script 1    •    Chapter_12_query_pl...    Replication_Cache_R... ×

▷ Run   Undo  ∨           Query plan    Connect to  ● packtadesqlpool        Use database  packtadesqlpool

 2   CREATE TABLE #temp(id int, table_name varchar(2000))
 3   INSERT into #temp (id,table_name)
 4   SELECT Row_number() OVER(Order by t.name) as id,  '[' + sch.[name] + '].[' + t.[name] + ']' AS table_name
 5       FROM sys.tables t
 6       JOIN sys.pdw_replicated_table_cache_state c
 7         ON c.object_id = t.object_id
 8       JOIN sys.pdw_table_distribution_properties p
 9         ON p.object_id = t.object_id
10       JOIN sys.schemas sch
11         ON t.schema_id = sch.schema_id
12     WHERE p.[distribution_policy_desc] = 'REPLICATE'
13       and c.state = 'NotReady'
14   SET @id = 1
15   Select @rowcount = count(*) from #temp
16   WHILE @id <=@rowcount
17   BEGIN
18   SELECT @rebuild_cache_qry = 'SELECT TOP 1 * FROM ' + table_name + ';', @table_name = table_name
19   FROM #temp
20   WHERE id = @id
21   EXEC sp_executesql @rebuild_cache_qry;
22   Print 'Replication Cache of ' + @table_name + ' is being rebuilt'
```

Figure 12.15 – Rebuilding the replication cache

6. Rerun the `SELECT` script in *step 3* that queried the **sys.pdw_replicated_table_cache_state** DMV. Notice that `dbo.supplier` will be in a **Ready** state:

```
▷ Run   Undo  ∨    Publish    Query plan    Connect to  ● packtadesqlpool        Use database  packtadesqlpool

 9   SELECT '[' + sch.[name] + '].[' + t.[name] + '];' AS table_name, c.[state] , p.[distribution_policy_desc]
10       FROM sys.tables t
11       JOIN sys.pdw_replicated_table_cache_state c
12         ON c.object_id = t.object_id
13       JOIN sys.pdw_table_distribution_properties p
14         ON p.object_id = t.object_id
15       JOIN sys.schemas sch
16         ON t.schema_id = sch.schema_id
17     WHERE p.[distribution_policy_desc] = 'REPLICATE'
18     ORDER BY c.[state], table_name
19
20   Select s.supplier_name sum(total_cost)
```

Results   Messages

View   Table   Chart   ↦ Export results ∨

Search

table_name	state	distribution_policy_desc
[dbo] [supplier];	Ready	REPLICATE

Figure 12.16 – The state after rebuilding

7. Rerun the SELECT queries in *step 4* to check whether **BroadcastMoveOperation** has been removed from the query plan:

Figure 12.17 – The query plan after rebuilding

How it works...

The replication table cache moves into the **NotReady** state when an INSERT/UPDATE/DELETE statement or when the table is modified using a **Data Definition Language** (DDL) statement (ALTER/CREATE/Truncate Table statements). The replication table can be moved back to the **Ready** state by issuing a SELECT query against the table. The Replication_Cache_Rebuild.SQL script finds the replicated tables in a **NotReady** state and fires a SELECT query against each of them, which rebuilds the replication cache.

Once the table has been rebuilt, a copy of the table is stored in all the compute nodes of the dedicated SQL pool instance, so the query engine doesn't need to broadcast the replicated table to all the nodes during query processing. As a result, the broadcast move operation in the query plan, which was observed when the replicated table was in the **NotReady** state, is not seen once the replicated table is in the **Ready** state.

After loading data into a replicated table, it is always recommended to run the `Select top 1 *` `from <table name>` command against the table to keep the replication cache ready. You may also schedule the `Replication_Cache_Rebuild.SQL` script to run regularly to keep the replication cache of all replicated tables in the **Ready** state.

Configuring result set caching in Azure Synapse dedicated SQL pool

Result set caching is a feature in Synapse dedicated SQL pools that caches the result of the query in the control node of the Synapse dedicated SQL pool instance. In other words, the first time a query is executed, the result of the query is stored in the control node of the Synapse dedicated SQL pool instance. The next time, when the same query is executed and if the underlying data hasn't changed, the query engine will quickly return the result directly from the cache in the control node without reading the data from any of the compute nodes of the Synapse dedicated SQL pool instance. Unlike other common memory-based caches, the result set cache is persisted even after the Synapse dedicated SQL pool instance has been restarted.

In this recipe, we will learn how to turn on result set caching and how to verify if the cache is being used.

Getting ready

Create a Synapse Analytics workspace, as explained in the *Provisioning an Azure Synapse Analytics workspace* recipe of *Chapter 8, Processing Data Using Azure Synapse Analytics*.

Complete the *Loading data into a dedicated SQL pool using PolyBase and T-SQL* recipe of *Chapter 10, Building the Serving Layer in Azure Synapse SQL Pool*, to create a dedicated SQL pool named **packtadesqlpool**, an external table named **dbo.ext_transaction_tbl**, and a dedicated SQL pool table named **dbo.transaction_tbl**. Alternatively, to create the **dbo.ext_transaction_tbl** and **dbo.transaction_tbl** tables, you may use the `Create_External_Table.SQL` script available at `https://github.com/PacktPublishing/Azure-Data-Engineering-Cookbook-2nd-edition/tree/main/chapter10`.

How to do it...

Follow these steps to configure the result set cache and check whether a query is using the cache:

1. Scale up the **packtadesqlpool** instance to DWU500, as described in *steps 1* to *3* of the *How to do it...* section of the *Analyzing a query plan and fixing table distribution* recipe.

2. Select **packtadesqlpool** under **Use database** and copy and paste the following script in the same query window:

```
Declare @request_id nvarchar(32),@session_id nvarchar(32)
Select
t1.sid, AVG(t2.total_cost),max(t1.order_count) as max_
ord_cnt, t2.transaction_date
from dbo.transaction_tbl t1 inner join dbo.transaction_
tbl t2
on t1.tid = t2.tid
Group by t1.sid,t2.transaction_date;


Select @request_id = req.request_id,@session_id = req.
session_id
from sys.dm_pdw_exec_requests req
where req.command like 'Select
t1.sid, AVG(t2.total_cost),max(t1.order_count) as max_
ord_cnt%'
and req.start_time between dateadd(ss,-
30,getdate()) and getdate()


Select  req.request_id, result_cache_hit, req.command,
 req.total_elapsed_time as total_query_elapsed_time
 from sys.dm_pdw_exec_requests req
   where req.request_id = @request_id and req.session_
id = @session_id


Select req_steps.command, req_steps.location_type,req_
steps.step_index,req_steps.operation_type
From sys.dm_pdw_request_steps req_steps
WHERE req_steps.request_id = @request_id
order by req_steps.step_index
```

The result is shown in the following screenshot. Under **Select** on the **Results** pane, select **Query 2**. This query provides the result from the **sys.dm_pdw_request_steps** DMV. Under **location_type**, we can see that the query involved processing and moving the data from a compute node:

Figure 12.18 – Query steps

Select **Query 1** under **Select** on the **Results** pane:

Figure 12.19 – Exec_Requests

Query 1 returns the result from **sys.dm_pdw_exec_requests**. The `Result_Cache_hit` column has a value of `-1`, which indicates that the result set cache was not used. We can also see that the query took **1718** milliseconds to run.

3. In the same query window, select **master** under **Use database** and copy and paste the following script. Click the **Run** button. The script will turn on result set caching:

    ```
    ALTER DATABASE packtadesqlpool SET RESULT_SET_CACHING ON
    ```

    As shown in the following screenshot, the `ALTER DATABASE` command turns on result set caching:

Figure 12.20 – Turning on result set caching

4.   Select **packtadesqlpool** under **Use database** and copy and run the script you executed in *step 2*:

Figure 12.21 – Loading the cache

Switch to **Query 1** under **Select** in the **Results** pane. The `result_cache_hit` column has a value of 0, implying that the result set cache was missed. The query didn't use the result set cache as it was fired for the first time since the result set cache was turned. The cache is being loaded at the first run.

5.   Rerun the script you executed in *step 2*. As shown in the **location_type** column of **sys. dm_pdw_request_steps**, the query didn't hit the compute node at all and was fully executed from the control node:

Figure 12.22 – Control node execution

6. Switch to **Query 1** under **Select** in the **Results** pane. The `result_cache_hit` column has a value of 1, implying that the result set cache was used successfully. The query also ran significantly quickly at **46** milliseconds compared to **1718** milliseconds earlier when result set caching was not turned on:

Figure 12.23 – Cache used successfully

7. Connect to the master database and execute the following script to scale the Synapse dedicated SQL pool down to DW100:

```
ALTER DATABASE packtadesqlpool
MODIFY (SERVICE_OBJECTIVE = 'DW100c');
```

How it works...

As we can see, the result set cache significantly reduces the query processing time by caching the results in the control node. However, the result set cache is not used when the underlying table or data changes. When the data changes, the corresponding result set is removed from the cache. The cache is loaded with the updated result set when the first SELECT query runs against the updated data.

Hence, result set caching may not be an ideal option for write-heavy environments with massive **Extract, Transform, Load** (**ETL**) tasks, and in environments with several queries where the resulting size is bigger than 1 GB. In such environments, result set caching may increase the load on the control node as lots of effort by the engine is required to keep the cache updated; this can result in detrimental performance. Result set caching is ideal for environments that have significant reporting workloads with SELECT queries making up over 70% of the workload.

There are also other scenarios where the query engine won't use result set caching, even if it is turned on. A few scenarios are when the query uses non-deterministic functions such as `getdate` or when the result set's size is over 10 GB. The list of unsupported scenarios for result set caching can be found at `https://docs.microsoft.com/en-us/azure/synapse-analytics/sql-data-warehouse/performance-tuning-result-set-caching#whats-not-cached`. When a query doesn't use the result set cache, a zero or a negative value is returned by the **sys.dm_pdw_exec_requests** DMV's **result_cache_hit** column. Based on the value returned by the **result_cache_hit** column, we can find the reason why the query missed the result set cache. Information about the **result_cache_hit** column's value and its implication can be found at `https://docs.microsoft.com/en-us/sql/relational-databases/system-dynamic-management-views/sys-dm-pdw-exec-requests-transact-sql?view=aps-pdw-2016-au7#remarks`.

Configuring longer backup retention for a Synapse SQL database

By default, backups of a Synapse SQL dedicated pool database are only available for 7 days. In other words, you can retrieve versions of the database that are no older than 7 days. Having a backup retention period of 7 days in most environments will not be sufficient. Therefore, in this recipe, we will learn how to retain backups with a retention period longer than 7 days.

Getting ready

Create a Synapse Analytics workspace, as explained in the *Provisioning an Azure Synapse Analytics workspace* recipe of *Chapter 8, Processing Data Using Azure Synapse Analytics*.

Create a Synapse dedicated SQL pool named **packtadesqlpool**, as described in *steps 1 to 3* of the *How to do it...* section of the *Loading data into a dedicated SQL pool using PolyBase and T-SQL* recipe of *Chapter 10, Building the Serving Layer in Azure Synapse SQL Pool*.

Create an Azure Automation account named **azadeautomation**, as described in *steps 3 to 4* of the *How to do it...* section of the *Provisioning and configuring a wake-up script for a serverless SQL database* recipe of *Chapter 5, Configuring and Securing Azure SQL Database*.

How to do it...

In this recipe, we will perform the following tasks to achieve backup retention longer than 7 days:

- Configure an Azure Automation account with the `Az.Accounts` and `Az.Synapse` modules installed and managed identity enabled

- Create a runbook in the Azure Automation account and connect to the Synapse workspace using a managed identity

- Run a PowerShell script from the runbook that does the following:

  - Takes a backup of the Synapse SQL dedicated pool database

  - Creates a new database using the backup created

  - Pauses the newly created database

Follow these steps:

1. Log into `portal.azure.com`, go to **All resources**, and search for `azadeautomation`. Look for the **Modules** menu under **Shared resources**. Click on **Browse gallery** and import the following modules in the order specified here:

   - `Az.Accounts`

   - `Az.Synapse`

   While importing, select the runtime version as 7.1 (or above) for both modules. Wait for `Az.Accounts` to be imported before you start importing `Az.Synapse`. For detailed instructions on importing modules, please refer to *steps 5* to *7* of the *How to do it…* section of the *Provisioning and configuring a wake-up script for a serverless SQL database* recipe of *Chapter 5, Configuring and Securing Azure SQL Database*.

2. In the **azadeautomation** account, go to **Identity** under **Account Settings**. Click on the **On** button under **Status** and click **Save**. Turning on **Identity** will enable **managed identity** authentication for your Azure Automation account. Using managed identity, your Azure automation account can connect to the Azure resources without you having to specify your user ID and password. Connecting using the **managed identity** of **azadeautomation** will only work for scripts deployed inside the Azure Automation account, making managed identity a very safe method of authentication:

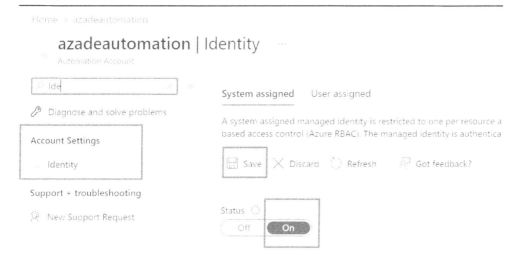

Figure 12.24 – Turning on managed identity

Click **Yes** when prompted for confirmation:

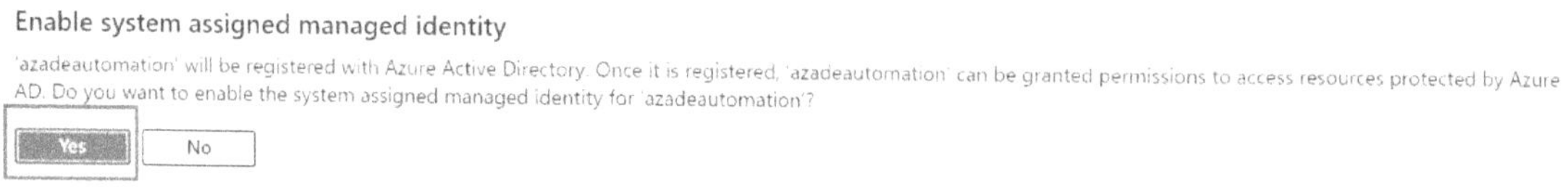

Enable system assigned managed identity
'azadeautomation' will be registered with Azure Active Directory. Once it is registered, 'azadeautomation' can be granted permissions to access resources protected by Azure AD. Do you want to enable the system assigned managed identity for 'azadeautomation'?
Yes
No

Figure 12.25 – Confirming managed identity creation

3. Click on **Azure role assignments**:

System assigned
User assigned
A system assigned managed identity is restricted to one per resource and is tied to the
based access control (Azure RBAC). The managed identity is authenticated with Azure
Save
Discard
Refresh
Got feedback?
Status
Off
On
Object (principal) ID
8d8309bc-783d-47dc-9b0c-d7f81eefb524
Permissions
Azure role assignments

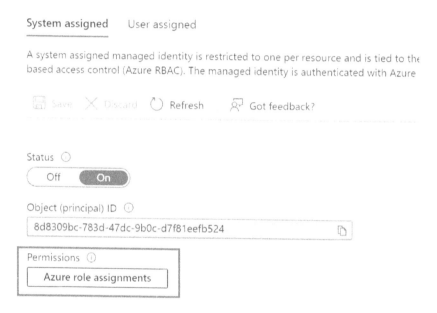

Figure 12.26 – Azure role assignments – managed identity

4. Click on + **Add role assignment (Preview)**. Set **Scope** to **Subscription** and **Role** to **Reader** and click **Save**:

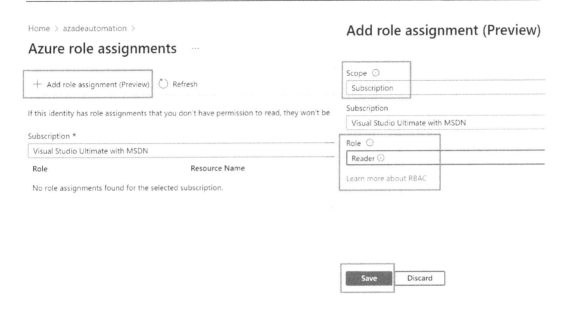

Figure 12.27 – Azure role assignment (Preview) – Subscription and Reader

5. Similarly, add another role assignment. Set **Scope** to **Resource Group**, **Resource Group** to **PacktADESynapse** (the resource group where the Synapse Analytics workspace resides), and **Role** to **Contributor** and click **Save**. Once done, the **Azure role assignments** page should reflect this, as shown in the following screenshot. It can take a few minutes for the roles to be listed. These roles are required for the managed identity of the Azure Automation account to connect to the subscription and back up and restore the Synapse dedicated SQL pool database:

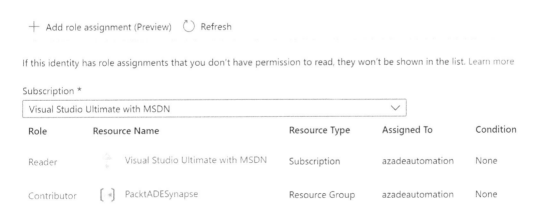

Figure 12.28 – The Azure role assignments page

6. In **azadeautomation**, the Azure Automation account you have created, go to the **Process Automation** section and click on **Runbooks**. Then, click **Create a runbook**:

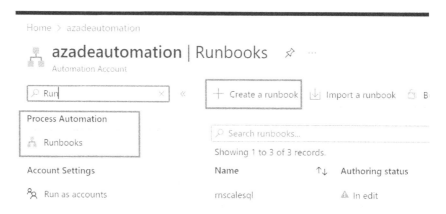

Figure 12.29 – Create a runbook

7. Set the runbook's **Name** to SynapseBackup, **Runbook type** to **PowerShell**, and **Runtime version** to **7.1 (preview)**:

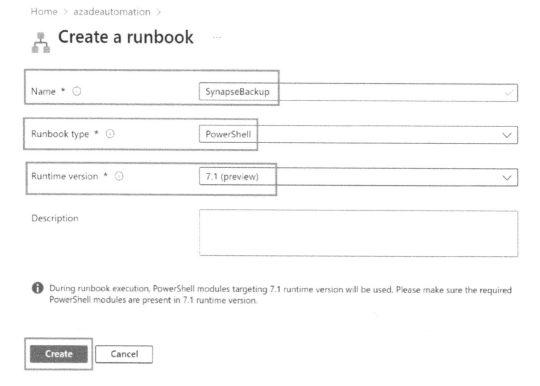

Figure 12.30 – The Create a runbook page

8. Copy and paste the following script in the runbook in sequence:

```
#Set Variables
$ResourceGroupName = "PacktADESynapse"
$SynapseAnalyticsWorksapce = "packtadesynapse"
$DatabaseName = "packtadesqlpool"
$label = $DatabaseName + (Get-Date -Format "yyyyMMdd")
```

Assign the appropriate values for the $ResourceGroupName, $SynapseAnalyticsWorksapce, and $DatabaseName variables. The script will create a backup database; the name of the backup database will be the original database name with the date suffixed to it. The $label variable will contain the new database name.

Copy and paste the following commands into the runbook:

```
#Login using Managed identity
$AzureContext = (Connect-AzAccount -Identity).context
$AzureContext = Set-AzContext -SubscriptionName
$AzureContext.Subscription -DefaultProfile $AzureContext
```

In the preceding code, we have the following:

- Connect-AzAccount -Identity initiates a connection to the Azure subscription using managed identity

- The Set-AzContext command sets the subscription against which the scripts are supposed to run

Copy and paste the following lines of code into the runbook:

```
$pool = Get-AzSynapseSqlPool -ResourceGroupName
$ResourceGroupName -WorkspaceName $SynapseAnalytics
Worksapce -Name $DatabaseName
$databaseId = $pool.Id -replace "Microsoft.Synapse",
"Microsoft.Sql" `
    -replace "workspaces", "servers" `
    -replace "sqlPools", "databases"


New-AzSynapseSqlPoolRestorePoint -WorkspaceName
$SynapseAnalyticsWorksapce -Name $DatabaseName
-RestorePointLabel $label
# Get the latest restore point
$restorePoint = $pool | Get-AzSynapseSqlPoolRestorePoint
| Select-Object -Last 1
# Restore to same workspace with source SQL pool
```

```
$restoredPool = Restore-AzSynapseSqlPool -FromRestorePoint
-RestorePoint $restorePoint.RestorePointCreationDate
-TargetSqlPoolName $label -ResourceGroupName $pool
.ResourceGroupName -WorkspaceName $pool.WorkspaceName
-ResourceId $databaseId -PerformanceLevel DW100c
# Pause the restored database
Suspend-AzSynapseSqlPool -WorkspaceName
$SynapseAnalyticsWorksapce -Name $label
```

In the preceding code, we have the following:

- The `Get-AzSynapseSqlPool` command gets details about the database that needs to be backed up and restored, assigned to a variable.

- The `New-AzSynapseSqlPoolRestorePoint` command creates a restore point. This restore point will be used to create a backup of the database.

- `Restore-AzSynapseSqlPool` creates a new database using the restore point created. The new database will contain the data as of the restore point and serve as the backup database. For example, if the restore point is created at 10 P.M., the backup database will contain the state of the database as of 10 P.M.

- `Suspend-AzSynapseSqlPool` will pause the new database that was created as the backup database needs to be running if and only if required. Pausing the new backup database keeps the cost of maintaining the backup solution minimal.

Click the **Save** button and then **Publish**:

Figure 12.31 – PowerShell script

9. Click **Start** to start executing the runbook:

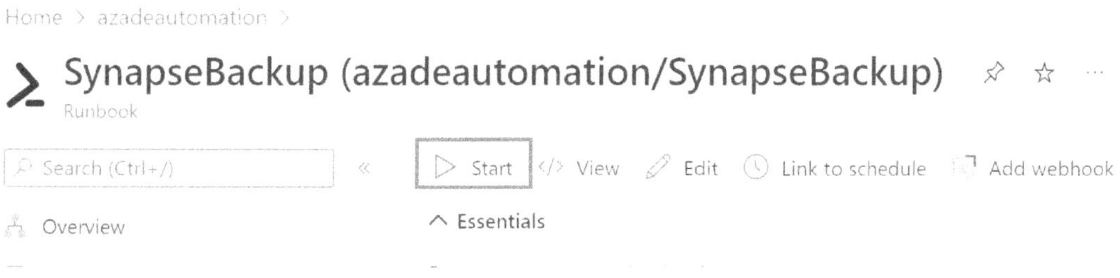

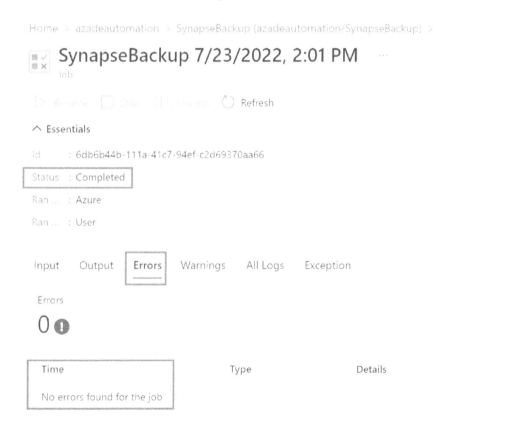

Figure 12.32 – Start execution

10. Wait until the runbook's status changes to **Completed**. Ensure no errors are listed:

Figure 12.33 – Run status

11. Go to **All resources** in the Azure portal and search for `packtadesqlpool`. You will find a database called **packtadesqlpool<yyyymmdd>**, as shown in the following screenshot (**<yyyymmdd>** will be replaced with the date when you ran the script):

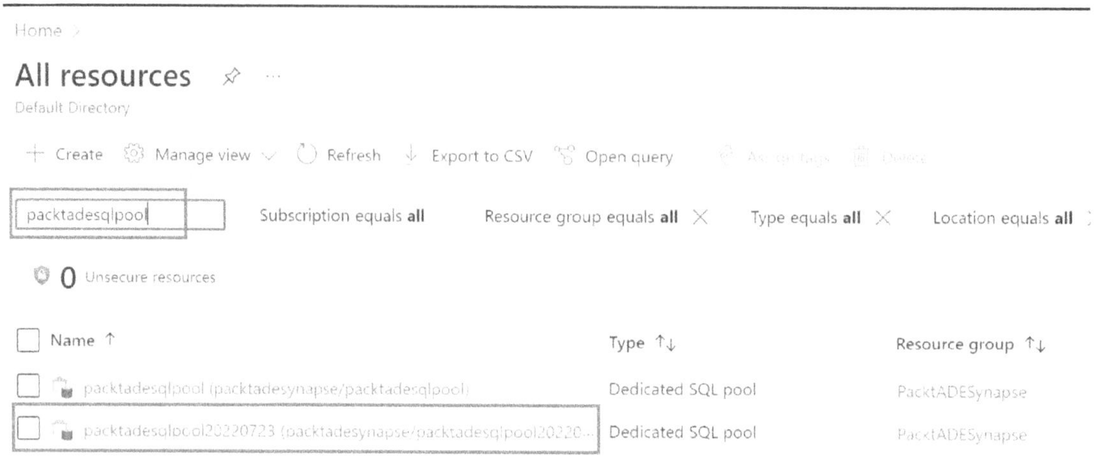

Figure 12.34 – Backup database result

How it works...

To create more copies of the database, you may schedule the runbook as per your preferred frequency (daily, weekly, and so on). The steps to schedule a runbook were provided in *steps 13* to *18* of the *How to do it...* section of the *Provisioning and configuring a wake-up script for a serverless SQL database* recipe of *Chapter 5, Configuring and Securing Azure SQL Database*. As the backup database is paused by the script, the cost incurred would be the storage cost due to the size of the backup instance. The total additional cost of the solution would depend on the storage cost and the number of copies of the database you retain. The longer the retention period, the higher the cost. You can use the `Remove-AzSynapseSqlPool` PowerShell command to remove older copies of the database that are not required as per your retention policy.

Auto pausing Synapse dedicated SQL pool

As of July 2022, unlike Azure SQL Database and Azure Databricks, the ability to automatically pause the database is not available in Synapse dedicated SQL pool databases. It would be useful to detect inactive periods in a Synapse dedicated SQL pool database and automatically pause it. This recipe will showcase a method that detects whether the Synapse dedicated SQL pool database was inactive for 30 minutes and pauses it if so.

Getting ready

Create a Synapse Analytics workspace, as explained in the *Provisioning an Azure Synapse Analytics workspace* recipe of *Chapter 8, Processing Data Using Azure Synapse Analytics*.

Create a Synapse SQL dedicated pool database named **packtadesqlpool**, as described in *steps* to *3* of the *How to do it...* section of the *Loading data into dedicated SQL pool using PolyBase and T-SQL* recipe of *Chapter 10, Building the Serving Layer in Azure Synapse SQL Pool*.

Create an Azure Automation account named **azadeautomation**, as described in *steps 3 and 4* of the *How to do it...* section of the *Provisioning and configuring a wake-up script for a serverless SQL database* recipe of *Chapter 5, Configuring and Securing Azure SQL Database*.

How to do it...

In this recipe, we will perform the following tasks to pause the Synapse dedicated SQL pool database when it is inactive for 30 minutes:

- Configure an Azure Automation account with the `SqlServer`, `Az.Accounts`, and `Az.Synapse` modules installed and managed identity enabled
- Add a credential to the Azure Automation account to connect to a dedicated SQL pool database
- Create a runbook in the Azure Automation account and execute a PowerShell script from the runbook that performs the following tasks:

  - Checks, using Synapse DMVs, whether there have been no active transactions for the last 30 minutes
  - Pauses the database if the database is inactive

Follow these steps:

1. Log into `portal.azure.com`, go to **All resources**, and search for `azadeautomation`. Look for the **Modules** menu under **Shared resources**. Click on **Browse gallery** and import the following modules in the order specified here:

   - `Az.Accounts`
   - `Az.Synapse`
   - `SqlServer`

   While importing, select the runtime version as 7.1 (or above) for both modules. Ensure that `Az.Accounts` has been imported before importing `Az.Synapse`. For detailed instructions on importing modules, please refer to *steps 5 to 7* of the *How to do it...* section of the *Provisioning and configuring a wake-up script for a serverless SQL database* recipe of *Chapter 5, Configuring and Securing Azure SQL Database*.

2. In the **azadeautomation** account, perform the following tasks:

    - Enable a system-managed identity account in the **Identity** section

    - Grant the following permissions:

        - **Reader** for **Azure Subscription**

        - **Contributor** for the **PacktADESynapse** resource group

    - Create an empty runbook called `SynapseAutoPause` on version 7.1

    For detailed screenshots of the preceding tasks, please refer to *steps 2* to *7* of the *How to do it...* section of the *Configuring longer backup retention for a Synapse SQL database* recipe.

3. Go to **Credentials** in the **Shared resources** section of the **azadeautomation** account and create a credential named `SynapseCred`. Specify the username as **sqladminuser** and the password as **PacktAdeSynapse123**. For detailed instructions and screenshots on a similar task, please refer to *steps 8* and *9* of the *How to do it...* section of the *Provisioning and configuring a wake-up script for a serverless SQL database* recipe of *Chapter 5, Configuring and Securing Azure SQL Database*.

    This credential contains the SQL database user ID and password to be used to connect to the Synapse dedicated SQL pool database; these were specified while creating the Azure Synapse Analytics workspace.

4. Paste the following script:

    ```
    $ResourceGroupName = "PacktADESynapse"
    $SynapseAnalyticsWorksapce = "packtadesynapse"
    $DatabaseName = "packtadesqlpool"
    $instanceName = $SynapseAnalyticsWorksapce + ".sql.
    azuresynapse.net"
    ```

    Assign the appropriate values for the `$ResourceGroupName`, `$SynapseAnalyticsWorksapce`, and `$DatabaseName` variables.

    Copy and paste the following code into the runbook:

    ```
    #Login using Managed identity
    $AzureContext = (Connect-AzAccount -Identity).context
    $AzureContext = Set-AzContext -SubscriptionName
    $AzureContext.Subscription -DefaultProfile $AzureContext
    ```

    In the preceding script, we have the following:

    - `Connect-AzAccount -Identity` initiates a connection to the Azure subscription using managed identity

- The Set-AzContext command sets the subscription against which the scripts are supposed to run

Copy and paste the following script into the runbook:

```
$Query = "
 select count(*) as request_count from sys.dm_pdw_exec_
requests req inner join sys.dm_pdw_exec_sessions ss on
ss.session_id = req.session_id
 where
 (req.status in ('Running','Suspended') or (
req.submit_time > DATEADD(minute, -30, GETDATE()) or
req.start_time > DATEADD(minute, -30, GETDATE()) or req.
end_time >
DATEADD(minute, -30, GETDATE())))
 and req.[label] not like 'SynapseAutoPause Job' and ss.
app_name not in ('Internal') OPTION (LABEL =
'SynapseAutoPause Job')"
```

Here, we are creating a variable called $Query that contains the query based on the **sys.dm_pdw_exec_requests** and **sys.dm_pdw_exec_sessions** DMVs to count the number of active queries/requests in the last 30 minutes. The script takes into account any query that was running, started, or finished within the last 30 minutes. The script excludes system sessions and any connection from the SynapseAutoPause runbook. If the count of requests is equal to zero, then the database is said to be inactive and it qualifies to be paused.

Copy and paste the following code into the runbook:

```
$pool = Get-AzSynapseSqlPool -ResourceGroupName
$ResourceGroupName -WorkspaceName
$SynapseAnalyticsWorksapce -Name $DatabaseName

if ($pool.Status -like 'paused' )
{
Write-Output "Synapse SQL DB is already paused"
}
```

The Get-AzSynapseSqlPool command gets the current status of the Synapse dedicated SQL pool database. If the database is already paused, no action is needed from the script and it prints a message stating **Synapse SQL DB is already paused**.

Copy and paste the following code into the runbook:

```
else
{
    $result = invoke-sqlcmd
```

```
-ServerInstance $instanceName
-Database $DatabaseName -Credential
$SynapseCred -Query $Query -Encrypt
    if ($result.request_count -eq 0)
 {
     $msg = "SQL Pool Database " +
 "$DatabaseName + " being paused as no
active transctions found"
    Write-Output  $msg
    Suspend-AzSynapseSqlPool -WorkspaceName
$SynapseAnalyticsWorksapce -Name $DatabaseName
     $msg = "paused azure synapse sql pool
- " + $DatabaseName
    Write-Output $msg
 }
 else
 {
    $msg = $DatabaseName + " cant be paused
as there are active transactions"
    Write-Output $msg
 }


 }
```

In the preceding code, we have the following:

- The invoke-sqlcmd command executes the query we prepared to check whether the database is inactive. The result of the query is assigned to a variable named $result.

- The if ($result.request_count -eq 0) condition checks if 0 user queries were running in the last 30 minutes to establish whether the database was inactive. If the condition is found to be true, the Suspend-AzSynapseSqlPool command is fired to pause the database.

- If the if ($result.request_count -eq 0) condition is evaluated to false, it implies there is at least one active query from any user in the last 30 minutes, and the script will exit without pausing the database.

Click the **Save** button and then **Publish**:

Figure 12.35 – Publishing the auto pause script

5. Let's test the script. In the Azure portal, go to **All resources** and search for `PacktADESynapse`, the Synapse Analytics workspace. Open **Synapse Studio**. Click on the notebook-like icon on the left-hand side of the screen. This will take you to the **Develop** section. Click the + button at the top and click on **SQL script**:

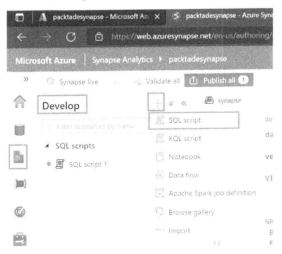

Figure 12.36 – New SQL script

6. Select **packtadesqlpool** under **Connect to**. Insert a simple query such as `Select GETDATE()`, as shown in the following screenshot, and click **Run**:

Figure 12.37 – Running the SQL script

7. Don't run any other query on the **packtadesqlpool** database. Wait for 30 minutes or longer and go back to the **azadeautomation** account in the Azure portal. Then, click on **Runbooks** under **Process Automation** and click the **SynapseAutoPause** runbook:

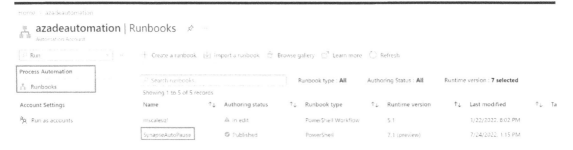

Figure 12.38 – Opening the runbook

8. Click on the **Start** button to start the runbook. Click **Yes** when prompted for confirmation to start the runbook:

Figure 12.39 – Starting the runbook

9. Verify that the runbook has finished running without any errors:

Figure 12.40 – Starting the runbook

10. Go to **All resources** in the Azure portal and search for `packtadesqlpool`, the Synapse dedicated SQL pool database. Click on it. Observe its status and notice that the database's status is **Paused**. The database was paused by the runbook as there was no activity for 30 minutes:

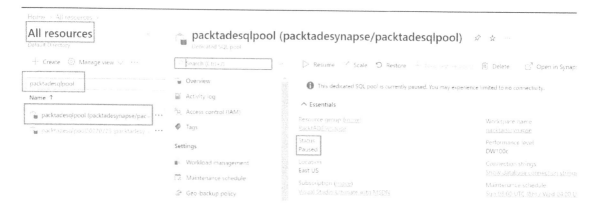

Figure 12.41 – Paused database

How it works...

In this recipe, we created a runbook called **SynapseAutoPause** in our Azure Automation account to run a PowerShell script that pauses the Synapse dedicated SQL pool database if the database is inactive for 30 minutes. To test the runbook, we ran a simple SQL script against the Synapse dedicated SQL pool database and left the database quiet for 30 minutes. After 30 minutes, we ran the runbook and verified that the database has been paused. The runbook can be scheduled at a suitable frequency to check if the database is inactive and pause it to reduce the cost incurred from the Synapse dedicated SQL pool. The steps to schedule a runbook were provided in *steps 13* to *18* of the *How to do it...* section of the *Provisioning and configuring a wake-up script for a serverless SQL database* recipe of *Chapter 5, Configuring and Securing Azure SQL Database*. The script that was used in this recipe is suitable for development, **user acceptance testing** (UAT) environments, and projects where the Synapse dedicated SQL pool may not need to be running all the time.

Optimizing Delta tables in a Synapse Spark pool lake database

As covered in the *Processing data using Spark pools and lake databases* recipe of *Chapter 8, Processing Data Using Azure Synapse Analytics*, a lake database allows you to store processed data in Delta tables, which are powered by Parquet files. Delta tables are very suitable for storing processed data that can be consumed by reporting solutions such as Power BI.

To achieve optimal performance in Delta tables, it is essential to evenly distribute the data among the Parquet files and purge the unwanted ones. The `OPTIMIZE` command helps optimally distribute the data among Parquet files, while the `VACUUM` command purges redundant Parquet files from the Azure Data Lake filesystem. The `OPTIMIZE` and `VACUUM` commands need to be executed regularly on the lake database so that you have optimal performance for the queries run against Delta tables.

In this recipe, we will be writing a script that can scan all Delta tables, optimize them, and vacuum redundant Parquet files.

Getting ready

To get started, log into `https://portal.azure.com` using your Azure credentials.

Create a Synapse Analytics workspace, as explained in the *Provisioning an Azure Synapse Analytics workspace* recipe of *Chapter 8, Processing Data Using Azure Synapse Analytics.*

Create a Spark pool cluster, as explained in the *Provisioning and configuring Spark pools* recipe of *Chapter 8, Processing Data Using Azure Synapse Analytics.*

Download the `transaction-tbl.csv` file from `https://github.com/PacktPublishing/Azure-Data-Engineering-Cookbook-2nd-edition/tree/main/chapter10`. In the Synapse Analytics workspace, create a folder named `files` in the data lake account attached to it. Upload the `transaction-tbl.csv` file to the `files` folder. For detailed screenshots of a similar task, follow *steps 1 to 4* in the *How to do it...* section of the *Analyzing data using a serverless SQL pool* recipe of *Chapter 8, Processing Data Using Azure Synapse Analytics.*

How to do it...

Follow these steps to run a script that can optimize and vacuum all delta tables:

1. First, let's create a lake database and a few Delta tables. Download the `Create_Delta_Table. ipynb` notebook from `https://github.com/PacktPublishing/Azure-Data-Engineering-Cookbook-2nd-edition/blob/main/chapter12`.

2. In **Synapse Studio**, click on the **Develop** button (the notebook-like button) on the left. Click on the + symbol, select **Import**, and select the `Create_Delta_Table.ipynb` file that we downloaded:

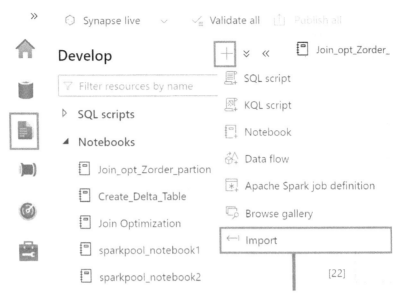

Figure 12.42 – Import

3. Go to the **Create_Delta_Table** notebook. Select **packtsparkpool** under **Attach to** and click **Run all**. The notebook will create a lake database called **lake_db** and two tables called **lake_db.transaction_tbl_t1** and **lake_db.transaction_tbl_t2**. Once the tables have been created, the notebook will perform three INSERT statements in the **lake_db.transaction_tbl_t1** table:

Figure 12.43 – Creating a Delta table

4. Click on the + symbol and create a new notebook:

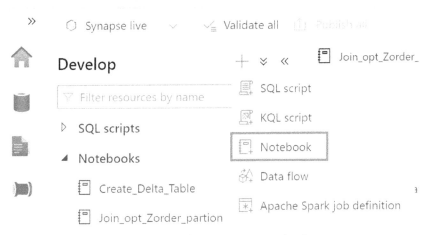

Figure 12.44 – Creating a new notebook

5. Select **packtsparkpool** under **Attach to**. Copy the following script and click the **Run** (small triangle) button:

```sql
%%sql
Describe detail lake_db.transaction_tbl_t1;
Describe detail lake_db.transaction_tbl_t2;
```

Scroll to the right and observe the **numfiles** and **sizeinBytes** columns. Notice that although the **lake_db.transaction_tbl_t1** and **lake_db.transaction_tbl_t2** tables are similar in size, the **lake_db.transaction_tbl_t1** table contains 18 files as we ran some additional insert queries to insert a few rows:

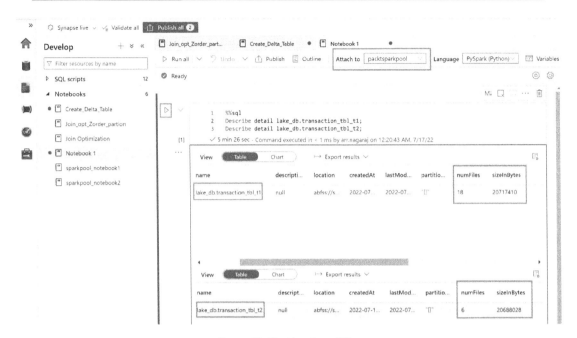

Figure 12.45 – Number of files

6. Add a new cell by hovering your mouse below the result set area of the previous cell and clicking the + **Code** button. Copy and paste the following script and click the **Run** button. This script runs the VACUUM and OPTIMIZE commands for all the tables in the database:

```
try:
    database_name = "lake_db"
    tables = spark.sql(f"SHOW TABLES FROM `{database_
name}`").select("tableName").collect()
    tables = [(row.tableName) for row in tables]
    for table_name in tables:
        spark.sql(f"OPTIMIZE `{database_name}`.`{table_
name}`")
        spark.sql(f"VACUUM `{database_name}`.`{table_
name}` RETAIN 720 HOURS")
except Exception as ex:
    raise Exception(f"error:`{str(ex)}`")
```

The script loops through all the tables in the **lake_db** database and executes the OPTIMIZE and VACUUM commands. By default, Delta tables retain previous versions of the tables for up to 30 days, which is useful for audit purposes. Executing the VACUUM command deletes previous versions of the table; older version data may not be retrievable after the VACUUM command is executed. To prevent loss of older version data, you can add the RETAIN 720 HOURS option after the VACUUM `{database_name}`.`{table_name}` command. The RETAIN 720 HOURS option will ensure only the versions that are older than 30 days from the current version of the table are deleted. You may customize the retention hours based on your organization's data retention requirements:

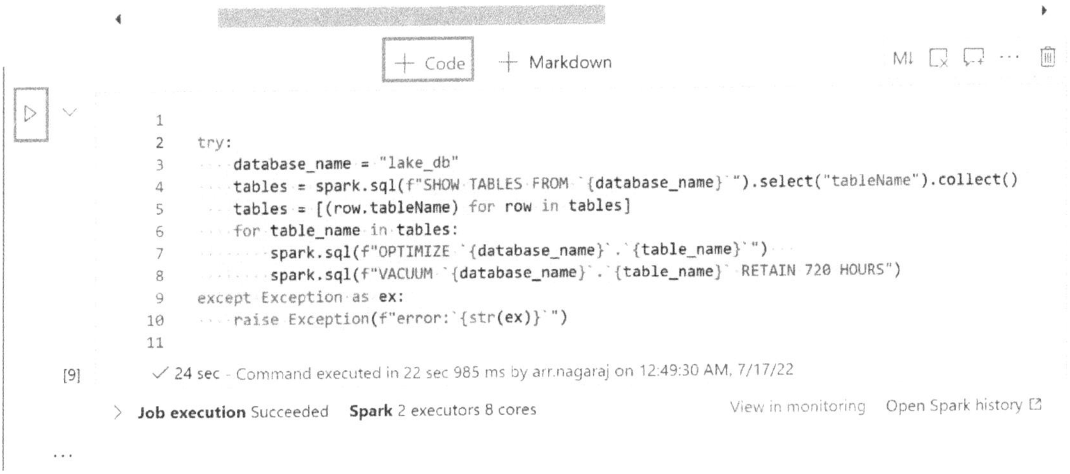

```
1
2    try:
3        database_name = "lake_db"
4        tables = spark.sql(f"SHOW TABLES FROM `{database_name}`").select("tableName").collect()
5        tables = [(row.tableName) for row in tables]
6        for table_name in tables:
7            spark.sql(f"OPTIMIZE `{database_name}`.`{table_name}`")
8            spark.sql(f"VACUUM `{database_name}`.`{table_name}` RETAIN 720 HOURS")
9    except Exception as ex:
10       raise Exception(f"error:`{str(ex)}`")
11
```

[9]   ✓ 24 sec - Command executed in 22 sec 985 ms by arr.nagaraj on 12:49:30 AM, 7/17/22

> **Job execution** Succeeded   **Spark** 2 executors 8 cores   View in monitoring   Open Spark history

...

Figure 12.46 – Vacuuming and optimizing all tables

7. Rerun the Describe detail <database_name>.<table_name> script that you ran in *step 5*. Notice that both tables now only have one Parquet file since the OPTIMIZE command has optimized the data distribution:

Figure 12.47 – Post-optimization

How it works...

VACUUM and OPTIMIZE are fundamental commands for maintaining the health of Delta tables. The script uses a simple for loop to iterate through all the tables and perform the OPTIMIZE and VACUUM operations. The script can easily be saved as a notebook and scheduled via integration pipelines to regularly optimize the Delta tables and maintain them in a good state.

Optimizing query performance in Synapse Spark pools

There are several methods you can use to optimize the performance of queries in a lake database, such as caching, indexing, partitioning, Z-ordering, data skipping, and using query hints. This recipe will showcase the following two methods to optimize the performance of a query:

- **Z-ordering:** Z-ordering helps the Spark engine easily locate columns with the same value
- **Partitioning:** Partitioning will partition the Delta lake table into smaller chunks, creating subfolders in the data lake storage account for each distinct value on the partitioned column

Getting ready

To get started, log into `https://portal.azure.com` using your Azure credentials.

Create a Synapse Analytics workspace, as explained in the *Provisioning an Azure Synapse Analytics workspace* recipe of *Chapter 8, Processing Data Using Azure Synapse Analytics*.

Create a Spark pool cluster, as explained in the *Provisioning and configuring Spark pools* recipe of *Chapter 8, Processing Data Using Azure Synapse Analytics*.

Download the `transaction-tbl.csv` file from `https://github.com/PacktPublishing/Azure-Data-Engineering-Cookbook-2nd-edition/tree/main/chapter10`. In the Synapse Analytics workspace, create a folder named `files` in the data lake account attached to it. Upload the `transaction-tbl.csv` file to the `files` folder. For detailed screenshots for a similar task, follow *steps 1* to *4* in the *How to do it...* section of the *Analyze data using a serverless SQL pool* recipe of *Chapter 8, Processing Data Using Azure Synapse Analytics*.

How to do it...

Follow these steps to optimize a query for reading Delta tables:

1. Download the `Optimize_Delta_Queries.ipynb` notebook from `https://github.com/PacktPublishing/Azure-Data-Engineering-Cookbook-2nd-edition/blob/main/chapter12`. Import the notebook into Synapse Studio, as we did in *steps 1* and *2* in the *How to do it...* section of the *Optimizing Delta tables in a Synapse Spark Pool lake database* recipe.

2. Go to the `Optimize_Delta_Queries` notebook. Select **packtsparkpool** under **Attach to** and click **Run all**:

Figure 12.48 – Importing the notebook

3. The first six cells do the following:

    • Read the file and create two tables called `transaction_tbl_f1 transaction_tbl_f2`:

      • `transaction_tbl_f1` contains approximately 25 million rows

      • `transaction_tbl_f2` contains approximately 500,000 rows

4. Go to the last cell in the notebook. This cell runs a SQL script that does the following:

    • Joins the `transaction_tbl_f1` and `transaction_tbl_f2` tables on the `tid` column

    • Filters by the `pid` column

    • Groups the result by the `tid` column

    The query joins the mid-sized table, `transaction_tbl_f2`, which contains 500,000 rows with the large table, `transaction_tbl_f1`, which contains 25 million rows. The query takes 37 seconds to run. In this recipe, we will explore options to optimize this query:

Figure 12.49 – Query execution time – not optimized

5. Download the `Optimize_Delta_Queries_Zorder_Partition.ipynb` notebook from https://github.com/PacktPublishing/Azure-Data-Engineering-Cookbook-2nd-edition/blob/main/chapter12. Import the notebook into Synapse Studio, as you did in *steps 1* and *2* in the *How to do it...* section of the *Optimizing Delta tables in a Synapse Spark Pool lake database* recipe. Go to the `Optimize_Delta_Queries_Zorder_Partition` notebook. Select **packtsparkpool** under **Attach to** and click **Run all**:

Figure 12.50 – Running the optimized notebook

6. Observe the result in the first cell, as shown in the following screenshot:

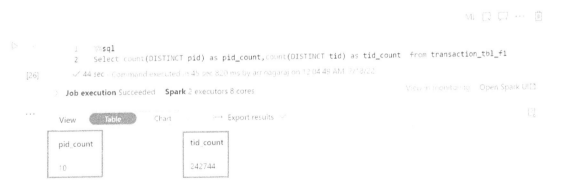

Figure 12.51 – Distinct row count

We know that the `pid` column will contain 10 distinct values and that the `tid` column will contain approximately 250,000 distinct values. A column participating in the WHERE clause or JOIN condition is a good candidate for partitioning. However, the column that's used for partitioning should not have too many distinct values as it will create too many files and subfolders. Also, there should be at least 1 million rows per partition. We know that **transaction_tbl_f1** contains 25 million rows and that there are only 10 distinct values for the `pid` column in the **transaction_tbl_f1** table. So, for each value of `pid`, we should have around 2.5 million rows on average. Observe the following query that we want to tune:

```
Select t1.pid, sum(t2.order_count)
FROM transaction_tbl_f1 t1
inner join transaction_tbl_f2 t2 on t1.tid = t2.tid
WHERE t1.pid between 3 and 7
Group by t1.pid
```

Here, we can see that the `pid` column is participating in the WHERE clause, so the `pid` column would make a good fit for the partition column as it has fewer distinct values and is participating in a WHERE clause.

7. Observe the second cell in the notebook. This call creates a table called **transaction_tbl_opt_f1** that's been partitioned by the `pid` column using **transaction_tbl_f1**. The **transaction_tbl_opt_f2** table has been created without partitioning from **transaction_tbl_f2** as it is a smaller table that only contains 500,000 rows; partitioning it would result in small subfolders, which is not recommended:

```
1   %%sql
2
3   CREATE TABLE transaction_tbl_opt_f1
4   USING DELTA PARTITIONED BY (pid)
5   AS
6   SELECT * FROM transaction_tbl_f1;
7   CREATE TABLE transaction_tbl_opt_f2
8   USING DELTA
9   SELECT * FROM transaction_tbl_f2;
```

✓ 5 min 6 sec - Command executed in 1 ms by arr.nagaraj on 12:18:10 AM, 7/18/22

Figure 12.52 – Creating partitioned tables

8. The **transaction_tbl_opt_f1** and **transaction_tbl_opt_f2** tables are joined using the `tid` column. Columns participating in the join condition can be optimized by **Z-ordering**. Z-ordering helps locate columns with the same value in different Parquet files. In our query, as we need to join **transaction_tbl_opt_f1** and **transaction_tbl_opt_f2** with the `tid` column, Z-ordering both the tables using the `tid` column will help the Delta engine identify the matching rows easily. Z-ordering is only effective when you have significant distinct values in the column; therefore, the `tid` column will be a good fit for Z-ordering as it contains approximately 250,000 rows. Cell 3 in the notebook performs Z-ordering, as shown in the following screenshot:

+ Code
+ Markdown
%%sql
Optimize transaction_tbl_opt_f1 zorder by (tid);
Optimize transaction_tbl_opt_f2 zorder by (tid);
7 min 45 sec - Command executed in < 1 ms by arr.nagaraj on 12:25:56 AM, 7/18/22
View
Table
Chart
Export results
name
value
numFilesBefore
2170
numBytesBefore
6418144533
numFilesAfter
10

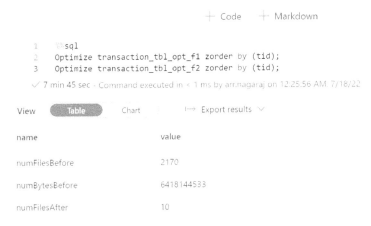

Figure 12.53 – Z-ordering

9. Cell 4 of the notebook reruns the queries but now with the **transaction_tbl_opt_f1** and **transaction_tbl_opt_f2** tables as they have been partitioned and Z-ordered. The query was completed in 26 seconds compared to the earlier run, which took 37 seconds:

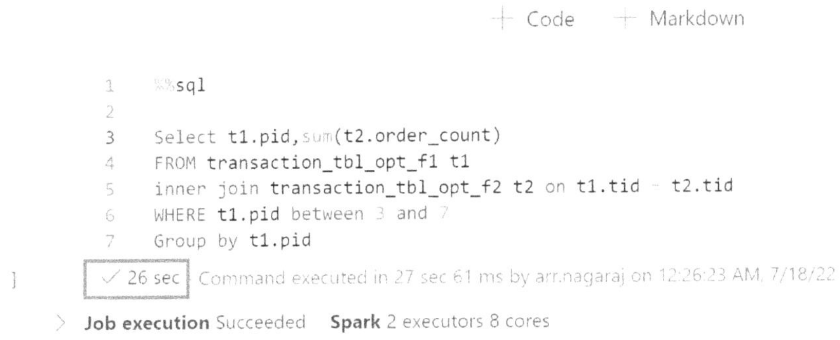

+ Code
+ Markdown
%%sql
Select t1.pid,sum(t2.order_count)
FROM transaction_tbl_opt_f1 t1
inner join transaction_tbl_opt_f2 t2 on t1.tid = t2.tid
WHERE t1.pid between 3 and 7
Group by t1.pid
26 sec
Command executed in 27 sec 61 ms by arr.nagaraj on 12:26:23 AM, 7/18/22
Job execution Succeeded
Spark 2 executors 8 cores

Figure 12.54 – The optimized result

How it works...

Partitioning the **transaction_tbl_opt_f1** table using `pid` organizes the table with a subfolder for each value of `pid`. Instead of all Parquet files in one folder named `transaction_tbl_opt_f1`, the Parquet files were placed in 10 subfolders inside the `transaction_tbl_opt_f1` folder in the data lake, which makes it easier for the Spark engine to read only the folders containing Parquet files that satisfy the **t1.pid** filter condition between 3 and 7. In Synapse Studio, click on the storage icon (the cylinder-like symbol on left). Then, click on the **Linked** tab, expand **packtadesynapse**, click on **synapse (Primary)**, and navigate to the `synapse | workspaces | packtadesynapse | warehouse | transaction_tbl_opt_f1` folder. The folder should look as follows:

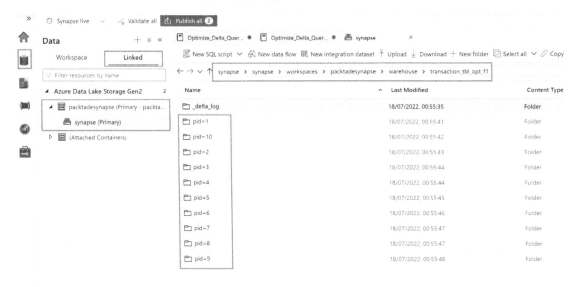

Figure 12.55 – subfolders – partitioning

Z-ordering helps identify the matching rows that have the same value for the `tid` column in both tables. Z-ordering needs to be scheduled regularly for it to be effective.

There are other techniques to optimize Delta lake query performance, such as indexing, caching, and many more. You may refer to `https://docs.microsoft.com/en-us/azure/synapse-analytics/spark/apache-spark-performance` to explore other techniques.

13

Monitoring and Maintaining Azure Data Engineering Pipelines

12 chapters and 500+ pages into this book should have given you a fair understanding of building Data Factory/Synapse integration pipelines to ingest, process, and store your data. The next key aspect of Azure data engineering is to monitor and maintain the pipelines that you've developed.

In this chapter, we will cover the following recipes:

- Monitoring Synapse integration pipelines using Log Analytics and workbooks
- Tracing SQL queries for Synapse dedicated SQL pools to Synapse integration pipelines
- Provisioning a Microsoft Purview account and creating a data catalog
- Integrating a Synapse workspace with Microsoft Purview and tracking data lineage
- Applying Azure tags using PowerShell to multiple Azure resources

After completing these recipes, you will be familiar with integrating Synapse workspaces with Log Analytics, monitoring Synapse pipeline performance and failures, using labels to trace queries fired by Synapse pipelines on dedicated SQL pools, tracking data lineage using Microsoft Purview, Purview integration with Synapse, and bulk tagging resources using PowerShell.

Technical requirements

For this chapter, you will need the following:

- A Microsoft Azure subscription
- PowerShell 7 and above
- Follow the *Install the Azure Az PowerShell module* instructions at `https://docs.microsoft.com/en-us/powershell/azure/install-az-ps?view=azps-8.0.0`

Monitoring Synapse integration pipelines using Log Analytics and workbooks

One of the key tasks after developing complex Synapse integration pipelines is to monitor them for aspects such as long-running activities, large data movements, and common errors. In this recipe, we will integrate a Synapse workspace with Azure Log Analytics and deploy a ready-made workbook to monitor the pipelines. In this recipe, at a high level, we will be performing the following tasks:

- Import a pipeline from a template. The pipeline's execution data will be monitored using an Azure Log Analytics workspace and workbook.
- Integrate a Synapse workspace with Azure Log Analytics.
- Deploy a workbook for monitoring the Synapse integration pipeline.

Getting ready

To get started, log into `https://portal.azure.com` using your Azure credentials and perform the following steps:

1. Create a Synapse Analytics workspace, as explained in the *Provisioning an Azure Synapse Analytics workspace* recipe of *Chapter 8*, *Processing Data Using Azure Synapse Analytics*.

2. Download the `transaction-tbl.csv` file from `https://github.com/PacktPublishing/Azure-Data-Engineering-Cookbook-2nd-edition/tree/main/chapter10`. In the Synapse Analytics workspace, create two folders named CSV and `Parquet` in the data lake account attached to it. Upload the `transaction-tbl.csv` file to the CSV folder. For detailed screenshots of a similar task, follow *steps 1* to *4* in the *How to do it...* section of *Analyzing Data using a Serverless SQL pool* recipe in *Chapter 8*, *Processing Data Using Azure Synapse Analytics*.

3. Create a Synapse dedicated SQL pool named `packtadesqlpool`, as described in *steps 1 to 3* in the *How to do it...* section of the *Loading data into a dedicated SQL pool using PolyBase and T-SQL* recipe of *Chapter 10, Building the Serving Layer in Azure Synapse SQL Pool.*

4. Create a Log Analytics workspace (if you haven't done so already), as explained in *steps 1* and *2* in the *How to do it...* section of the *Configuring a Log Analytics workspace for a Synapse dedicated SQL pool* recipe of *Chapter 11, Monitoring Synapse SQL and Spark Pools.*

How to do it...

Follow these steps to monitor Synapse pipelines using a Log Analytics workbook:

1. Let's integrate the Log Analytics workspace with Synapse. Log into `portal.azure.com`, go to **All resources**, and search for `packtadesynapse`, the Synapse Analytics workspace you created previously. Find **Diagnostic settings** under the **Monitoring** section on the left and click on **Add diagnostic setting**:

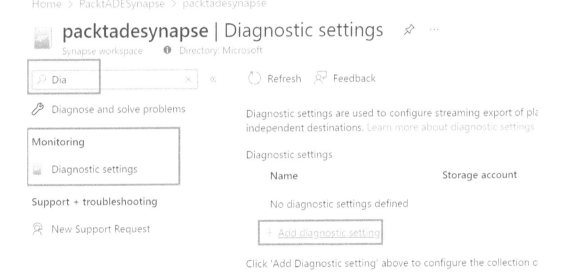

Figure 13.1 – Configuring the Log Analytics workspace

2. Set **Diagnostic setting** to `IntegrationPipeline`. Select **Integration Pipeline Runs**, **Integration Activity Runs**, and **Integration Trigger Runs** under **Categories**. Select the **Send to Log Analytics workspace** option under **Destination details** and select **PacktADELogAnalytics** under **Log Analytics workspace**. Then, click **Save**:

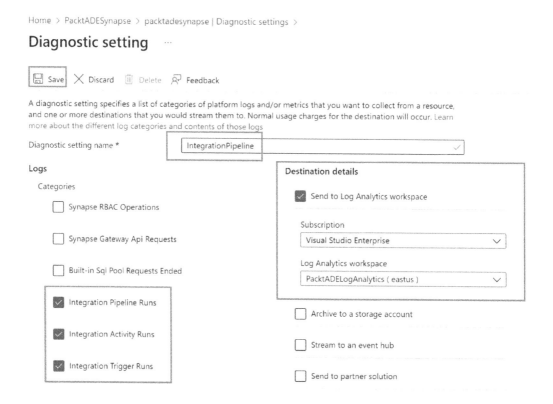

Figure 13.2 – Saving the Log Analytics workspace

3. The next step is to import a sample pipeline. We will be monitoring the sample pipeline's execution and diagnostics data via the Log Analytics workspace. The sample pipeline will read the `transaction-tbl.csv` file, load it into the **packtadesqlpool** database's table using the **Copy activity** task, perform transformations using the **Data flow** task, and finally save the result as Parquet files in the `Parquet` folder in the data lake account. Follow these steps to import a pipeline:

    *   Download the `SynapsePipeline.zip` file from `https://github.com/PacktPublishing/Azure-Data-Engineering-Cookbook-2nd-edition/tree/main/chapter13`.

    *   Open **Synapse Studio**. Click on the integration pipeline icon (the pipe-like symbol on the left). Click on + and then **Import from pipeline template**:

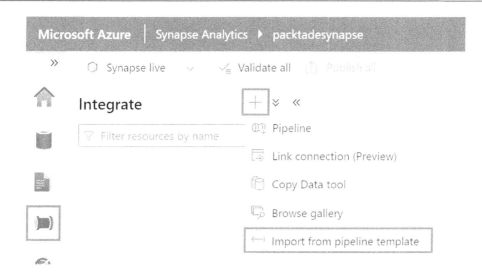

Figure 13.3 – Import from pipeline template

- Select the `SynapsePipeline.zip` file you downloaded. Then, select **packtadesynapse-WorkspaceDefaultSQLServer** for the first **Linked service** box and then **packtadesynapse-WorkspaceDefaultStorage** for the second **Linked service** box for the data lake connection, as shown in the following screenshot. Then, click **Open pipeline**:

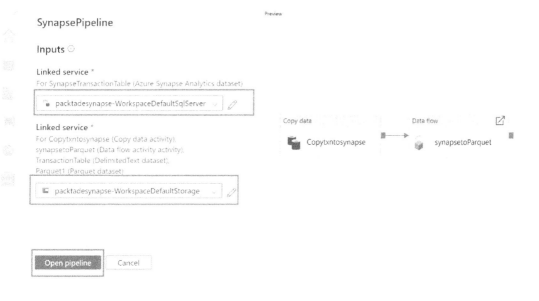

Figure 13.4 – Import pipeline – setting a linked service

- Click on the **Publish all** button:

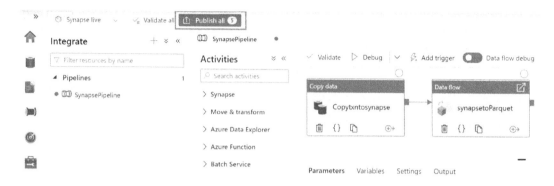

Figure 13.5 – Publishing the pipeline

- Click on **Add trigger** and select **Trigger now** to execute the pipeline. Optionally, run the pipeline two or three times if you wish to have more diagnostic/execution data for the Log Analytics workspace:

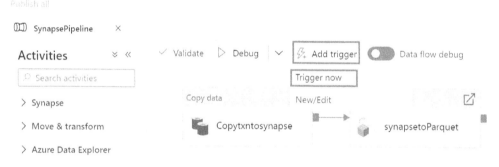

Figure 13.6 – Executing the pipeline

4. Download the `SynapsePipeline.workbook` file from `https://github.com/PacktPublishing/Azure-Data-Engineering-Cookbook-2nd-edition/blob/main/chapter13`.

5. Go to **All resources** and search for `PacktADELogAnalytics`, the Log Analytics workspace. Under the **General** section, click on **Workbooks**. Use the contents of `SynapsePipeline.workbook` and create a new workbook, as explained in *steps 2 to 5* in the *How to do it…* section of the *Creating workbooks in a Log Analytics Workspace to visualize monitoring data* recipe of *Chapter 11, Monitoring Synapse SQL and Spark Pools*. Once done, select **Log Analytics Subscription** as your subscription, **Log Analytics workspace** as **PacktADELogAnalytics**, and **Synapse/Datafactory** as **packtadesynapse**. Then, click **Done Editing**:

Figure 13.7 – Pipeline monitoring workbook

6. Click on the **Save** button and use `SynapsePipeline` as the workbook's name:

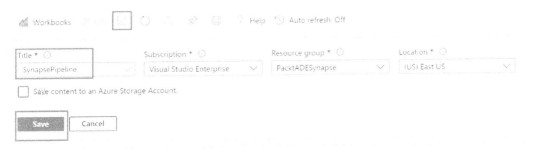

Figure 13.8 – Saving the pipeline

7. **Overview Dashboard** specifies the average duration of pipelines in the workspace and insight into the successes/failures of pipelines:

Figure 13.9 – Pipeline monitoring dashboard

8. The **Pipeline Details** tab provides useful insights into information such as error trends, count of error messages by type, and recent failure details. However, as our workspace has been just created, it doesn't have enough data to display, hence why it's empty:

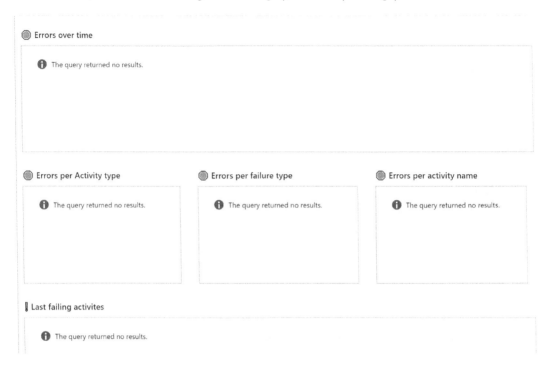

Figure 13.10 – Pipeline monitoring – error details

Refer to `https://github.com/microsoft/Azure_Synapse_Toolbox/blob/master/Collateral/Screenshots/SynapsePipelineWorkbook1.png` for an example dashboard where failures were recorded.

How it works...

A Log Analytics workspace, once integrated with an Azure Synapse workspace for the Integration Activity/Trigger/Pipeline runs categories, collects execution data for all integration pipelines in the Synapse workspace. The workbook presents the data collected by Log Analytics in a visually interactive way. The workbook that was used in this recipe can be found at `https://github.com/microsoft/Azure_Synapse_Toolbox/tree/master/Monitor_Workbooks`, the official GitHub page maintained by Microsoft for Synapse.

The best part about the solution is that even when new additional pipelines are added to the Synapse workspace, the workbook in Log Analytics will automatically reflect the execution data of new pipelines. Monitoring for error trends and pipeline duration becomes extremely critical when you have hundreds of pipelines, with each pipeline containing several tasks and complex workflows. This workbook will be very handy in identifying errors and performance issues and addressing them proactively.

Tracing SQL queries for dedicated SQL pool to Synapse integration pipelines

In the *Using Kusto queries to monitor SQL and Spark pools* recipe of *Chapter 11, Monitoring Synapse SQL and Spark Pools*, we explored using Kusto queries and a Log Analytic workspace to find expensive queries in a dedicated SQL pool. However, in data engineering projects, finding the expensive queries in a dedicated SQL pool alone wouldn't be sufficient as you need to find the details about the integration pipeline that fired the query. To do this, we need to find a way to correlate the Log Analytics data from the integration pipelines and a dedicated SQL pool.

Fortunately, **Copy activity** in an integration pipeline automatically adds a label to the SQL query it uses to copy the data. We can easily identify the pipeline and activity name from the label attached to the SQL query in the dedicated SQL pool. However, other activities, such as data flows and SQL stored procedure tasks, don't automatically append labels to queries, making it harder to trace the pipeline details from the query. A workaround for other activities, such as data flows, is to manually append labels to the queries fired inside them so that they can be traced back to the activity/pipeline.

In this recipe, we will use Kusto queries and a Log Analytics workspace to trace the pipeline details for queries from a **Copy activity** and showcase manually appending labels to queries in a data flow activity.

Getting ready

To get started, log into `https://portal.azure.com` using your Azure credentials. Then, follow these steps:

1. Complete all the steps listed in the *Getting ready* section of the *Monitoring Synapse integration pipelines using Log Analytics and workbooks* recipe.

2. Link the Log Analytics workspace to a dedicated SQL pool, as explained in *steps 3 to 5* in the *How to do it…* section of the *Configuring a Log Analytics workspace for a Synapse dedicated SQL pool* recipe of *Chapter 11, Monitoring Synapse SQL and Spark Pools*.

3. Complete *steps 1 to 3* in the *How to do it….* section of the *Monitoring Synapse integration pipelines using Log Analytics and workbooks* recipe to configure Log Analytics for integration pipelines and to execute a sample pipeline.

How to do it...

After completing the *Getting ready* section, you would have executed **SynapsePipeline**, which contains a **Copy** activity and **Data flow** activity. First, we will trace the pipeline and activity details of queries fired from the Copy activity.

Follow these steps to find the pipelines and activities details for queries fired from the Copy activity:

1. In the Azure portal, go to **All resources** and search for PacktADELogAnalytics, the Log Analytics workspace, and click on it. Click on **Logs**. Then, close the **Queries** popup using the **close** button at the top right:

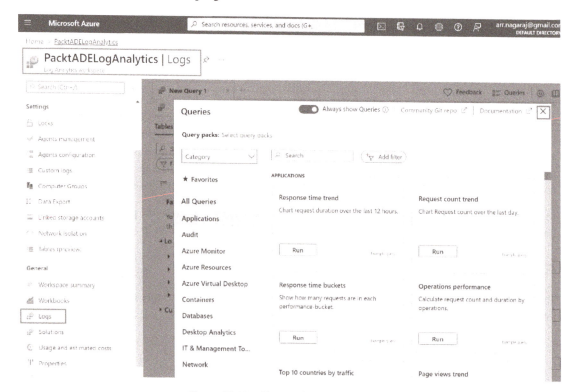

Figure 13.11 – Closing the Queries popup

2. In the query window, paste the following script:

```
let SQLQueries =
SynapseSqlPoolExecRequests
| where  StartTime between (datetime(2000-05-20)..
TimeGenerated)
| where Label !="health_checker" and Label contains "ADF"
| where Status contains "Running"
| where _ResourceId endswith DatabaseName
```

```
| extend duration_sec = datetime_diff("second",
TimeGenerated,StartTime)
| summarize Query_duration_sec = max(duration_
sec),StartTime = min(StartTime), Command =
any(Command),Label = any(Label),ResourceClass =
any(ResourceClass),QueryPlan = any(ExplainOutput),Status
= any(Status),Source = any(SourceSystem) by RequestId;
```

The preceding script finds the queries on a dedicated SQL pool that have labels that contain the word ADF. Copy the activity in the Synapse integration pipeline; Azure Data Factory will automatically add a label in "ADF Activity: <Activity ID>" format. The Activity ID is a unique identifier that you can use to identify whether an activity has been executed inside a Synapse integration pipeline or a Data Factory pipeline.

Append the following script to the same query window:

```
let PipelineActivity =
SynapseIntegrationActivityRuns
| extend Label = strcat("ADF Activity ID:","
",ActivityRunId),activity_duration_sec = datetime_diff
("second", TimeGenerated,Start)
| summarize activity_duration_sec=max(activity_duration_
sec), PipelineName = any(PipelineName) , ActivityName =
any(ActivityName),PipelineRunId = any(PipelineRunId),
ActivityType=any(ActivityType),
EffectiveIntegrationRuntime = any(EffectiveIntegration
Runtime) by Label;
```

The preceding script gets the pipeline's name, run ID, activity name, and the activity IDs of the pipelines that were executed. Append the following script to the same query window:

```
SQLQueries
| join kind = leftouter PipelineActivity on Label
| project RequestId,StartTime,Query_duration_
sec,Command,PipelineName,ActivityName,Label,PipelineRunId
```

The preceding script combines the SQL query details and pipeline/activity details using the activity ID, which is stored in the **Label** column. Select the time range to **Last 24 hours**, select all the queries you've pasted, and hit the **Run** button:

Figure 13.12 – Activity and pipeline details

As you can see, the script provides the pipeline's name and the name of the Copy activity that fired the query on the dedicated SQL pool. Using **PipelineRunId**, we can find the exact execution of the pipeline that fired the query.

As explained initially, data flows don't automatically add labels; they need to be appended manually to queries from data flow. Next, we will learn how to manually add a label to trace the activity's details.

3. Go to **Synapse Studio**. Click on the develop icon (the notebook-like icon), click on the + button, and select **SQL script**. Then, connect to **packtadesqlpool**, copy and paste the following script, and execute it:

```
CREATE PROC [dbo].[get_transactiontable_df] AS

Select * from dbo.transactiontable
Option (LABEL = 'ADF: SQLtoParquet - Dataflow')
```

The preceding script creates a stored procedure that contains a SELECT query that reads all the rows from **dbo.transactiontable**. The SELECT query uses an OPTION clause that adds the label to the query:

Figure 13.13 – Stored procedure creation

4. Expand **Data flows** and click on the **SQLtoParquet** data flow, Then, click on **source1**, then **Source options**. Select **Stored procedure** under **Input** and set **Schema name** to **dbo**. Finally, set **Stored procedure** to **get_transactiontable_df**, the name of the stored procedure you created in *step 3*. Then, click on **Publish all**:

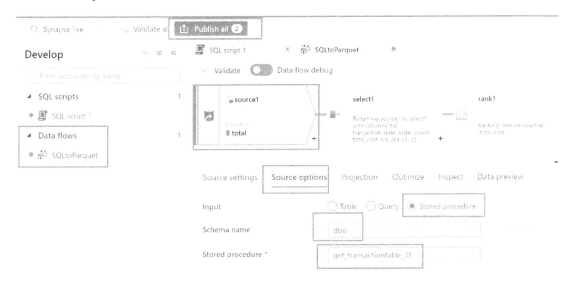

Figure 13.14 – Stored procedure input

Previously, we used **Table** as the input option, which obtained all the rows from **dbo. transactiontable**. We have set **Stored procedure** to **Input** here, which allows us to manually append the label and fetch all the rows from the table.

5. Click on **Integration Pipeline Icon**, go to **SynapsePipeline**, and execute **SynapsePipeline** by clicking on **Add trigger** | **Trigger now**:

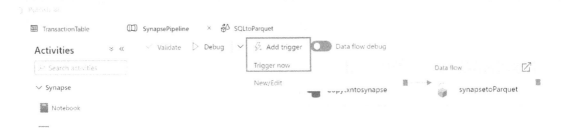

Figure 13.15 – Triggering the pipeline

6. Rerun the KQL script you used in *step 2* in the Azure Log Analytics workspace. Now, the query from the data flow is also listed along with the query from the Copy activity. The label we added in the stored procedure is listed in the result set. Using that label, we can identify that the query was fired from the data flow:

Figure 13.16 – Data flow labels listed

How it works...

Integrating a Synapse integration pipeline and a dedicated SQL pool in the same Log Analytics workspace allows us to easily correlate the diagnostic data from both services. Query labels allow us to tag queries so that we can locate them during database monitoring and troubleshooting tasks. The Copy activity automatically adds labels to queries, which helps us identify the queries fired by it easily. Other activities don't automatically add labels to queries. Hence, for the data flow activity, we manually added a label using a stored procedure and tracked the query in the Log Analytics workspace. It is recommended that you label the queries while developing data flows as it will make monitoring easier, especially in a complex environment that contains hundreds of pipelines.

Provisioning a Microsoft Purview account and creating a data catalog

Microsoft Purview is a data governance service that allows you to manage all your data assets in one single place. Using Microsoft Purview, you can build data catalogs to maintain various information about data assets such as dataset details, schema details of the dataset, data source details, dataset classification, owners of the dataset, and many more.

In this recipe, we will provision a Purview account and build a basic data catalog. After that, we will register a resource group and scan the resources/assets in the resource group. By doing this, Purview will collect and store the metadata information about all the data assets in the resource group and build a basic data catalog.

Getting ready

To get started, log into `https://portal.azure.com` using your Azure credentials. Then, follow these steps:

1. Create a Synapse Analytics workspace, as explained in the *Provisioning a Azure Synapse Analytics workspace* recipe of *Chapter 8, Processing Data Using Azure Synapse Analytics*.

2. Download the `transaction-tbl.csv` file from `https://github.com/PacktPublishing/Azure-Data-Engineering-Cookbook-2nd-edition/tree/main/chapter10`. In a Synapse Analytics workspace, create two folders named `CSV` and `Parquet` in the data lake account attached to it. Upload the `transaction-tbl.csv` file to the `CSV` folder. For detailed screenshots for a similar task, follow *steps 1 to 4* in the *How to do it...* section of the *Analyzing data using a Serverless SQL pool* recipe of *Chapter 8, Processing Data Using Azure Synapse Analytics*.

3. Create a Synapse-dedicated SQL pool and name it `packtadesqlpool`, as described in *steps 1 to 3* in the *How to do it…* section of the *Loading data into a dedicated SQL pool using PolyBase and T-SQL* recipe of *Chapter 10, Building the Serving Layer in Azure Synapse SQL Pool.*

4. Complete *step 3* in the *How to do it…* section of the *Monitoring Synapse integration pipelines using Log Analytics and workbooks* recipe to deploy a sample pipeline called **SynapsePipeline** and execute it.

How to do it...

Follow these steps to provision a Purview account, register services, and scan assets:

1. First, create a Purview account. Go to `portal.azure.com` and click on **Create a resource**. Search for `Microsoft Purview` and click on the **Create** button:

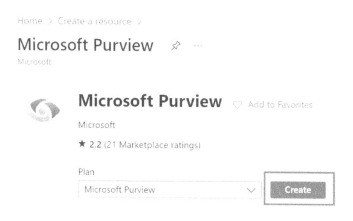

Figure 13.17 – Purview – Create a resource

2. Set **Resource group** to `PacktADESynapse` and **Microsoft Purview account name** to `packtadepurview`. Then, click **Review + Create**:

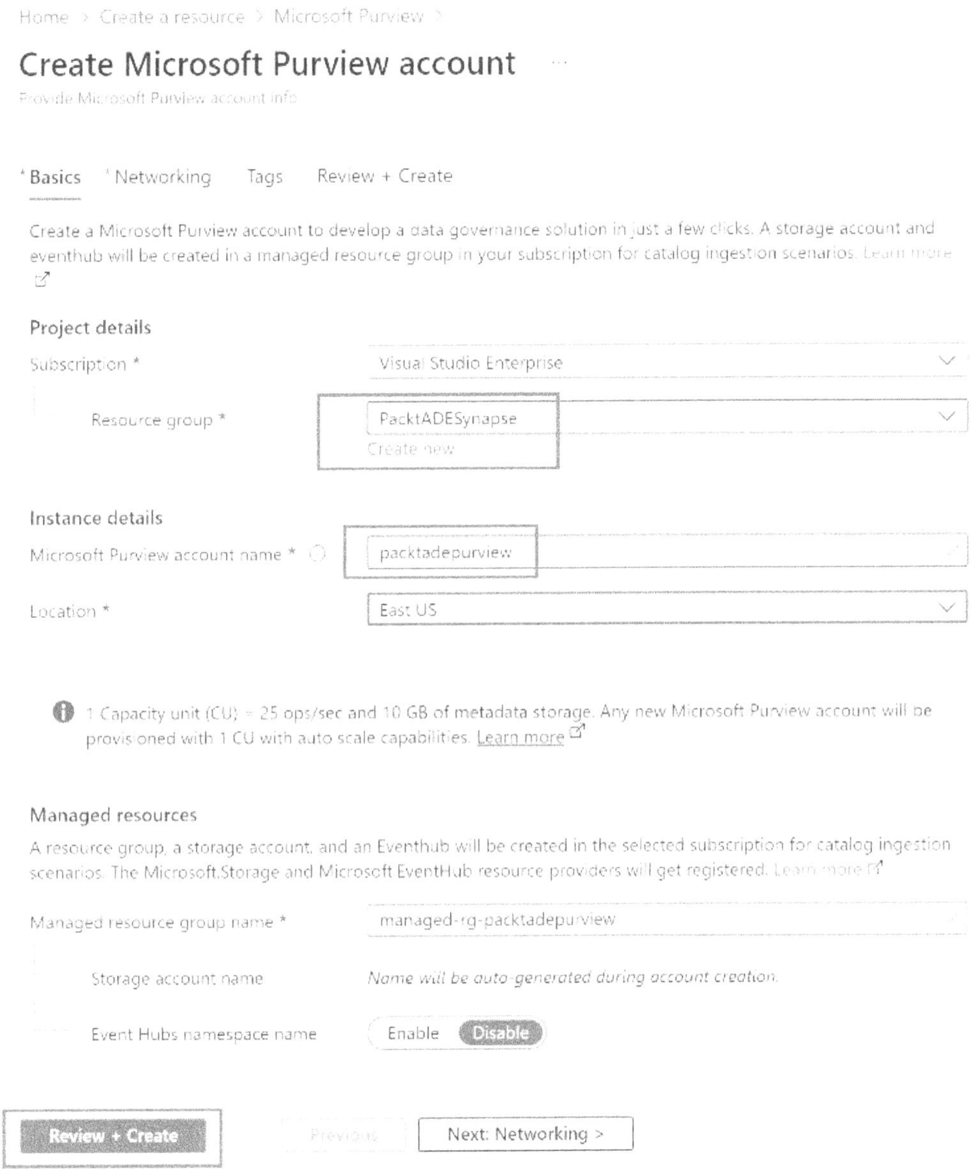

Home > Create a resource > Microsoft Purview >

Create Microsoft Purview account

Provide Microsoft Purview account info

* Basics   ' Networking   Tags   Review + Create

Create a Microsoft Purview account to develop a data governance solution in just a few clicks. A storage account and eventhub will be created in a managed resource group in your subscription for catalog ingestion scenarios. Learn more

Project details

Subscription *   Visual Studio Enterprise

Resource group *   PacktADESynapse
                                        Create new

Instance details

Microsoft Purview account name * ○   packtadepurview

Location *   East US

ℹ 1 Capacity unit (CU) = 25 ops/sec and 10 GB of metadata storage. Any new Microsoft Purview account will be provisioned with 1 CU with auto scale capabilities. Learn more

Managed resources

A resource group, a storage account, and an Eventhub will be created in the selected subscription for catalog ingestion scenarios. The Microsoft.Storage and Microsoft.EventHub resource providers will get registered. Learn more

Managed resource group name *   managed-rg-packtadepurview

Storage account name   *Name will be auto-generated during account creation.*

Event Hubs namespace name   Enable   **Disable**

Review + Create   Previous   Next: Networking >

Figure 13.18 – Provisioning Purview

3. Once the Purview account has been provisioned, go to it. Click on **Open Microsoft Purview Governance Portal**:

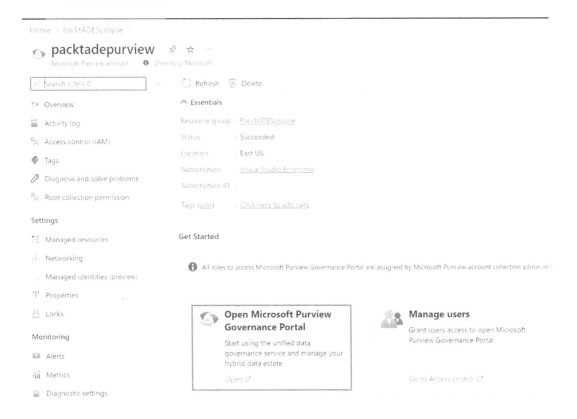

Figure 13.19 – Open Microsoft Purview Governance Portal

4. Click on the **Data map** icon (the second icon on the left) and click **Register**:

Figure 13.20 – Registering resources in Purview

5. Select **Azure MULTIPLE** and click **Continue**:

Register sources

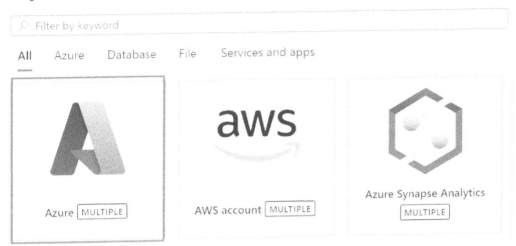

Figure 13.21 – Azure MULTIPLE

6. Provide PacktADESynapse as the name (this is the same name as the resource group to be scanned as it's easier to identify). Set both your subscription name and resource group name to PacktADESynapse. Then, click **Register**:

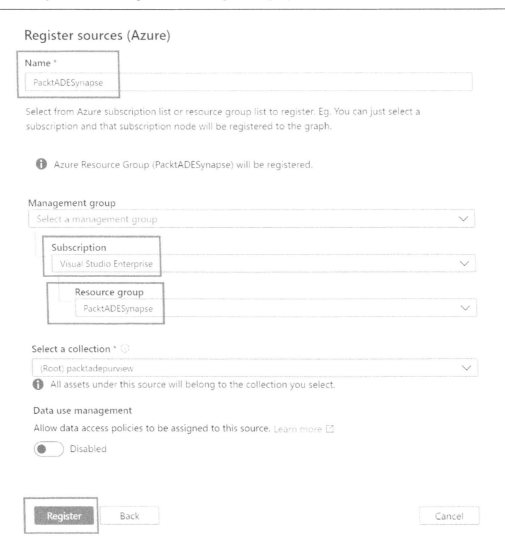

Figure 13.22 – Registering a resource group in Purview

7. Now that we have registered the resource group, we must scan the resources in the resource group using Microsoft Purview so that Purview can collect data about the data assets in the resource group. To do this, we need to add a **Purview managed identity** account to the following roles:

    • The **reader** role to the **packtadesynapse** workspace

    • The **Storage blob reader** role to the Azure Data Lake account

    • The **db_datareader** role to the dedicated SQL pool database

Use the following PowerShell script to add the roles for the Synapse workspace and Azure Data Lake to a Purview managed identity account. Assign the appropriate values for the $Resourcegroup, $Synapse_ws_name, $storage_name, and $purview variables:

```
$Resourcegroup = "PacktADESynapse"

$Synapse_ws_name = "packtadesynapse"

$storage_name = "packtadesynapse"

$purview = "packtadepurview"

$role = Get-AzADServicePrincipal -DisplayName $purview

$storage = Get-AzStorageAccount -ResourceGroupName
$Resourcegroup -Name $storage_name

New-AzRoleAssignment -ObjectId $role.id
-RoleDefinitionName "Storage Blob Data Reader" -Scope
$storage.id

$synapse = Get-AzSynapseWorkspace -ResourceGroupName
$Resourcegroup -Name $Synapse_ws_name

New-AzRoleAssignment -ObjectId $role.id
-RoleDefinitionName "Reader" -Scope $synapse.id
```

Upon executing the script, you will get the following output:

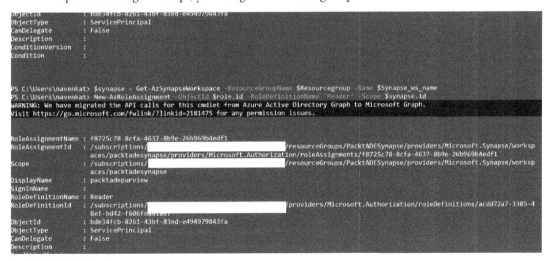

Figure 13.23 – Granting permissions to PowerShell

8. Go to **Synapse Studio**. Click on the **develop** icon (the notebook-like icon). Then, click the + button and select **SQL script**. Connect to **packtadesqlpool**. Copy and paste the following script:

```
CREATE USER [packtadepurview] FROM EXTERNAL PROVIDER
GO
EXEC sp_addrolemember 'db_datareader', [packtadepurview]
GO
```

Hit the **Run** button. This script will add a Purview managed identity to the **db_datareader** role on the dedicated SQL pool database:

Figure 13.24 – Granting the db_datareader permission

9. Go to the **packtadepurview** Purview account and open **Microsoft Purview Governance Portal**. Click on the **Data Map** icon (the second icon on the left). Then, click the **New scan** button under **PacktADESynapse**:

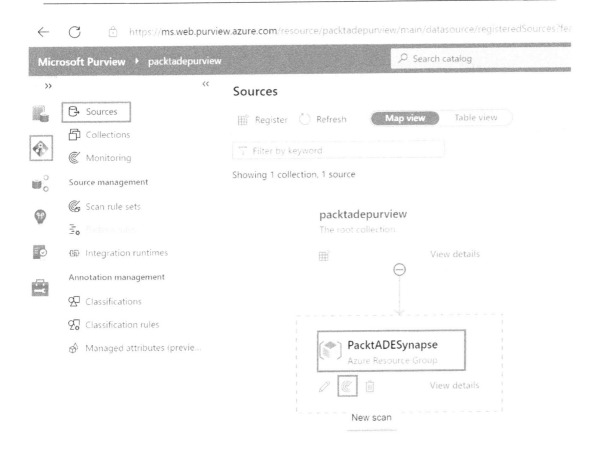

Figure 13.25 – Scanning resources

10. Select **Microsoft Purview MSI (system)** under **Credential** and click **Continue**:

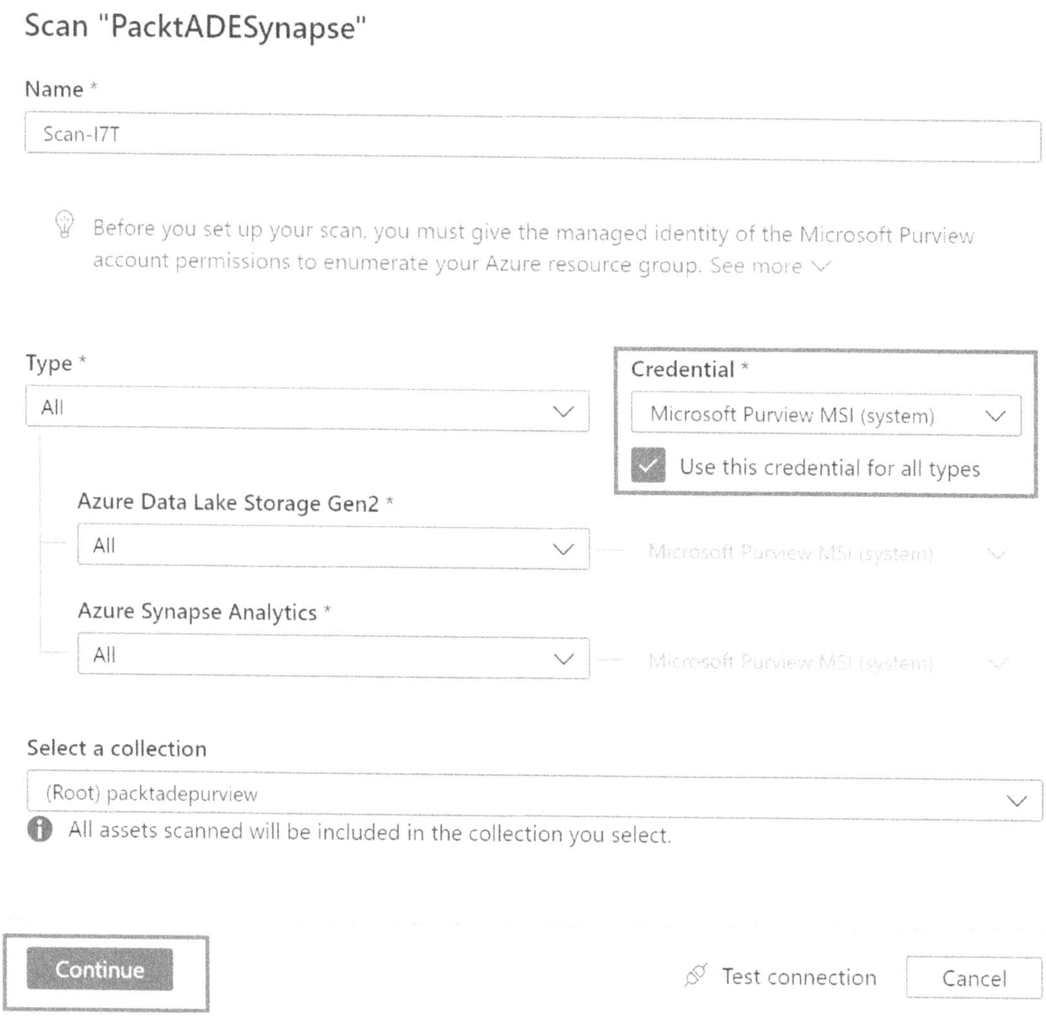

Scan "PacktADESynapse"

Name *

Scan-I7T

💡 Before you set up your scan, you must give the managed identity of the Microsoft Purview
     account permissions to enumerate your Azure resource group. See more ∨

Type *

All   ∨

Credential *

Microsoft Purview MSI (system)   ∨

☑ Use this credential for all types

Azure Data Lake Storage Gen2 *

All   ∨   Microsoft Purview MSI (system)   ∨

Azure Synapse Analytics *

All   ∨   Microsoft Purview MSI (system)   ∨

Select a collection

(Root) packtadepurview   ∨

ℹ All assets scanned will be included in the collection you select.

Continue   🔗 Test connection   Cancel

Figure 13.26 – Setting a credential

11. Ignore the **Select a scan rule set** window for now and click **Continue**:

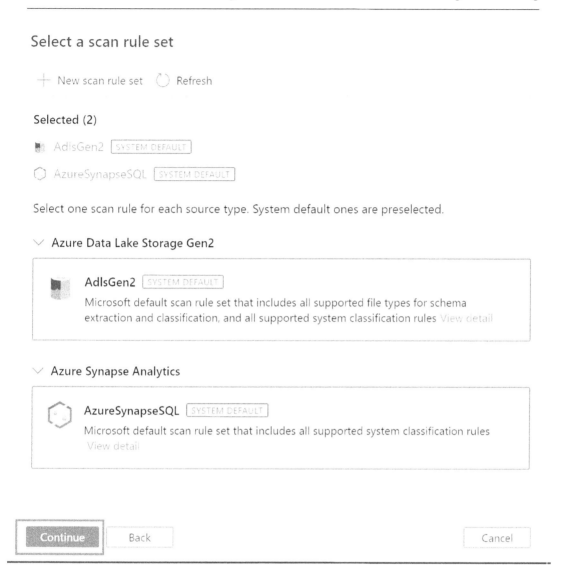

Figure 13.27 – Skipping Select a scan rule set

12. Set the scan schedule to **Once** instead of **Recurring**. Setting the scan schedule to **Recurring** will run a resource scan at a specified time and ensure the Purview data catalog retains the updated data. For testing purposes, it is sufficient to run the scan once:

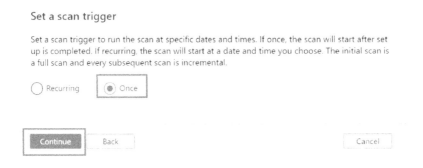

Figure 13.28 – Scanning the schedule

13. Click on **Save and Run** to run the scan:

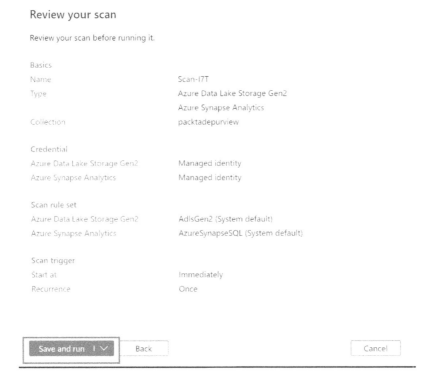

Figure 13.29 – Confirming the scan

14. Click on **View details** under **PacktADESynapse** to check the status of the scan:

Figure 13.30 – Scan details

15. Once the run completes, you will notice that the scan has discovered assets from the resource group. Ignore the scan failure report as the error message will be related to a serverless SQL database. We can ignore it as we don't have any objects in this recipe related to a serverless SQL database. The run typically takes around 15 to 20 minutes:

Figure 13.31 – Scan results

16. Click on **Collections** at the top, then **packtadepurview**, and then **Assets**:

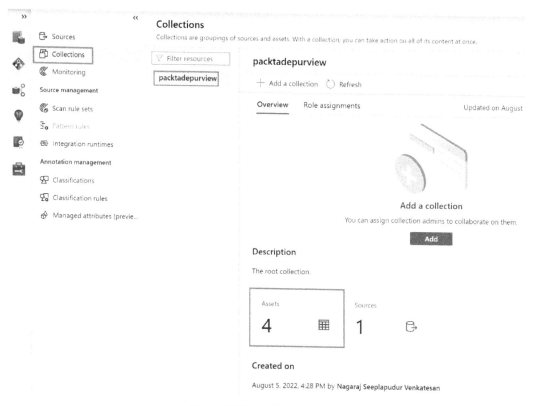

Figure 13.32 – Collection details

17. The assets it has discovered will be listed here. Click on **transactiontable**:

Assets in collection

⊞ Move

▽ Filter by keyword   (Source type : **all**)   (Instance : **all**)   ▽ Clear all filters

Showing 1 - 4 of 4 items

	Name ↑	Source type
☐	🛢 dbo	Azure Synapse Analytics
☐	packtadesqlpool	Azure Synapse Analytics
☐	packtadesynapse.azuresynapse.net:443	Azure Synapse Analytics
☐	⊞ transactiontable	Azure Synapse Analytics

Figure 13.33 – Assets list

18. Click on the **Schema** tab to see the columns in the table and their data types:

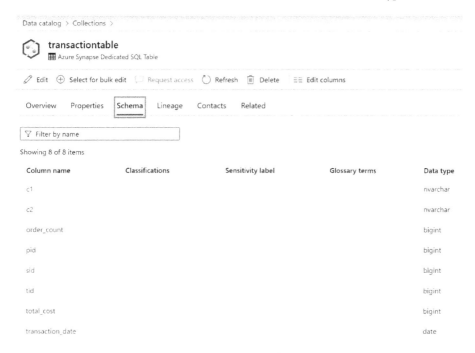

Figure 13.34 – Assets details

19. Click on the **Related** tab. The objects that are related to **transactiontable** are displayed here:

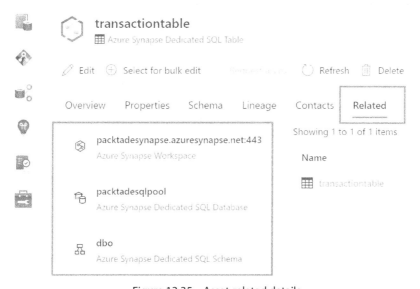

Figure 13.35 – Asset-related details

How it works...

The Purview account's managed identity was granted permissions on the assets in the resource group and Purview's scan was able to record all the metadata information about the assets in the resource group. Microsoft Purview creates a data catalog where you can easily locate a table or a column and find information about it, as well as list objects related to it.

You can learn more about Purview by going to the links and playing with the options on the Purview governance portal. Check out `https://docs.microsoft.com/en-us/azure/purview/use-azure-purview-studio` to learn more about the Purview governance portal.

Integrating a Synapse workspace with Microsoft Purview and tracking data lineage

One of the key features of Microsoft Purview is tracking data lineage. Microsoft Purview offers end-to-end tracking of data lineage, using which you can track the life cycle of a particular data asset – for example, if data is loaded from a CSV file into a dedicated SQL pool table and then further processed and stored in a Parquet file. Purview will be able to represent the various phases (CSV file -> SQL table -> Parquet file) of data processing graphically and automatically. Tracking data lineage in data engineering projects is critical as you need to have a bird's-eye view of the various processing phases of a data asset and its properties.

As you know, a Synapse workspace holds all the data engineering pipelines, which do all the data movement and processing. By integrating a Synapse workspace with Microsoft Purview, we can track data lineage with minimal effort.

In this recipe, we will integrate a Synapse workspace with a Microsoft Purview account and track data lineage.

Getting ready

Complete all the steps listed in the *Getting ready* section of the *Provisioning a Microsoft Purview account and creating a data catalog* recipe.

You must also complete *steps 1 to 13* in the *How to do it...* section of the *Provisioning a Microsoft Purview account and creating a data catalog* recipe.

How to do it...

Follow these steps to integrate a Synapse workspace with your Microsoft Purview account and track data lineage:

1. Log into `portal.azure.com`, go to **All resources**, and search for `packtadesynapse`. Open **Synapse Studio**. Click on the **Manage** icon (the briefcase-like icon). Click on **Microsoft Purview**, then **Connect to a Purview account**:

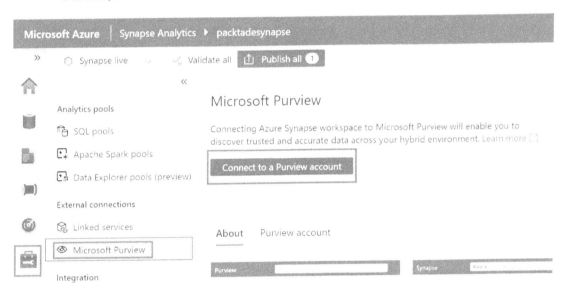

Figure 13.36 – Connect to a Purview account

2.   Set **Purview account name** to `packtadepurview` and click **Apply**:

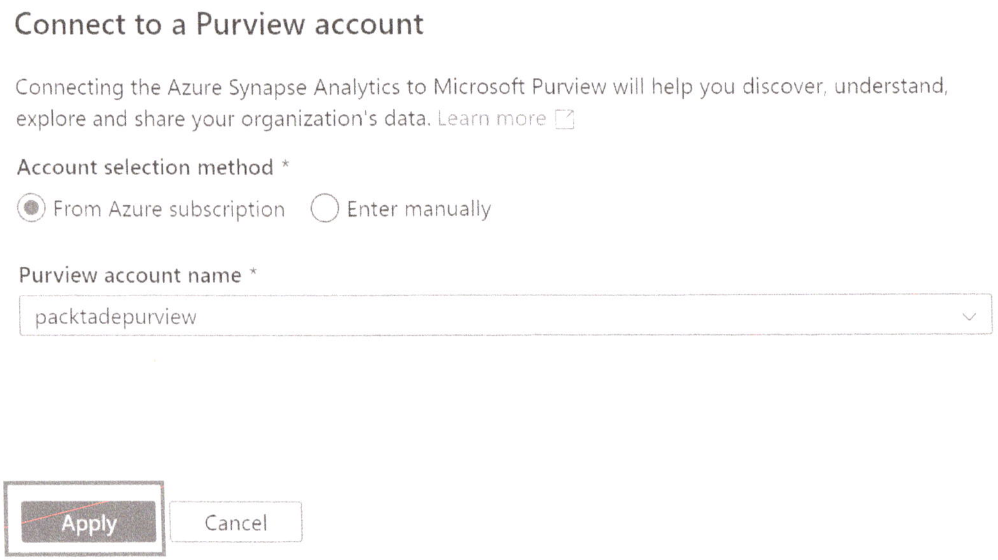

Connect to a Purview account

Connecting the Azure Synapse Analytics to Microsoft Purview will help you discover, understand, explore and share your organization's data. Learn more ☐

Account selection method *

⦿ From Azure subscription   ◯ Enter manually

Purview account name *

| packtadepurview   ⌄ |

Apply   Cancel

Figure 13.37 – Integrating Synapse with Purview

3.   Click on the **Purview account** tab and ensure the connection status is **Connected**, as shown here:

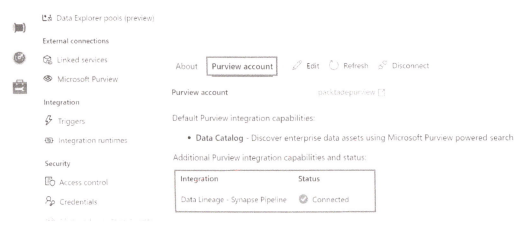

Figure 13.38 – Synapse integration confirmation

4. Click on the **Integrate** icon (the pipe-like icon on the left). Expand **Pipelines** and click **SynapsePipeline**. Then, click **Add trigger**, then **Trigger now**:

Figure 13.39 – Triggering the pipeline

5. The pipeline that was executed performs the following tasks:

- Reads a CSV file called `transaction-tbl.csv` from **packtadesynapse**, an Azure Data Lake account

- Loads the file to a table called `transactiontable` in **packtadesqlpool**, a dedicated SQL pool database that uses a Copy activity in the integration pipeline

- Reads `transactiontable`, adds a column called `rank_by_cost`, and copies the rows in Parquet format to a folder named `Parquet` in the **packtadesynapse** data lake account

Using the Purview integration, we can track the lineage from the data lake account to the `Parquet` folder.

Go to the **Search** area at the top of the window. Type `order_count`; Synapse will automatically prompt the **order_count** column. Click on the suggestion. **order_count** is a column that exists in the CSV file, Synapse table, and the destination Parquet file:

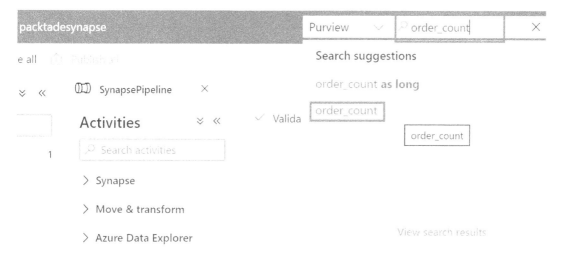

Figure 13.40 – The search column

6. All the assets that contain the **order_count** column will be listed. Click on **transactiontable**:

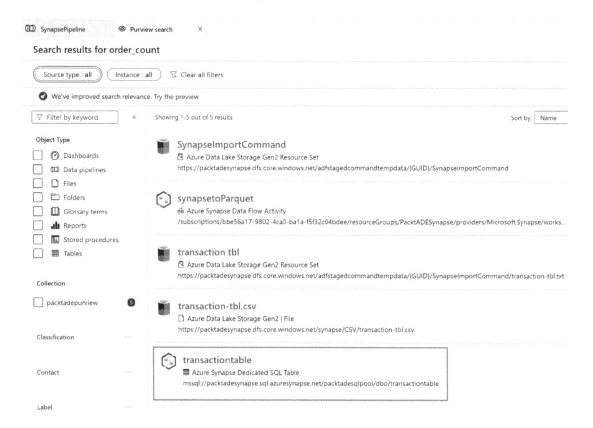

Figure 13.41 – Selecting an asset

7. Click on the **Lineage** tab. The **Lineage** tab shows the data asset's complete life span. It visually shows how the data moved from the data lake account to **transactiontable** in the Synapse-dedicated pool database via the Copy activity and then to the Parquet folder via a data flow. Hover your mouse over each block to learn more about the processing phase:

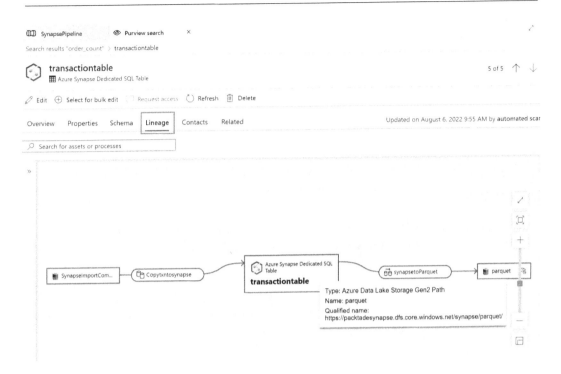

Figure 13.42 – Data lineage

If the Parquet file's destination is missing for you, go to the **synapsetoParquet** data flow, click on **source 1**, click on the **Source options** tab, and ensure **Input** is set to **Table**, as shown in the following screenshot. Make sure that you deselect the **Stored procedure** option for **Input**. Once set, publish the data flow, rerun the pipeline, and check the lineage after a few minutes:

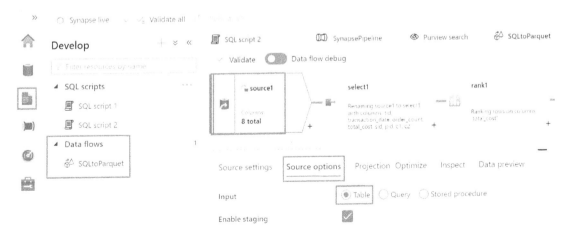

Figure 13.43 – Setting the data flow input

How it works...

Integrating a Synapse workspace with a Purview account helps us track data lineage easily. After each pipeline run, Synapse automatically updates the Purview account with the data lineage details. Data lineage for new pipelines and data flows is automatically updated to the Purview account. There is no need to perform a Purview scan via the Purview governance portal (as you did in the *Provisioning a Microsoft Purview account and creating a data catalog* recipe) to keep the lineage updated.

Applying Azure tags using PowerShell to multiple Azure resources

Azure tags help us organize the Azure resources we have created in Azure data engineering projects. Tags add key-value pairs to resources, which can help us easily categorize the resources and identify them. For example, adding a tag with the key set to `Environment` and the value set to `production` for resources can help us identify all production environment resources in our tenant. You can add multiple tags to the same resource too. For example, you may have two tags with `Environment` and `Project` as keys. `Environment` helps identify the environment that the resource belongs to, while `Project` helps identify which project the resource is part of.

It is always good practice to add tags at the time of resource creation. However, in some scenarios, you may need to add tags to several resources after they have been created. In this recipe, we will use a PowerShell script to tag several resources in one go.

Getting ready

You may perform this recipe in any Azure subscription with resources created.

How to do it...

In this recipe, we will add two tags – **Project** and **Classification** – to resources that match the following conditions:

- Resources starting with `Packt` – for example, resources named **packtadesqlpool**.
- All resources in resource groups whose names start with `Packt` – for example, all resources in a resource group named **PacktADESynapse**.

Follow these steps to add tags using PowerShell:

1. Connect to your Azure subscription using the `connect-Azaccount` command. Copy and paste the following command into the PowerShell console and run it:

```
$Tags = @{"Project"="Packt"; "Classification"="Public"}
$tr = get-Azresource | Where-Object {$_.Name -like
"packt*" -or $_.ResourceGroupName -like "packt*"}
  foreach ($Res in $tr) {
        Set-AzResource -ResourceGroupName $Res.
ResourceGroupName -Name $Res.Name -ResourceType $Res.
ResourceType -Tag $tags -Force
      }
```

This script will add the two tags with the following values:

- `Project` as the key and `Book` as the value. This implies that resources that match our condition belong to the project named `Book`.

- `Classification` as the key and `Public` as the value. This implies that resources that match our condition don't contain sensitive data and can be classified as public or non-sensitive.

The script will have a lengthy output. It should look similar to the following:

Figure 12.44 – Publish Auto Pause Script

2. Let's use the Azure portal and explore the use of the Azure tags we've created. Go to the Azure portal and click on **All resources**. Then, click on **Add filter**. Notice that the tags we created are listed **under** Tags. Select **Classification** under the **Filter** option:

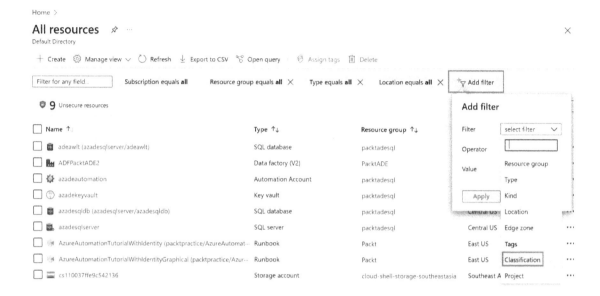

Figure 12.45 – Add filter

3. Select **Public** as the value and click the **Apply** button:

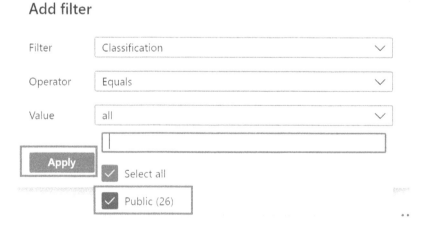

Figure 12.46 – Applying the filter

4. You will see that the resources we have tagged are listed:

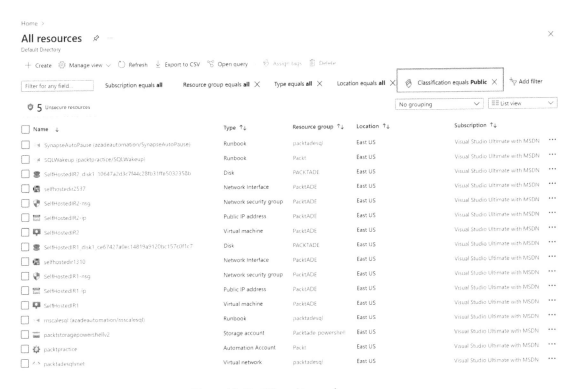

Figure 13.47 – Filtered tagged resources

How it works...

PowerShell script helps search through the subscription and apply the necessary tags. Let's look at what each command does:

- `get-Azresource | Where-Object {$_.Name -like "packt*" -or $_.ResourceGroupName -like "packt*"}`: This command filters for resources starting with `Packt` or that belong to resource groups starting with `packt`

- `foreach ($Res in $tr)`: This command loops through all the resources

- `Set-AzResource -ResourceGroupName $Res.ResourceGroupName -Name $Res.Name -ResourceType $Res.ResourceType -Tag $tags -Force`: This command applies the necessary tags

Even though they seem simple, Azure tags are an extremely powerful feature when you're managing complex projects that contain thousands of resources. For example, Azure tags are very useful in preparing Azure billing reports, where you need to find the monthly cost of the resources that belong to a particular project. Azure billing reports have a column containing tags that have been created, which we can use to filter (as we did in **All resources** in the Azure portal) and track the cost of a particular project easily.

Index

Packt.com

Subscribe to our online digital library for full access to over 7,000 books and videos, as well as industry leading tools to help you plan your personal development and advance your career. For more information, please visit our website.

Why subscribe?

- Spend less time learning and more time coding with practical eBooks and Videos from over 4,000 industry professionals

- Improve your learning with Skill Plans built especially for you

- Get a free eBook or video every month

- Fully searchable for easy access to vital information

- Copy and paste, print, and bookmark content

Did you know that Packt offers eBook versions of every book published, with PDF and ePub files available? You can upgrade to the eBook version at packt.com and as a print book customer, you are entitled to a discount on the eBook copy. Get in touch with us at customercare@packtpub.com for more details.

At www.packt.com, you can also read a collection of free technical articles, sign up for a range of free newsletters, and receive exclusive discounts and offers on Packt books and eBooks.

Other Books You May Enjoy

If you enjoyed this book, you may be interested in these other books by Packt:

Data Engineering with Alteryx

Paul Houghton

ISBN: 9781803236483

- Build a working pipeline to integrate an external data source
- Develop monitoring processes for the pipeline example
- Understand and apply DataOps principles to an Alteryx data pipeline
- Gain skills for data engineering with the Alteryx software stack
- Work with spatial analytics and machine learning techniques in an Alteryx workflow
- Explore Alteryx workflow deployment strategies using metadata validation and continuous integration
- Organize content on Alteryx Server and secure user access

Simplifying Data Engineering and Analytics with Delta

Anindita Mahapatra

ISBN: 9781801814867

- Explore the key challenges of traditional data lakes
- Appreciate the unique features of Delta that come out of the box
- Address reliability, performance, and governance concerns using Delta
- Analyze the open data format for an extensible and pluggable architecture
- Handle multiple use cases to support BI, AI, streaming, and data discovery
- Discover how common data and machine learning design patterns are executed on Delta
- Build and deploy data and machine learning pipelines at scale using Delta

Packt is searching for authors like you

If you're interested in becoming an author for Packt, please visit `authors.packtpub.com` and apply today. We have worked with thousands of developers and tech professionals, just like you, to help them share their insight with the global tech community. You can make a general application, apply for a specific hot topic that we are recruiting an author for, or submit your own idea.

Share your thoughts

Now you've finished *Azure Data Engineering Cookbook, Second Edition*, we'd love to hear your thoughts! Scan the QR code below to go straight to the Amazon review page for this book and share your feedback or leave a review on the site that you purchased it from.

`https://packt.link/r/1-803-24678-2`

Your review is important to us and the tech community and will help us make sure we're delivering excellent quality content.

www.ingramcontent.com/pod-product-compliance
Lightning Source LLC
Chambersburg PA
CBHW060635060326
40690CB00020B/4406